RATTLING

the

CAGE

RATTLING

the

CAGE

Toward Legal Rights
for Animals

STEVEN M. WISE

FOREWORD BY JANE GOODALL

A Merloyd Lawrence Book

PERSEUS PUBLISHING

Cambridge, Massachusetts

A CIP catalog record for this book is available from the Library of Congress.
ISBN: 0-7382-0437-4

Perseus Publishing is a member of the Perseus Books Group

Text design by Jeff Williams
Set in 11-point Palatino by the Perseus Books Group

1 2 3 4 5 6 7 8 9 10—03 02 01 00
First paperback printing, December 2000

Perseus Publishing books are available at special discounts for bulk purchases in the U.S. by corporations, institutions, and other organizations. For more information, please contact the Special Markets Department at HarperCollins Publishers, 10 East 53rd Street, New York, NY 10022, or call 1–212–207–7528.

Find us on the World Wide Web at http://www.perseuspublishing.com

For Jerom
a person, not a thing

CONTENTS

FOREWORD

by Jane Goodall

Rattling the Cage is an important book, an exciting book. It will be welcomed by everyone who is concerned about the well-being of animals, by those who are, as I am, kept awake by grim mental images of the abuse inflicted on other animals by humans. I was honoured when Steve Wise asked me to write this introduction, for I believe that *Rattling the Cage*, thanks to all the long years of research that went into the writing, will make an impact, and leave its mark on the process of law. I see it as a major stepping-stone along a road that is gradually leading to a new legal relationship between humans and other sentient, sapient life forms.

Steve Wise is a law school professor. He is also an accomplished animal-rights legal scholar and one of the world's most prominent animal-rights lawyers. Steve and his wife, Debi, defend a variety of animal species across the United States and advise those who defend animals around the world. In writing *Rattling the Cage*, Steve has used his experience in both science and law to great advantage, and he has a trial lawyer's knack for telling a good story. He explains, for example, why it matters so much today whether an ox who gored a passerby on a road in the Middle East four thousand years ago was Babylonian or Hebrew. And why, four hundred years ago, an early animal advocate stood up for barley-eating rats in a French courtroom. And, most surprisingly, why John Quincy Adams would thunder on the floor of the U.S. House of Representatives that he would present petitions to the Congress from horses or dogs if they asked him to.

In many ways this book can be seen as the animals' Magna Carta, Declaration of Independence, and Universal Declaration of

Rights all in one. And it is timely. Twenty—even ten—years ago, Steve would have been out on a limb, ridiculed by his colleagues and largely ignored by the lay public. But attitudes toward animals have changed. Very few scientists today believe that nonhuman animals are simply mindless machines, collections of stimuli and responses. Of course, it would be convenient to believe that this was true, that there was a basic and fundamental difference between ourselves and the rest of the animal kingdom. Then we could do unpleasant things to them without any feelings of guilt. But this is scarcely an option today, when there have been so many descriptions of incredibly complex social behavior and so many examples of intelligent behavior from so many careful field studies on a whole variety of animal species. Our thirty-nine years with wild chimpanzees at Gombe, for example, has taught us much about these relatives of ours, each with his or her own unique personality. They share so many of our behaviors. They form close affectionate bonds with each other that may persist through a life of sixty or more years; they feel joy and sorrow and despair, mental as well as physical suffering; they show many of the intellectual skills that until recently we believed unique to ourselves; they look into mirrors and see themselves as individuals—they have consciousness of "self." Admittedly, chimpanzees are capable, as are we, of acts of brutality. But they also demonstrate empathy, compassion, altruism, and love. Should not beings of this sort have the same kind of legal rights as those we grant to human infants or the mentally disabled, who also cannot speak for themselves?

This book outlines how legal changes for animals, once thought impossible (and there were very few who even bothered to consider this at all), can actually happen. Then we shall have another way of fighting the injustice that is still perpetuated on animals of all kinds—by science, agribusiness, the pharmaceutical industry, the live-animal traders, and so on. If only such a change in law could have happened in time to save Jojo, Jade, and Dick.

Jojo was the first adult male chimpanzee whom I met in a medical research laboratory—which was, of course, in the basement, with no windows. JoJo was, like the other nine adult males who shared space with him, confined in a five-foot by five-foot cage. There were thick steel bars between JoJo and me. And there were bars on either side of him, and above and below. His view of the world was utterly distorted by thick steel bars. He had one motor tyre in his cell, and a drinking spout. He had been born in the African forest; he had spent more than ten years in the lab. Then there was Jade. For more than seven years she had earned her living by attending up-market birthday parties, and other such social events. When I met her, she was about eight years old and her teeth had been pulled. She was dressed in human clothes. She had been brought, as an "ambassador for her species," to a fund-raising dinner. When we met, I greeted her in chimp style, and after that, from the opposite end of the vary large table, she gave her toothless grin every time I caught her eye. I knew she wanted to come over; I knew she had been disciplined to remain in her place, slurping soft foods. I desperately avoided looking at her, filled with anger that she was thus exploited. At the end of the meal, she was allowed to come and clamp her arms around me and breathe her sadness into my neck. And there was Dick. When I met him he lived in a small zoo cage with a cement floor. He had a female with him, but clearly had little time for her. He sat in the corner endlessly tapping each finger of his left hand in turn with the index finger of his right hand, in time to the rhythmic opening and shutting of his mouth. In the cage next to him, a lone male gorilla endlessly vomited into his hand and reingested the vomit. Dick was the first captive chimpanzee to whom I made a commitment—I would work to try to better his condition, and that of countless other captive apes around the world.

Since I met Dick in 1956, ethical concerns about our treatment of animals are surfacing everywhere; there are, for example, groups of physicians, surgeons, psychologists, veterinarians—and lawyers—who protest abuse of animals and lobby for change. There are more than seven thousand different animal rights/welfare groups in the United States alone, and there are increasing numbers of people speaking out against intensive or factory farming of food animals, against trapping, hunting, exotic animals in circuses, movies and advertising, puppy mills—and on and on. And there is growing concern for animal welfare in all parts of the world, including the developing world.

The trouble is, while those guilty of cruelty may be prosecuted, often successfully—Steve has saved hundreds of animals' lives in the courtrooms—in the legal sense, animals are regarded as "things," as mere objects that can be bought, sold, discarded, or destroyed at an owner's whim. Only when animals can be regarded as "persons" in the eyes of the law will it be possible to give teeth to the often-fuzzy laws protecting animals from abuse. *Rattling the Cage* explains how legal rights for animals can help to stop so much of the abuse that, today, goes unpunished.

As Steve hints in his last chapter, this book represents a first step toward seeking legal rights for other animal species, rights modified in appropriate ways for different kinds of animals. Chimpanzees along with bonobos are our closest living relatives, differing from us in structure of DNA by only just over 1 percent. This makes these apes "our sibling species"—thus it is fitting that they should be the first to acquire rights, as surely they will—in the eyes of the law. In 1996, Steve and I made a presentation to the Senior Lawyers' Division of the American Bar Association, explaining just why justice demands that we extend fundamental legal rights to chimpanzees. This book makes the same point, in huge detail and in clear language, so that lawyers and judges and law professors—indeed everyone everywhere—will be able to follow the argument. So that in the end the machinery of the law can be changed in favour of the great apes.

It will be too late for JoJo, Jade, and Dick—they are gone. Yet still I think of them, and I feel deep shame; shame that we, with our more sophisticated intellect, with our greater capacity for understanding and compassion, deprived them of freedom, stole from them the dim greens and browns; the soft gray light of that African forest, the peace of afternoon; when the sun flecks through the canopy and small creatures rustle and flit and creep among the leaves. Deprived them of the freedom to choose, each day, how they would spend their time, and where and with whom. Deprived them of the sounds of nature, the gurgling of streams, murmuring wind in the branches, of chimpanzee calls that ring out so clear, and rise up through the tree tops and drift away in the hills. Deprived them of their comforts, the soft leafy floor of the forest, the springy, leafy branches from which sleeping nests can be made. But it is not too late for hundreds of others who are, as this book goes to press, languishing in man-made prisons.

Steve Wise has marshaled the facts and presented them at a crucial point in Western history—at the end of a millennium. Let us hope that as we enter the twenty-first century, the new and more enlightened attitude concerning our own moral relationship with the rest of the animal kingdom will be reflected in appropriate changes in the legal system. Certainly this book *Rattling the Cage*, will give the process a mighty shove. Thank you, Steve.

ACKNOWLEDGMENTS

Anyone who writes a book that tries to sweep across 4,000 years then burrow deeply into the minds of two species of nonhuman animals requires a lot of help. That is why I owe a debt to the law professors, lawyers, judges, and philosophers who were kind enough to review portions of this book, or earlier articles, or to debate issues that helped me to clarify my thinking: Taimie Bryant, David Favre, Thomas G. Kelch, Sarah Luick, Richard Posner, Stuart Shanker, Peter Singer, Eliot Sober, L.W. Sumner, Paul Waldau, Carl Wellman, and David Wolfson. I warmly thank Daniel Coquillette, who put me on speaking terms with legal history almost 25 years ago and gave me a model for how a great teacher teaches while he was doing it.

My debt is enormous to the many scientists from whom it has been my rare privilege to learn. Despite having never met a lawyer of the species, jus animal, and sometimes suspicious of the entire genus, they quickly realized that I wanted to learn as much as possible about the magnificent apes with whom they share their lives or study with such interest and empathy. Some opened their laboratories, read drafts of chapters on the minds of chimpanzees and bonobos, good-naturedly put up with pestering for the results of their latest research, and welcomed me as friend: Sally Boysen, Roger Fouts, Sue Savage-Rumbaugh and Duane Rumbaugh. Others fortunate enough not to have to endure my visits to their laboratories sent me reprints of their articles, patiently answered my unending questions about the cognition of chimpanzees, bonobos, and human children, and gently pointed out my most egregious errors: Juan Carlos Gómez, Donald R. Griffin, Marc D. Hauser, Michael A. Huffman, Evangeline Lillard, Tetsuro Matsuzawa, David Premack, Daniel Shillito, Michael Tomasello, Frans de Waal, and Richard Wrangham. The master himself, Morris Goodman, enthusiastically gave me a primer on primate taxonomy. Charles Sedgwick, D.V.M. kindly relayed the story of the accidental death of Jamal at the Chimpanzees of Mahale Mountains exhibit at the Los Angeles Zoo, while Deborah Blum, who knows every primatologist in the world, graciously reviewed several chapters with her sharp journalist's eye. My friend, Jane Goodall, whose pathbreaking work and immunity to jet lag has allowed the wonders of the chimpanzee to be revealed to humankind, penned an overly-generous Foreword. Thanks, Jane, whichever continent you might be on. And a special thanks to Rachel Weiss, who made me understand what it was like for Jerom to wither and perish at the Yerkes Regional Primate Research Center; at least he did not die alone or unloved.

I am fortunate to have Merloyd Lawrence as my friend and editor at Perseus Books. Her personal commitment to attaining rights for chimpanzees and bonobos rivals my own. As an editor, she cuts without wounding. I thank my agent, Charles Everitt, experienced in every possible aspect of book publishing, who thought he saw the potential for a good book then tirelessly doubled as a second editor to try to realize it.

My greatest debt is, of course, to my wife and law partner, Debi Slater-Wise. At work she is the scourge of anyone who dares try harm a nonhuman animal and a merciless editor of her husband's work. At home hers is the hand that guides the family rudder, steering three children, Roma and those twin soldiers of entropy, Siena and Christopher, towards secure, happy, and productive lives. Nunc scio quid sit amor.

Finally, my deepest thanks to Fred Gates, who made everything possible.

1

The Problem with Being a Thing

It is difficult, to handle simply as property, a creature possessing human passions and human feelings . . . while on the other hand, the absolute necessity of dealing with property as a thing, greatly embarrasses a man in any attempt to treat it as a person.

—Frederick Law Olmsted, traveling in the American South before the Civil War[1]

Jerom's Story

Jerom died on February 13, 1996, ten days shy of his fourteenth birthday. The teenager was dull, bloated, depressed, sapped, anemic, and plagued by diarrhea. He had not played in fresh air for eleven years. As a thirty-month-old infant, he had been intentionally infected with HIV virus SF2. At the age of four, he had been infected with another HIV strain, LAV-1. A month short of five, he was infected with yet a third strain, NDK. Throughout the Iran-Contra hearings, almost to the brink of the Gulf War, he sat in the small, windowless, cinder-block Infectious Disease Building. Then he was moved a short distance to a large, windowless, gray concrete box, one of eleven bleak steel-and-concrete cells 9 feet by 11 feet by 8.5 feet. Throughout the war and into Bill Clinton's campaign for a second term as president, he languished in his cell. This was the Chimpanzee Infectious Disease Building. It stood in the Yerkes Regional Primate Research Center near grassy tree-lined Emory University, minutes from the bustle of downtown Atlanta, Georgia.

Entrance to the chimpanzee cell room was through a tiny, cramped, and dirty anteroom bursting with supplies from ceiling to floor. Inside, five cells lined the left wall of the cell room, six lined the right. The front and ceiling of each cell were a checkerboard of steel bars, criss-crossed in three-inch squares. The rear wall was the same gray concrete. A sliding door was set into the eight-inch-thick concrete side walls. Each door was punctured by a one-half-inch hole, through which a chimpanzee could catch glimpses of his neighbors. Each cell was flushed by a red rubber fire hose twice a day and was regularly scrubbed with deck brushes and disinfected with chemicals. Incandescent bulbs hanging from the dropped ceiling provided the only light. Sometimes the cold overstrained the box's inadequate heating units, and the temperature would sink below 50°F.

Although Jerom lived alone in his cell for the last four months of his life, others were nearby. Twelve other chimpanzees—Buster, Manuel, Arctica, Betsie, Joye, Sara, Nathan, Marc, Jonah, Roberta, Hallie, and Tika—filled the bleak cells, living in twos and threes, each with access to two of the cells. But none of them had any regular sense of changes in weather or the turn of the seasons. None of them knew whether it was day or night. Each slowly rotted in that humid and sunless gray concrete box. Nearly all had been intentionally infected with HIV. Just five months before Jerom died of AIDS born of an amalgam of two of the three HIV strains injected into his blood, Nathan was injected with 40 ml of Jerom's HIV-infested blood.[2] Nathan's level of CD4 cells, the white blood cells that HIV destroys, has plummeted. He will probably sicken and die.

Sales Tax for Loulis

The biologist Vincent Sarich has pointed out that from the standpoint of immunology, humans and chimpanzees are as similar as "two subspecies of gophers living on opposite sides of the Colorado River."[3] Rachel Weiss, a young Yerkes "care-tech" who watched Nathan being injected with Jerom's dirty blood and saw Jerom himself waste away and die, wrote about what she had seen. During the time she cared for the chimpanzees of the Yerkes Chimpanzee Infectious Disease Building, Rachel learned firsthand that chimpanzees possess "passions" and "feelings" that, if not human, are certainly humanlike. It made them no less "difficult to handle simply as property." She stopped thinking of them as "property" and resigned from Yerkes shortly after Jerom's death.

Seventeen years before Jerom's death, the primatologist Roger Fouts encountered Loulis staring at him through the bars of another Yerkes cage. Loulis's mother was huddled in a corner. Four metal bolts jutted from her head. Fouts doubted that the brain research she had endured allowed her even to know that Loulis was her son. He plucked up the ten-month-old, signed the necessary loan papers, then drove Loulis halfway across the United States to his adopted mother.

Washoe was a signing chimpanzee who lived on an island in a pond at the Institute for Primate Studies in Norman, Oklahoma. Loulis did not want to sleep in Washoe's arms that first night and curled up instead on a metal bench. At four o'clock in the morning, Washoe suddenly awakened and loudly signed *"Come, baby."* The sound jerked Loulis awake, and he jumped into Washoe's arms.[4] Within eight days, he had learned his first sign. Eight weeks later, he was signing to humans and to the other chimpanzees in Washoe's family. In five months, Loulis, by now an accepted family member, was using combinations of signs. At the end of five years, he was regularly using fifty-one signs; he had initiated thousands of chimpanzee conversations and had participated in thousands more. He had learned everything he knew from the other chimpanzees, for no human ever signed to him.

As years passed, Fouts realized that Yerkes could call in its loan and put Loulis to the knife, as his mother had been. When Loulis was seventeen years old, Fouts sought to buy him outright. Yerkes agreed to sell for $10,000, which Fouts didn't have. After strenuous efforts, he raised that amount. But at the last second, a hitch developed. Ten thousand dollars was Loulis's purchase price. As if Yerkes were selling Fouts a desk or chair, Fouts was charged another 7.5 percent in Georgia sales tax.[5]

The scientists who injected Jerom and Nathan kept the baker's dozen chimps imprisoned in a dungeon, and invaded the brain of Loulis's mother and the administrators who collected sales tax for Loulis believed that chimpanzees are things. But they didn't know why. Rachel Weiss and Roger Fouts show that we can come to believe—as they do—that chimpanzees are persons and not just things.

DEMOLISHING A WALL

For four thousand years, a thick and impenetrable legal wall has separated all human from all nonhuman animals. On one side, even the most trivial interests of a single species—ours—are jealously guarded. We have assigned ourselves, alone among the million animal species, the status of "legal persons." On the other side of that wall lies the legal refuse of an entire kingdom, not just chimpanzees and bonobos but also gorillas, orangutans, and monkeys, dogs, elephants, and dolphins. They are "legal things." Their most basic and fundamental interests—their pains, their lives, their freedoms—are intentionally ignored, often maliciously trampled, and routinely abused. Ancient philosophers claimed that all nonhuman animals had been designed and placed on this earth just for human beings. Ancient jurists declared that law had been created just for human beings. Although philosophy and science have long since recanted, the law has not.

This book demands legal personhood for chimpanzees and bonobos. Legal personhood establishes one's legal right to be "recognized as a potential bearer of legal rights."[6] That is why the Universal Declaration of Human Rights, the International Covenant on Civil and Political Rights, and the American Convention on Human Rights nearly identically state that "[e]veryone has the right to recognition everywhere as a person before the law."[7] Intended to prevent a recurrence of one of the worst excesses of Nazi law, this guarantee is "often deemed to be rather trivial and self-evident"[8] because no state today denies legal personhood to human beings. But its importance cannot be overemphasized. Without legal personhood, one is invisible to civil law. One has no civil rights. One might as well be dead.

Throngs of Romans scoot past the gaping Coliseum every day without giving it a glance. Athenians rarely squint up at their

Parthenon perched high on its Acropolis. In the same way, when we encounter this legal wall, it is so tall, its stones are so thick, and it has been standing for so long that we do not see it. Even after litigating for many years on behalf of nonhuman animals, I did not see it. I saved a handful from death or misery, but for most, there was nothing I could do. I was powerless to represent them directly. They were things, not persons, ignored by judges. But I was butting into something. Finally I saw that wall.

In Chapters 2 through 4, we will see how it was built by the Babylonians four thousand years ago, then strengthened by the Israelites, Greeks, and Romans, and buttressed again by early Christians and medieval Europeans. As one might expect, its mortar is now cracked and stones are missing. It may appear firm and sturdy, but its intellectual foundations are so unprincipled and arbitrary, so unfair and unjust, that it is crumbling. It has some years left, but it is so weak that one good book could topple it. This is meant to be that book.

In Chapters 5, 6, and 7, I hope to convince you that equality and liberty, the two most powerful legal principles and values of which Western law can boast, demand the destruction of that wall. But there are about 1 million species of animals. Many of them, say, beetles and ants, should never have these rights. So the wall must be rebuilt. But how? In Chapter 8, I will show you that the hallmark of the common law, which is the judge-made law of English-speaking peoples, is flexibility. It abhors thick high legal walls, except when they bulwark such fundamental interests as bodily integrity and bodily liberty, and prefers sturdy dividers that can be dismantled and re-erected as new discoveries, morality, and public policy dictate.

WHY CHIMPANZEES AND BONOBOS?

Chimpanzees and bonobos (sometimes referred to as "pygmy chimpanzees") are kidnapped for use as biomedical research subjects or as pets or in entertainment. They are massacred for their meat to feed "the growing fad for 'bush meat' on the tables of the elite in Cameroon, Gabon, the Congo, the Central African Republic, and other countries," so that their hands, feet, and skulls can be displayed as trophies, and for their babies.[9] Thousands are jailed around the world in biomedical research institutions like Yerkes or are imprisoned in decrepit roadside zoos or chained alone and lonely in private dwellings. When the last century turned, there were 5 million wild chimpanzees in Africa.[10] We don't know the number of bonobos because they weren't then considered a species separate from chimpanzees. But it was probably about half a million. By 1998, only 200,000 chimpanzees remained, perhaps as few as 120,000, and maybe 20,000 bonobos.[11] One of the world's most prominent bonobo experts, Takayoshi Kano, believes that less than 10,000 bonobos may have survived.[12] Thousands of chimpanzees and bonobos are slaughtered every year. They are nearing annihilation.[13]

In Chapters 9 and 10, you will get a close look at the kinds of creatures these apes are and how similar their genes and brain structures are to ours. You will learn about the scientific revolt that has broken out as an increasing number of scientists demand they be tucked into the genus *Homo* with us. We will peel back the layers of their minds and try to understand what is known about how they feel and what they think; why they are conscious and self-conscious; how they understand cause and effect, relationships among objects, and even relationships among relationships; how they use and make tools; how they can live in societies so complex and fluid that they have been dubbed "Machiavellian"; how they deceive and empathize, count simple numbers and add fractions, treat their illnesses with medicinal plants, communicate with symbols, understand English and use sign or lexigram languages, and how they might know what others think. We will compare what we think we know about their minds with what we think we know about ours.

I didn't choose to describe the plights of Jerom and Nathan and the rest of the Yerkes chimpanzees because they are not the worst known examples of legal chimpanzee abuse. That dubious prize probably goes to the notorious SEMA, Inc., renamed Diagnon, located in Rockville, Maryland. Sometime in 1986, a nauseated employee tipped off the True Friends, a band of animal-rights activists, who broke into the lab and videotaped what was happening inside. AIDS-infected baby chimpanzees were housed alone in what SEMA called "isolettes," metal cubes 40 inches high, 31 inches deep, and 26 inches wide, each of which contained a small window. Inside, the babies rocked and rocked as would the emotionally starved or the mentally ill.[14]

I hope you will conclude, as I do in Chapter 11, that justice entitles chimpanzees and bonobos to legal personhood and to the fundamental legal rights of bodily integrity and bodily liberty—now. Kidnapping them, selling them, imprisoning them, and vivisecting them must stop—now. Their abuse and their murder must be forbidden for what they are: genocide.

2

Trapped in a Universe That No Longer Exists

A Peculiar Universe

The ocean tides were designed to move our ships in and out of ports. Horses and oxen exist just to work our fields. Apes and parrots were produced to entertain us. Pigs were created for us to eat. Slaves live for the sake of their masters. The human races were placed on separate continents so they would not mix. Nature has marked Chinese as inferior to whites. Women are made for men. Blacks lie so far below whites on the scale of created beings that they have no rights that whites are bound to respect. Each of these claims has been made.

More than 2,300 years ago, Aristotle wrote that men were by nature superior to women and that slaves lived for the sake of their masters.[1] A century later, the Greek Stoic Chrysippus claimed that horses and oxen existed so they could labor for us and that "as for the pig, that most appetizing of delicacies, it was created for no other purpose than slaughter, and god, in furnishing our cuisine, mixed soul in with its flesh like salt."[2] Unfortunately, most of these opinions cannot be ascribed to the Ancients. It was said in 1582 that apes and parrots were put on earth just to make us laugh.[3] The Abbé Pluche instructed us on the purpose of the tides in 1732.[4] Half a decade before the American Civil War, the California Supreme Court barred Chinese witnesses from testifying against whites in court because they were a race "whom nature has marked as inferior and who are incapable of progress or intellectual development beyond a certain point."[5] Two years later, the U.S. Supreme Court wrote in the infamous *Dred Scott* case that at the time of the American Revolution, blacks were thought to exist so far below whites in the scale of created beings that they had no rights that whites were bound to respect.[6] In 1965, a Virginia judge upheld a statute that forbade marriages between people of different races because "Almighty God created the races white, black, yellow, malay, and red, and he placed them on separate continents . . . The fact that He separated the races shows that he did not intend for the races to mix."[7]

These beliefs about the purposes of horses and oxen, women and races, ocean tides and pigs, parrots, slaves, and apes may appear to be unconnected. But they are not. Each believer hears the universe tick in a particular unchanging way. The ancient Greeks heard one sound. Medieval Christians heard the same. But virtually no modern scientist or philosopher does. No evidence supports the existence of this peculiar universe. Yet the belief that nonhuman animals are somehow made for us lies at the root of what the law says we can do to them today. How did nonhuman animals become trapped in this obsolete universe?

"MEN CAN MAKE USE OF THE BEASTS FOR THEIR OWN PURPOSES WITHOUT INJUSTICE"

Archaeologists tell us that humans first exploited nonhuman animals thousands of years before they knew how to record it. How they justified this, whether they felt they had to, or whether they had the mental capacities to understand what they were doing is unknown. But we do know that thousands of years ago our use of nonhuman animals came to be seen, as everything came to be seen, as part of the sacred and thereby subject to the will of the gods. Thus, in blind Homer's *Iliad*, "the will of Zeus was done" the moment that Agamemnon and Achilles began to quarrel.[8] Nothing happened by accident.[9]

The Greek poet Hesiod, an eighth-century B.C. contemporary of Homer, sang that animals devoured each other because they had not been given the sense of right and wrong. Zeus had given the law of justice to humans alone.[10] This world began to shift thanks to a trio of early- to mid-sixth-century B.C. Milesian philosophers, Thales, Anaximander, and Anaximenes, who made most of the Nobel Prizes and almost all of Western technology possible. They invented science and philosophy.[11] And so the first faint glimmerings of a universe that danced not on divine strings but to the tune of the cause and effect of physical laws began to emerge.

Our story begins to get interesting, as many stories do, when we encounter a second trio of Greek philosophers, Socrates, Plato, and Aristotle, who often passed ideas from one to the next like Tinker to Evers to Chance.[12] Hardly any Greeks of their time thought anything happened by accident.[13] Socrates' friend Xenophon tells us that the old philosopher believed that nonhuman animals existed just for us; even the sun crossed the heavens for our sake.[14] Plato invented a "principle of plenitude," by which every conceivable form that could exist in the universe did.[15] Aristotle envisioned all nature as having been designed like a ladder.

These ideas fused into the "Great Chain of Being." It became, in Professor Owen Lovejoy's words, "one of the half-dozen most potent and persistent presuppositions in Western thought. It was, until not much more than a century ago, probably the most widely familiar conception of the general *scheme* of things, of the constitutive pattern of the universe."[16] It worked like this: An infinite number of finely graded forms were arranged along the ladder. Creatures who were barely alive occupied the lowest rungs. Above them ranged the sentient beings, conscious, perhaps able to experience. Rational beings inhabited higher rungs, with the most rational human beings on the highest rungs that could be assigned to beings with physical bodies. Above them, looming incredibly high, dwelled an infinite number of spiritual and divine beings.[17] The lower-rung dwellers were designed to serve the higher-rung dwellers, for they generated more heat, had souls made from better stuff, and were more perfect.[18]

Plato identified three kinds of souls. Animals and plants had the primitive one that lay below our navels. The mortal soul that was located in the chest and belly of animals listened to reason and passion. It was irrational, though it allowed nonhuman animals and slaves—those humans who were most like beasts—the ability to comprehend simple things like "my master is coming toward me." The immortal soul, however, resided only in human heads. This was the seat of reason and our connection to the divine.[19] But not all humans had it. Children lacked it. Slaves lacked it. Certainly, nonhuman animals lacked it. Some adults received it late, others never at all.[20] Aristotle thought there were five souls. Plants and animals shared the simple nutritive soul. All animals shared the appetitive, locomotive, and sensory souls. But only humans possessed the rational soul that allowed them to reason, listen to reason, think, and believe. And virtually all of them had it.[21]

But classical Greece was not revolutionary Philadelphia and Aristotle was no Thomas Jefferson. All "men" were most emphatically *not* created equal. Aristotle thought that one portion of the rational soul was actually used to reason. Greek men, who had it in full, occupied the topmost possible rung. Greek women were a bit colder, slightly deficient in reason and in the sense of justice, and so a little less perfect.[22] Children had it, though they couldn't reason even as well as their mothers.[23] But another part of the rational soul only appreciated how others reasoned.[24] The natural slave (generally non-Greeks) had this part. He had some sense of reason, in that he could appreciate how his master reasoned. But he could not reason himself.[25] "[T]he slave is a living tool and the tool is a lifeless slave," Aristotle said, and he advised his most famous student, Alexander the Great, "to be a hegemon [leader] to the Greeks and a despot to the barbarians [anyone not Greek], to look after the former as after friends and relatives, and to deal with the latter as with beasts and plants."[26] Justice, law, and even friendship reasoning beings owed only to each other. Considering how the Greeks dealt with "beasts and plants," it was advisable to have others believe that you could reason.

The great thing about being more perfect was that the less perfect always acted "for the sake of" you. In Aristotle's world, rain didn't just fall. It fell "for the sake of" the corn. Corn grew "for the sake of" nonhuman animals. In turn, animals existed for the sake of man, the tame for use and food, the wild, if not all, at least the greater part of them, for food and the provisions of clothing and various instruments. Now if nature makes nothing incomplete, and nothing in vain, the inference must be that she has made all animals for the sake of man.[27]

This was "teleological anthropocentrism."[28] The "teleological" part means that everything in nature was imbued with a purpose. Everything had a goal. "Anthropocentrism" means that the world was designed for us.

The Great Chain of Being prescribed an unchanging and tidy universe in which each form occupied its appropriate, necessary, and permanent place that had been designed just for it in the natural hierarchy.[29] It was, in the words of Harvard paleontologist Stephen Jay Gould "explicitly and vehemently antievolutionary."[30] Nothing progressed, nothing regressed, nothing shifted rungs. No species ever went extinct; none could be created.

Aristotle forged many intellectual molds in science, ethics, taxonomy, politics, psychology, and philosophy. Some were not broken for hundreds, even thousands, of years. One of them was the syllogism. He virtually invented it ("Socrates is a man; all men are mortal; therefore Socrates is mortal"). Whether intentional or not, Aristotle's own place on the Great Chain of Being illustrated a syllogism. It was this: "Greek males occupy the top rung of the Great Chain of Being; I am a Greek male; therefore I occupy the top rung." Over the centuries, it generalized to this: "Only groups to which I belong occupy the top ring; I belong to those groups; therefore I occupy the top rung." It has remained in constant use in determining who has what rights. We'll call it "Aristotle's Axiom," and it is an axiom because no one ever, *ever*, assigns a group to which he or she belongs to any place in a hierarchy of rights other than the top. Mel Brooks nicely summarized Aristotle's Axiom in his movie *The History of the World, Part One:* "It's great to be the king!"

These hierarchies are created in two ways. One group either pushes every other group below by force or threat of force or persuades the others that they belong on the lower rungs. Soldiers like the first way; philosophers, legal writers, taxonomists, and priests prefer the second. The problem for nonhuman animals is that they can neither fight nor write. Well, they can fight a little, and sometimes do very well one-on-one. But they are uniformly terrible at organized warfare against humans, and we are excellent at slaughtering them. That is why until humans learn to fight for them or write for them, nonhuman animals will never have any rights.

Although blinded by teleological anthropocentrism, the Greeks were not blind. They could see that nonhuman animals (and slaves) were not literally "lifeless tools." They were alive. They had senses and could perceive. But Aristotle compared them to "automatic puppets."[31] He said they did not know *that* they perceived or, contradicting Plato, they did not believe *what* they perceived. So while both human and nonhuman animals could *perceive* the sun as being small, only humans could *believe* that it was large. Nonhuman animals could also feel pleasure and pain. They could learn, remember, and experience, for experiences were just a chain of memories.[32] In short, Aristotle denied them the abilities to reason, to possess intellect, thought, and belief.[33] Because they lacked these abilities, they had no emotions, even if they sometimes acted as if they did. They were oblivious to justice and injustice, to good and bad, even to their own welfare.[34] Both nonhuman animals and rebellious slaves could be hunted in a "just war." Aristotle wrote that "the art of war is a natural art of acquisition, for the art of acquisition includes hunting, an art which we ought to practice against wild beasts, and against men who, though intended by nature to be governed, will not submit; for war of such a kind is naturally just."[35]

Shortly after Aristotle died, a third trio of Greek philosophers began to take the stage. Zeno, and later Cleanthes and Chrysippus, taught in the famous "Painted Stoa" on the north side of the Athenian agora. Their "Stoic" philosophy penetrated deeply into Greek, then Roman, thought and exerted a tremendous influence upon ethics, science, and law for more than five hundred years.[36] The Stoics sometimes disagreed with Aristotle. They took the first important steps toward equality by widening some of the rungs on the Great Chain. A natural equality among humans existed, they insisted. Everyone could reason enough to understand natural law. Human masters were not superior to slaves, nor was a husband superior to his wife.[37] Justice was owed to both.

But the Stoics shared Aristotle's belief that the universe operated according to a divine plan and that some had been created to benefit others, for example, "plants for the support of animals, animals for the support and service of man, the world for the benefit of gods and men."[38] And though Chrysippus argued that "[n]o human is a slave by nature," he said that "irrational animals take the place of slaves."[39] In the first century B.C., the Roman Stoic Cicero wrote that

> as Chrysippus cleverly put it, just as a shield-case is made for the sake of a shield and a sheath for the sake of a sword, so everything else except the world was created for the sake of some other thing; thus the corn and fruits produced by the earth were created for the sake of animals, and animals for the sake of man: for example the horse for riding, the ox for ploughing, the dog for hunting and keeping guard.[40]

Stoic fleas existed just to awaken slumbering humans, asses lived to bear human burdens, horses had the purpose of carrying humans, sheep provided human clothing, dogs guarded and protected humans, and mice stimulated human tidiness.[41] The modern way in which we use the word "stoic" to mean "unemotional" and "passionless" honors the Stoic belief that everything existed for the sake of the rational. As the most important part of the universe was the human being, so the most important capacity of the human being was the ability to reason. Irrational nonhuman animals were contemptible, beneath notice, and Stoic writers accordingly ignored them.[42]

Didn't nonhuman animals have even the tiniest drop of sweet reason? Not to Aristotle and not to a Stoic. Nonhuman animals were alive. They could perceive through their senses.[43] They could act on impulse. But they could do almost nothing more. The Stoics denied them many of the few capacities that Aristotle had granted them—every ability to perceive, conceive, reason, remember, believe, even experience—as we would deny those abilities to a computer.[44] Nonhuman animals could know nothing of the future. They could know nothing of the past. The Roman Stoic Seneca wrote that

> [t]he dumb animal grasps what is present by its sense. It is reminded of the past when it encounters something that alerts its senses. Thus the horse is reminded of the road when it is brought to where it starts. But in its stable, it has no memory of it, however often it has

been trodden. As for the third time, the future, that does not concern dumb animals.[45]

The Stoics thought that the ability to "assent" was necessary for beings to know that they perceived, as well as for memory, emotions, beliefs, and reason (though some Stoics later conceded the possibility that nonhuman animals might possess some very limited instinctual, self-conscious understanding of their own bodies that allowed them to know what could harm or help them).[46] Humans could assent. Nonhuman animals, limited to life, sensation, and impulse, could not.[47] Nor could they desire, know good, or learn from experience.[48] Their very voices were merely air struck by an "impulse," whereas human voices were directed by the mind.[49] One of Cicero's characters made it plain that nonhuman animals fared no better in the Stoic world than in the world of Aristotle.

[N]o right exists between man and beast. For Chrysippus well said, that all other things were created for the sake of men and gods, but that these exist for their own mutual fellowship and society, so that men can make use of the beasts for their own purposes without injustice.[50]

They existed apart from the community of reasoning beings. They were not entitled to Stoic justice.[51]

We leave the Greeks with a mention of one final and exceedingly influential Stoic idea. The Greeks had long believed that the workings of the universe betrayed a moral significance.[52] Stoics sought to live as closely as possible to nature, because it embodied universal, immutable, natural justice, and law.[53] As we will see in Chapter 6, natural law and its close cousin, natural rights, insist that the moral is inseparable from the legal. The Stoics, and especially Cicero, fashioned this idea into something close to how we understand it today. Cicero's writings about natural law were so crisp and timeless—"there will not be different laws at Rome as at Athens, or different law now and in the future, but one eternal and unchangeable law will be valid for all nations and at all times"—that they helped catalyze the development of important Western political institutions and law.[54] Nearly two thousand years later, these ideas influenced the American Declaration of Independence, the French Declaration of the Rights of Man, every American federal and state constitution, and numerous other national constitutions and international human rights declarations and treaties.

The Long Reign of the Great Chain

The Great Chain resonated in a Judeo-Christian world suckled on Genesis.[55] God had originally granted humans dominion "over the fish of the sea, and over the fowl of the air, and over every living thing that moveth upon the earth."[56] But the Flood was a turning point, for God had apparently changed His mind. He had forbidden the killing of nonhuman animals for food.[57] Now He drowned every animal not bobbing on the Ark. Rashi, a medieval Jewish scholar, explained why innocent nonhuman animals were drowned along with evil humanity: "[S]ince animals exist for the sake of man, their survival without man would be pointless."[58] After the Flood, God told Noah and his sons that

> the fear of you and the dread of you shall be upon every beast of the earth, and upon every fowl of the air, upon all that moveth upon the earth, and upon all the fishes of the sea; into your hand are they delivered, every moving thing that liveth shall be meat for you; even as the green herb have I given you all things.[59]

The rest, as they say, was history, but history sired by bad philosophy and dreadful science. It remained only for Christianity to absorb Greek teaching. It didn't seem an animal-friendly religion. In the Gospel of Mark, Jesus cast devils from a man into a herd of two thousand swine who ran "violently down a steep place into the sea . . . and were choked in the sea."[60] Saint Paul emphasized that God did not care for oxen.[61] It was Paul who stitched Stoic natural law into the fabric of Christianity.[62]

In the second century, the Catholic church Father Clement used the Great Chain of Being to help explain grades of evil. The lower a being was on the chain, "the least real, least good, least spiritual, most deprived of being, and consequently most evil" it was.[63] In the third century, the Christian theologian Origen tried to develop a Christian doctrine that animals were created for the use of humans.[64] But it was the earthy and practical Saint Augustine of Hippo (Lord, "Give me chastity . . . but not yet") who, in the fifth century, diverted the nearly parallel Christian and Stoic streams into a single flow.[65] With his teacher, Saint Ambrose, Augustine, "took natural law from Cicero, baptised it, and handed it on for preservation in the Church."[66] According to the "order of nature," the sentient animals rank higher than that which lacks sensation, the intelligent rank above the mass without intelligence, and the immortal rank above the mortal.[67] The Sixth Commandment, "Thou shalt not kill," protected only humans.

[S]ome attempt to extend this command even to beasts and cattle, as if it forbade us to take the life from any creature. But if so why not extend it also to the plants, and all that is rooted and nourished by the earth? For though this class of creatures have no sensation, yet they also are said to live, and consequently they can die; and therefore, if violence be done them, can be killed . . . Must we therefore reckon it a breaking of this Commandment, "Thou shalt not kill," to pull a flower? . . . when we say, Thou shalt not kill, we do not understand this of the plants, since they have no sensation, nor of the irrational animals that fly, swim, walk, or creep, since they are dissociated from us by their want of reason, and are therefore by the just appointment of the Creator subjected to us to kill or keep alive for our own uses; if so then it remains that we understand that commandment simply of man.[68]

Not only did Augustine accept Aristotle's five souls, but he dressed Jesus in the dogma of Chrysippus.[69] Augustine thought that mind, intelligence, language, ethics, and understanding lodged only in the rational soul and that existed only in humans.[70] Humans alone could memorize, deliberately recall, imagine, or know whether that what they perceived was true.[71] Nonhuman animals could recognize, remember, even instinctively understand what gave them pleasure and pain.[72] But they could "know" nothing.[73] They had no emotions. They could not reason or assent.[74] Humans and nonhuman animals could equally perceive what their senses transmitted, but only humans were conscious that they were actually perceiving.[75] Not only had Jesus driven devils into swine, but when he cursed the barren fig tree,

Christ himself shows that to refrain from the killing of animals and the destroying of plants is the height of superstition, for judging that there are no common rights between us and the beasts and trees, he sent the devils into a herd of swine and with a curse withered the tree on which he found no fruit.[76]

Unlike Aristotle, Augustine, and later, in the thirteenth century, Saint Thomas Aquinas even argued that wild animals were not entitled to be hunted only in a just war. In a "just war," human "right intention insists that charity and love exist even among enemies. Enemies must be treated as human beings with rights."[77] But nonhuman animals did not rise to the dignity of human enemies.[78] They could therefore be the targets of unrestricted, merciless, perpetual war. The perceived irrationality of nonhuman animals, which had excluded them from Greek justice and Stoic justice, now excluded them from Christian justice.[79] It still does.

"How extremely stupid not to have thought of that!"

Professor Robert Brumbaugh has written that "(f)rom Xenophon through Aristotle through the Stoic school, the preposterous idea of a world designed for human exploitation diffused quite thoroughly into Western common sense."[80] Professor Lovejoy believed it "one of the most curious movements of human imbecility."[81] Preposterous as it may be, imbecilic as it may now appear, it buried itself so deeply into philosophy, science, political science, and finally, the law, that it has proven exceedingly resistant to change.[82] It became a commonplace in the Middle Ages. Aquinas accepted it almost exactly the way Aristotle had proposed it seventeen centuries before.

Now all animals are naturally subject to man. This can be proved in three ways. First, from the order observed by nature. For just as in the generation of things we perceive a certain order of procession from the imperfect (thus matter is for the sake of form; and the imperfect form for the sake of the perfect), so also is there order in the use of natural things. For the imperfect are for the use of the perfect: as the plants make use of the earth for their nourishment, animals make use of plants, and man makes use of both plants and animals.

Therefore it is in keeping with the order of nature, that man should be master over animals. Hence the philosopher [Aristotle] says that the hunting of wild animals is just and natural, because man thereby exercises a natural right. Secondly, this is proved from the order of divine providence which always governs inferior things by the superior. Wherefore, since man, being made to the image of God, is above other animals, these are rightly governed by him . . . [83]

The Renaissance whirlwind dispersed many ancient conceptions of the physical world. Its first squalls blew through astronomy and physics. In 1543, Copernicus said that the sun and not the earth was the center of the universe. Galileo demanded we consider that "[w]e abrogate too much to ourselves if we suppose that the care of us is the adequate work of God, the end beyond which the divine wisdom and power do not extend . . . "[84] By the end of the seventeenth century, the universe was commonly thought to be without boundaries, even populated by creatures who lived on other worlds.[85] At the same time, European explorers were regularly chancing upon vast areas bursting with strange new plants and animals that had adapted to lands in which humans had never lived.[86] The Dutch naturalist Antonie van Leeuwenhoek discovered an unknown microscopic world teeming with uncountable numbers of immensely tiny organisms.[87] To the mind of Descartes,

[i]t is not probable that all things have been created for us in such a manner that God has no other end in creating them . . . Such a supposition would, I think, be very inept in reasoning about physical questions; for we cannot doubt that an infinite of things exist, or did exist though they have now ceased to do so, which have never been beheld or comprehended by man, and have never been of any use to him.[88]

Scientists began to think about physical phenomena as natural processes and to explain them in mechanical terms. Galileo's student Evangelista Toricelli described the physics of a suction pump not in Aristotelian terms of the water finding its "proper place" but as due to the weight of air.[89] Geological evidence steadily accumulated that the earth was vastly older than the few thousand years many had presumed and was discovered to have suffered violent upheavals.[90]

When fossils were discovered, it became obvious that species had lived and died long before human consciousness had awakened. Worse breaches so compromised the Great Chain of Being's beauty, symmetry, and equilibrium that they threatened to collapse the entire system. Valiant rearguard actions ensued. The eighteenth-century taxonomist Linnaeus devised an antievolutionary system of classification.[91] Eventually forced by facts to include some mechanism for natural change, he concluded that existing species merely formed hybrids but clung to the ancient idea that new species could never arise.[92] Others tried to salvage the Great Chain by "temporalizing" it to allow for the movement and progress of species.[93] Thus, the Chain of Being temporarily became an "Escalator of Being," upon which species could ascend.[94] But even this dodge caved in under the weight of the growing mass of contradictory facts.

Standing on the shoulders of Copernicus, Kepler, Galileo, and Newton, whose methods and insights into astronomy, physics, and chemistry made his discoveries possible, Darwin delivered the coup de grâce in *The Origin of Species*.[95] His theory of evolution by natural selection operated through a process of gradual change. Species multiplied either by splitting or evolving into new species when populations became isolated.[96] This idea struck many as simplicity itself. "How extremely stupid not to have thought of that!" the English biologist Thomas Huxley exclaimed.[97]

So the world had been designed not by God, but by Greeks. The universe was not static; it ceaselessly fluctuated. The structure of every creature could be explained as the gradual adaptation of species to change. Creatures were neither "higher" nor "lower." Instead, each adapted to his or her environment. Life had evolved not as rungs on a ladder but as a bristling bush. Biology had no more purpose than did chemistry or physics and didn't need one.[98] Darwin's world had no Design; it needed no Designer.[99] Today, 95 percent of the biologists within the elite National Academy of Sciences claim either to be atheists or agnostics.[100] Because all organisms had descended from a common ancestor, the Great Chain could not possibly exist.[101] The "grand master metaphor (that had) dominated, perverted, and obstructed European efforts to discover man's place in nature" was so thoroughly demolished that its proponents today are confined to a small number of fundamentalist theologies.[102]

Darwin's earthquake rumbled through not just science but also theology, philosophy, sociology, and, inevitably, political science and law. For hundreds of years, the privileged had used the Great Chain against agitators for human rights and "especially against all equalitarian movements."[103] Inequality had for so long been thought to be a cornerstone of nature that demands for equality could be powerfully criticized as subverting the laws of nature and of God.[104] The destruction of the Great Chain of Being kicked open the door to the acceptance of human equality. It also opened the human mind to the idea of the nonhuman mind.

But the Great Chain did not disappear. The heart of the curious and imaginary world of the Ancients that spurred its invention remains beating within the breasts of judges, animating the common law that regulates the modern relationships between human and nonhuman animals. Its intellectual foundations rotted away long ago. Yet, as we will see, it undermines the living common law's most fundamental principles.[105] How could one of the world's greatest revolutions have bypassed the common law? In Chapter 3, I will show how the Great Chain wormed into Roman law, then fixed itself first into English, then into American, common law. Understanding that it is there is the first step toward digging it out, freeing nonhuman animals from their outdated universe, and recognizing the long-delayed fundamental legal rights to which at least some of them have long been entitled.

3

The Legal Thinghood of Nonhuman Animals

THE OX THAT GORED

A yoked ox plows a Near Eastern field in the late afternoon of a spring day three—no, let's say four thousand years ago. Light rain has been falling on and off since mid morning. Thunder rumbles from deep within a huge bank of clouds that has blackened the western sky. It's quitting time. The farmer turns the ox toward home. Suddenly a blinding bolt of lightning sizzles, followed by a terrific clap of thunder. Deafened and surprised, the farmer pitches to the ground. The normally placid ox begins to lumber in panic toward the dirt road that passes the edge of the field. The farmer, after finally regaining his wits, yells to the ox to stop. To no avail. Large raindrops begin to pound the road, raising cloudlets of dust. A passerby sizes up the situation and tries to stop the frightened animal. Nearly blinded by the rain, he fails to see the ox lower his head and is gored in the abdomen. By the time the breathless farmer arrives, the ox is calm and the rain has nearly stopped. But the passerby lies motionless, blood seeping into the mud. The farmer's head sinks into his two brown hands.

What did this tragedy mean for the farmer? What did it mean for the goring ox? What does it mean for the rights of nonhuman animals today? Some of the thousands of clay tablets that have been found record actual ancient Near Eastern lawsuits, but none mentions a goring ox.[1] Neither does the Old Testament. Nevertheless, differing ancient Near Eastern solutions to the theoretical problem of the ox that gored will help us understand the modern law of nonhuman animals and why change is overdue.

"All Law Was Established for Men's Sake"

After reading Chapter 2, we should not be surprised to learn that ancient worlds marinated in the belief that everything in the universe was created for the sake of human beings also hatched the idea that, in the words of the Roman jurist Hermogenianus, "Hominum causa omne jus constitum" ("All law was established for men's sake").[2] But why should we care what an obscure Roman jurist said sixteen or seventeen centuries ago? Because the idea is very much with us today. A respected modern treatise on jurisprudence contains the identical words "Hominum causa omne jus constitum," followed by an explanation: "The law is made for men and allows no fellowship or bonds of obligation between them and the lower animals."[3]

Law—good, mediocre, and bad—tends to survive, borrowed from one age by another. This borrowing of law, whether consciously or unconsciously (as Oliver Wendell Holmes, Jr., the great American judge believed), has long been the primary workaday business of lawmakers.[4] Professor Alan Watson, an expert in comparing the law of different legal systems in different ages, tells us that "to a truly astounding degree the law is rooted in the past."[5] This makes sense. Borrowing law is simpler than constantly beginning anew. It provides continuity and stability. But when we borrow past law, we borrow the past. The law of a modern society often springs from a different time and place, perhaps even from a culture that may have believed in an entirely different cosmology or belief about how the universe works.[6] Legal rules that may have made good sense when fashioned may make little sense when transplanted to a vastly different time, place, and culture. Raised by age to the status of self-evident truths, ancient legal rules mindlessly borrowed may perpetrate ancient injustices that may once have been less unjust because we knew no better. But they may no longer reflect shared values and often constitute little more than evidence for the extraordinary respect that lawmakers have for the past.[7]

Early this century, the philosopher George Santayana famously claimed, "Those who cannot remember the past are condemned to repeat it."[8] Every legal rule has its tangled history. Sometimes that history has nothing to do with whether a borrowed law is just in a new and different context. Holmes explained that often "(s)ome ground of policy is thought of that which seems to explain it and to reconcile it with the present state of things: and then the rule adapts itself to the new reasons which have been found for it, and enters on a new career."[9] But this is not always so. Sometimes a rule does not embark upon a new career at all but becomes an anachronism stubbornly holding out in defiance of modern sensibilities. Whole books about these leftover laws have been written to amuse us. We think it's funny when a law enacted at the turn of the last century still requires someone to walk in front of an automobile with a lantern to warn unwary horsemen of its approach. We may be less amused to learn that other laws demand a belief in God in order to hold public office. Few legal rules are as doddering, or as unjust, as the legal thinghood of every nonhuman animal. Some untangling will be necessary to spur its overdue reconsideration.[10] Like Theseus in the palace of the Minotaur, we will follow its winding thread through the labyrinth of legal history. It will lead us to the most ancient legal systems known.

"The Ox Shall Be Stoned to Death"

The oldest known laws were written in cuneiform on Mesopotamian—specifically Sumerian—clay tablets about 3000 B.C.[11] Law may have existed earlier, but we know nothing of it. The Sumerians were the first to emerge from prehistory, because they invented writing. Mesopotamian judicial and legislative power was concentrated in the hands of a single ruler who enacted the law.[12] But its source was neither the ruler nor the gods but rather a "transcendent primordial force" that outranked the gods.[13] The gods did not so much reveal the law to the ruler as allow him to understand its source and enable him to make laws that harmonized with it.[14] The outstanding Near Eastern exception was the Old Testament law of the Israelites. They believed that law was written by the One God. Not even their king could escape it.[15]

All Near Eastern law, Mesopotamian and Israelite, recognized that humans could own nonhuman animals.[16] Cuneiform tablets refer to the ownership, express or implied, of sheep, donkeys, oxen, asses, pigs, goats, cattle, bees, and dogs. The earliest known "law codes" are the Laws of Ur-Nammu, king of Sumer and Akkad, of the third dynasty of Ur, dating from about 2100 B.C.,[17] the Lipit-Ishtar Law Code, of which only a portion is known, probably dating from the first half of the nineteenth century B.C.,[18] the Laws of Eshnunna, created about 1920 B.C.,[19] and the Laws of Hammurabi, the Babylonian king whose reign probably began about 1728 B.C.[20] These were likely not "law codes" in the modern sense of a compilation of statutes, and we do not know for certain to what degree they actually reflected existing law.[21] Their paragraphs generally began in the style of "If, Given that" and ended with instructions to take the action required.[22] We will focus only upon the small portions of the Laws of Eshunna and the Laws of Hammurabi that concern the goring ox.

The Pentateuch, or the first five books of the Old Testament, contains numerous "laws" in the ancient Near Eastern sense, all traditionally attributed to Moses. Josephus, the first-century Jewish historian, claimed that Moses had invented the word for "law" and was the first legislator.[23] However, neither the Mesopotamians nor the biblical Hebrews actually had a word for "law," and no evidence that Moses even lived exists outside the Pentateuch.[24] Within it, however, can be found seven Near Eastern law codes, four of which are independent.[25] When the Pentateuch emerged in its familiar form around the end of the fifth century B.C., it stitched together a Mosaic law (law attributed to Moses) actually written over perhaps five hundred years that reflected the legal thought of centuries before.[26] The very earliest of these laws are believed to have been written sometime between the fifteenth and thirteenth centuries B.C.[27]

Exodus, the second of the Five Books, contains the Covenant Code. It is the only set of Old Testament laws drawn similarly to the "If, Given that" mode of the earlier Mesopotamian cuneiform law codes.[28] As do the Laws of Eshunna and the Laws of Hammurabi, the Covenant Code meets the problem of the goring ox. Almost certainly this is no coincidence. The overall similarity of style and phraseology between the two Mesopotamian law codes and the Covenant Code about such an unusual subject makes it likely that all drew from the same well, if the Covenant Code did not outright copy much of the Mesopotamian codes. As we will see, Section 53 of the Laws of Eshunna and Exodus 21:32 are especially similar.[29] Here is how the three codes dealt with the problem of the fatally goring ox, translated by the eminent Near Eastern scholar Jacob J. Finkelstein.[30]

The Mesopotamians and Israelites tailored these solutions to the problem of the goring ox to reflect the differing ways in which they thought the world worked. The two cuneiform codes confined themselves to remedying injuries committed by an ox upon humans or their property.[32] The issue of the ox who fatally gored was tucked within sections of the Laws of Hammurabi that prescribed property settlements for damages inflicted by an ox.[33] But the silence of the cuneiform codes to the fate of the goring ox is, like the Holmesian dog (Sherlock, not Oliver Wendell) who failed to bark, deafening.

Why is it that the earlier cuneiform codes, which required the payment of money for wrongs inflicted by oxen, have a twentieth-century ring, whereas the later Covenant Code, with its draconian mandates to stone oxen to death and execute negligent ox owners, sounds primitive? The answer is that the touchstone of Mesopotamian, like modern, law was economics, not religion. Its main concern was protecting property and compensating for damages.[34] Mesopotamian cosmology did not view humans as qualitatively different from the rest of nature but as part of a continuum. Far from believing that the universe was made for humans, Mesopotamians thought that nature was complete before humans arrived. Human society was just one of many societies that existed in the universe.[35] Mesopotamians had not been ordered to tame nature.[36] Instead, humanity had been created merely to relieve gods of onerous physical labors; in sum, our less-than-exalted purpose was feeding and caring for the gods.[37]

Genesis, in contrast, charged the Israelites with subduing nature. We occupied the central place between God and nature.[38] A divine spark smoldered within every human. Created in the image of God, all humans were sacred. Humanity stood apart from, and was superior to, everything else in the natural world.[39] Any destroyer of sacred human life was to be severely punished and his or her trespasses upon the divine were not to be forgiven in this world.[40]

A. The Laws of Eshunna

sec. 53. If an ox has gored another ox and caused its death, the owners of the oxen shall divide between them the sale value of the living ox and the carcass of the dead ox.

sec. 54. If an ox was a habitual gorer, the local authorities having so duly notified its owner, yet he did not keep his ox in check and it then gored a man and caused his death, the owner of the ox shall pay two-thirds of a mina of silver (to the survivors of the victim).

sec. 55. If it gored a slave and caused his death, he shall pay fifteen shekels of silver.

B. The Laws of Hammurabi

sec. 250. If an ox, while walking along the street, gored a person and caused his death, no claims will be allowed in that case.

sec. 251. But if someone's ox was a habitual gorer, the local authority having notified him that it was a habitual gorer, yet he did not have its horns screened nor kept his ox under control, and that ox then gored a free-born man to death, he must pay one-half mina of silver.

sec. 252. If (the victim was) someone's slave, he shall pay one-third mina of silver (to the slave's owner).

C. The Covenant Code of Exodus

21:28. If an ox gores a man or woman to death, the ox shall be stoned to death, its flesh may not be eaten, but the owner of the ox is innocent.

21:29. But if the ox was previously reputed to have had the propensity to gore, its owner having been so warned, yet he did not keep it under control, so that it then killed a man or a woman, the ox shall be stoned to death, and its owner shall be put to death as well.

21:30. Should a ransom be imposed upon him, however, he shall pay as the redemption for his life as much as is assessed upon him.

21:31. Whether it (i.e., the ox) shall have gored a minor (lit. a son or a daughter) this same rule shall apply to him.

21:32. If the ox gores a slave or slavewoman, he must pay thirty shekels of silver to his owner, but the ox shall be stoned to death . . .[31]

> *21:35.* If an ox belonging to one man gores to death the ox of his fellow, they shall sell the live ox and divide the proceeds, and they shall divide the dead one as well.
>
> *21:36.* But if the ox was previously reputed to gore, and its owner had not kept it under control, he shall make good ox for ox, but will keep the dead one for himself.

The sections of the Covenant Code that concerned the ox who gored a human to death were not found among the laws concerning property settlements. They were nested among the laws that addressed crimes by one human against another.[41] Uniquely among the ancient law codes, Israelite law demanded capital punishment, and not merely money compensation, for murder.[42] Because humans were made in the image of God, the killing of a human by man or beast demanded the forfeiture of the killer's life.[43] Unlike Mesopotamian law, Israelite law made human life incommensurable with property of any kind. No death penalty was imposed for a property crime. But no amount of money could compensate for the taking of a human life.[44] In Finkelstein's words, the Covenant Code:

[t]he real crime of the ox is that by killing a human being—whether out of viciousness or by an involuntary motion—it has committed a *de facto* insurrection against the hierarchic order established by Creation: Man was designated by God "to rule over the fish of the sea, the fowl of the skies, the cattle, the earth, and all creatures that roam over the earth." . . . It was not merely that wrongs against the person are of greater gravity than wrongs against property. It is rather that the two realms belong to utterly different mental sets. Different scales are used to weigh the two wrongs, and the correlative measures prescribed are of two distinct qualitative orders.[45]

The ox killing of a human tore the fabric of the universe in a way that was different from any other. "A beast that kills a man destroys the image of God and must give a reckoning for it."[46] This "reckoning" mended that tear.

What was to become then of our hapless farmer? If he was lucky enough to be a Mesopotamian, it was time to dig up the family silver. But, if he was an Israelite, he had to determine, quickly and accurately, whether the unfortunate samaritan had been a slave. If so, like his Mesopotamian counterpart, he might buy his way out.[47] But if the dead man had been free and the goring ox had a reputation for ill-temper, he had to predict the chances of having a ransom imposed, as allowed by Exodus 21:30. If he didn't like those chances, he couldn't just go home to the missus ("I *told* thee to rid us of that nasty beast; *everybody* said so!"). It was pack the camel and head south of Beersheba or north of Dan, or perhaps east toward Persia, and fast, for Exodus 21:29 meant that one who allowed his ox of questionable reputation to kill a human was guilty of a homicide punishable by death, presumably carried out in an unpleasant manner.[48]

What was to become of the ox? If he was lucky enough to be a Mesopotamian ox, maybe nothing at all. But even execution in the usual way was not reckoning enough for the Israelite ox. Much later, the Mishnah, a law code created between the second century B.C. and the second century A.D., ordered that fatally goring oxen, along with domestic nonhuman animals used for bestial purposes and such wild human-killers as wolves, lions, bears, leopards, panthers, and serpents be "tried" before the Lesser Sanhedrin, a tribunal of twenty-three judges, then put to death.[49] But the Covenant Code demanded that the ox be killed by *stoning*.

Why stoning? Oxen were the most important animals in the Israelite economy. They were used for plowing and treading out corn, for draft and as beasts of burden, for food and religious sacrifices. The Israelites were not an especially inhumane people. Other parts of their law codes prohibited such cruelties as muzzling oxen while threshing corn and working them seven days a week.[50] The price of an ox often equaled that of a human slave, and his destruction could ruin any small owner lucky enough to be dunned for ransom.[51] No portion of the ox's body could ever be used. Why waste a perfectly good ox carcass, even if he had been ill-tempered? Something unusual was going on.

Death by stoning "in the biblical tradition and elsewhere in the ancient Near East, [wa]s reserved for crimes of a special character," such as the worship of foreign gods, child sacrifice, sorcery, blasphemy, and violation of the sabbath. These were seen to "'offend' the corporate community or were believed to compromise the most cherished values to the degree that the commission of the offense places the community itself in jeopardy."[52] They were "thought to strike at the moral and religious fibers which the community as a whole sees as defining its essence and integrity. Such crimes, in other words, amount to insurrections against the cosmic order itself."[53] Even the ox who killed a slave was stoned to death, for the slave was, after all, a human being.[54] The ox had to be obliterated from the consciousness of the community.[55]

This human-centered and religious cosmology and law of the Pentateuch swept aside the utilitarian and secular cosmology and law of Mesopotamia. When it merged with a Roman law threaded with Greek and Stoic ideas, it was to affect the legal thinghood of nonhuman animals profoundly.

GREEKS AND ROMANS

Little is known about how Greek law treated the ownership and acquisition of animals before 500 B.C.[56] But the writer Edith Hamilton's description of the status of human slaves in ancient Athens could double for animals as well. No one "ever gave a thought to slaves, no more in the West than in the East. Everywhere the way of life depended on them. One cannot say they were accepted as such, for there was no acceptance. Everyone used them; no one paid attention to them."[57] In the *Iliad* and the *Odyssey*, Homer treated "the ox almost as a unit of currency," with a cauldron valued at twelve oxen and a woman at four.[58] We know that lawsuits concerning domesticated animals were brought in ancient Athens, including one famous suit involving domesticated peafowl, and that animals were treated as legal things.[59] We think that the Athenians even held murder trials of animals and inanimate objects.[60] In contrast to the Mesopotamian "law codes," the early Greek codes were true law.[61] More important, they recognized rule by law, not as power but as justice.[62]

The outlines, structure, and details of Roman law have all had a dramatic impact upon much of Western law, but especially upon the law of property.[63] We instinctively think about law the way the Romans did. If we try to think about law any other way, we think we're thinking about something else. From the outset, the distinction between a person and a thing was critical. Gaius, a second-century Roman jurist, was the first to divide the law into the modern categories of persons, things, and actions.[64] A Roman "person" had rights, whereas a Roman "thing" was the object of the rights of a person.[65] All the beings that the Romans thought, at some time, lacked free will—women, children, slaves, the insane, and animals—were, at some time, or at all times, classified as things.[66] The legal right of a human to deprive animals (both wild and domesticated) of their lives and natural liberties was thought to be so natural and was so ingrained in Roman thought that it was always assumed and never justified.[67]

"A Natural State of Liberty"

The Romans wove their philosophy and theology deeply into law. Roman law was mostly legislation, and for many centuries, this legislation was based upon a natural law philosophy borrowed from the Stoics. The Romans called it the *jus naturale* and believed that it was derived from universal principles of natural justice and natural equity, not sheer power, and that it could be discovered through the use of reason.

In 533, two monumental works of Roman jurisprudence appeared, the Emperor Justinian's *Digest* and *Institutes*. The *Digest* was intended to be as free as possible from vagueness, contradictions, and incompatibilities with Christian teachings.[68] The *Digest* said that the "*jus naturale* is that which nature has taught to all animals; for it is not a law specific to mankind but is common to all animals—land animals, sea animals, and the birds as well."[69] Both the *Institutes* and *Digest* said that both wild animals and humans were born free by the law of nature and that human slavery was contrary to the law of nature and existed only by the *jus gentium*, which was the law that humans devised for themselves through their use of reason.[70]

This contradiction between the *jus naturale* and the *jus gentium* was the only one ever known to exist in Roman law. Professor David Brion Davis, an astute historian of human slavery, recognized that

> the institution of slavery has always been a source of conceptual contradiction . . . [that partly] arose from the impossibility of transforming a conscious being into a totally dependent and nonessential consciousness—one whose essence is to be the mere instrument and confirmation of an owner's will.[71]

This tension between the natural law, which abhorred slavery, and the law of nations, which often sanctioned it, was to last two thousand years.[72] It still exists in the law of nonhuman animals.

This link between human and nonhuman slavery was made early on. The Lex Aquilia, in force during the Roman republic (very roughly, the time before Christ), said that "[i]f anyone kills unlawfully a slave or servant-girl belonging to someone else or a four-footed beast of the class of cattle, let him be condemned to pay the owner the highest value that the property had attained in the preceding year."[73] At about the same time, the Edict of the Curule Aediles held sellers of slaves and domesticated animals strictly liable to buyers to whom defects had not been specifically declared.[74] The offspring of both nonhuman animals and slaves could be acquired through "accession," through which one's property naturally increased through birth.[75] Nineteenth-century American slavery was to draw some of its strength from these Judeo-Christian and Roman links between slaves and animals.[76]

If the denial of liberty to all animals, human and nonhuman, violated a natural law that applied equally to both, shouldn't the keeping of both in slavery have not equally violated natural law?[77] Alan Watson, whose point was that Justinian's texts did not consider human slavery immoral, has suggested that Roman jurists either didn't notice the contradictions or didn't consider the matter important enough to discuss.[78] Both seem unlikely since this was the only known instance of conflict between the *jus gentium* and *jus naturale* and because the Stoics had been loudly proclaiming human equality for centuries. Perhaps the jurists noticed but allowed the matter to remain unexamined because it contradicted their cosmology and religion. Better to accept the enslavement of those who were different than to scrutinize one's cherished beliefs. Or perhaps contradictions simply concerned them less than they concern us today.[79]

Later Roman law expressly incorporated the Stoic natural law idea that things were created for human use.[80] Both the *Digest* and the *Institutes* assumed that wild animals lived in a "natural state of freedom"[81] or "natural state of liberty."[82] Wild animals fell into the category of *res nullius*, things that belonged to no one until a human being "occupied" them, a category animals shared with unoccupied lands, abandoned property, precious stones, hidden treasure, and the property of an enemy captured in war.[83] So long as nonhuman animals remained unoccupied, they remained un-owned.[84]

The Roman *jus naturale* was no passing phase. As we will see, it merged with Saint Augustine's blend of Stoic philosophy and biblical cosmology to pass largely unchanged into the common law, first of England, then of the United States. It formed the loom upon which the Roman lawyers and jurisconsults wove their civil law and later generations their powerful idea of natural rights, upon which the American Declaration of Independence, every American constitution, many foreign constitutions, and much of modern international human rights law rests.[85] Ironically, it created both a powerful force for human liberty and animal slavery.

4

Border Crossings

THE HISTORICAL GULF BETWEEN HUMAN AND NONHUMAN ANIMALS

In 1522, the patience of French peasants with the barley-eating rats around Autun expired. A formal petition was drawn, charging the rats with crop gobbling. The magistrate presented it to the bishops' vicar and the gears of the Autun ecclesiastical court engaged. The judges summonsed the rats and solemnly appointed Bartholemy Chassenee as their defense counsel. Chassenee thought that animals should be tried with all the formalities and process that both ritual and complete fairness demanded.[1] So when the rats failed to appear at the appointed time, Chassenee argued that they not only inhabited a large number of villages but were scattered throughout the countryside. This raised a serious legal problem. How could a single summons notify village rats and country rats alike that proceedings had been commenced and where and when they should appear to answer the charges?

Another citation was issued that allowed the rats more time to put their affairs in order and complete the arduous journey (for a rat) to Autun. To ensure that word reached the ears of every rat, the court ordered the summons trumpeted from the pulpit of every church in every parish in which the rats were thought to live. Despite these exertions, not a rat appeared. Unruffled, Chassenée patiently explained to the judges just how long it took for rats to travel such distances. Obviously they needed more time. More time they received, but the rats still failed to attend.

The complaining citizens, Chassenée now charged, had done nothing to restrain their cats, who, everyone knew, relished nothing better than the defendants. His clients also believed, and reasonably so, that the complainants might be tempted to take the law into their own hands if they chanced to meet a defendant outside the court. The law allowed summonsees to ignore summonses that endangered them. The court fixed this problem. "Restrain your cats!" the judges warned the peasants, and presumably they did. But when the rats failed to appear a fourth time, the court, out of patience, defaulted them.[2]

Fierce anathemas and maledictions or, worse for the God-fearing, excommunications, were regularly pronounced throughout Europe, and occasionally in the Americas, from the end of the Dark Ages into the last century upon all manner of creatures. French, Italian, and Swiss weevils and locusts infested vineyards, French and German sparrows chattered during sermons, Italian caterpillars devoured orchards and gardens, and Brazilian termites ate forests of furniture. Along with Italian and French moles, Spanish rats, French snakes, flies, snails, and eels, Italian locusts, Swiss bloodsuckers, caterpillars, and beetles, French and Croatian grasshoppers, German and Swiss worms, Canadian turtledoves, and more, they were solemnly tried.[3] Humanity saw them as pests sent by God to punish sin, and it had no means but ecclesiastical trials to rid itself of them.[4]

An infestation of French weevils in St. Julien in 1546 was lifted when the local people repented of their sins, resolved to live more justly and pay their tithes, conducted public prayers, and held a series of high masses and processions of the host. But when the weevils returned forty years later, the ensuing ecclesiastical case became enmeshed in wranglings between counsel for the community, François Fay, and counsel for the weevils, Pierre Rembaud and Antoine Filliol. Everyone agreed that the divine hierarchy differentiated man from animal but disagreed on what that meant for the case. In despair, the townspeople set aside land for the weevils if they would only promise to leave the rest of the land alone. The prosecutor now claimed that the chosen land was fertile. But defense counsel said it was sterile and worthless. As compromise was impossible, a judicial decision was made in writing. But we will never know what it was, for the last page appears to have been eaten through—by weevils.[5] Even as they caused privation, barley-eating weevils, grape-eating locusts, and apple-eating caterpillars were said to be minor annoyances that God would remedy after mankind repented.

But nonhuman homicide and bestiality ranged well beyond annoyance; they upended the divine hierarchy itself. Another kind of trespass altogether, these matters were for Caesar.[6] The usual nonhuman perpetrators of homicides were not weevils or flies or locusts or moles or caterpillars. They were not even one of the diminishing number of wolves and other large European predators, though Gratiano, in *The Merchant of Venice*, makes it clear that the occasional wild animal killed and was punished for it.

> *Thy curish spirit*
> *Governed a wolf who, hanged for human slaughter,*
> *Even from the gallows did his fell soul fleet.*[7]

Most often they were the pigs who roamed the narrow streets of medieval villages. In 1266, a pig was burned to death at Fontaneux-aux-Roses, near Paris, convicted of eating a child.[8] In 1386, a sow was sentenced in the Norman city of Falaise to be mangled in the head and forelegs for having killed a child. Then she was hanged.[9] In 1394, a pig was convicted and hanged in Roumaygne for having murdered a baby.[10] In 1403, sows were executed for eating children in Mantes and Meullant.[11] In 1474, a pig, hung in Lausanne for having killed a human, was left to dangle from the noose as a warning to other pigs.[12] In 1499, a pig was hung near Chartres for killing an infant.[13] But not only pigs were executed. In 1314, a bull who had killed a man was hung in Moisy, France.[14] Dogs were executed for killing a Franciscan novice.[15] Echoing the goring ox, a cow was killed and buried, unflayed, for killing a Saxon woman in 1621 near Leipzig.[16]

It was not just animals but humans who were thought to be on the level of "beasts" who were treated this way. Jewish killers of Christians were hung upside down throughout Europe. In twelfth-century Burgundy, a homicide committed by "beasts or Jews" was punished by hanging, usually upside down.[17] Why upside down?

A beast or beastly human who killed a human reversed the ordained hierarchy and breached divine boundaries. Inversion set
the world right again.[18]

Perverted sexual couplings unhinged the universe in much the
same way. In Leviticus 18:23, God told Moses that no human could
"lie with any beast to defile thyself therewith; neither shall any
woman stand before a beast to lie down thereto; it is confusion." In
Deuteronomy 27:21, Moses cursed anyone who "lieth with any
manner of beast." In Exodus 22:19 and Leviticus 20:15 and 16, God
prescribed the penalty: death for the human and death for the
beast. Lord Coke, whom Shakespeare knew as chief justice of England, called this "the detestable, abominable sin, amongst Christians not to be named, committed by carnall knowledge against
the ordinance of the Creator and order of nature."[19] Throughout
Europe and even in colonial America, the perpetrators of acts "not
to be named" were punished in the secular courts, sometimes by
hanging, often by burning.[20] As with homicide, sexual relations
were often forbidden not just with beasts but with beastly humans.
In 1562, the Belgian jurist Damhouder wrote that sexual relations
with a Jew, Turk, or Saracen were punished by burning, "inso-
much as such persons in the eyes of the law and our holy faith differ in no wise from dogs."[21]

Examining Swedish bestiality trial records for the years 1630 to
1780, Jonas Liliequist found that between six hundred and seven
hundred Swedes were executed and that hundreds more were
flogged, sentenced to religious penalties, and forced to labor in
chains.[22] Nonhumans throughout Europe often shared their
abusers' fates.[23] Swedish court records revealed that hundreds of
cows, mares, sows, ewes, and bitches were executed as accom-
plices.[24] One eighteenth-century Swedish commentator implicitly
invoked the law of the goring ox when he wrote that the nonhuman victim was killed in order to obliterate all memory of the act.[25]
Probably the most famous exception was the she-ass of the village
of Vanvres. In 1750, one Jacques Ferron was surprised *in flagrante
delicto* and sentenced to death. The she-ass, however, was acquit-
ted after the prior of the convent in Vanvres and many of its
prominent citizens

signed a certificate saying they had known the said she-ass for four
years, and that she had always shown herself to be virtuous and

well-behaved both at home and abroad and had never given occasion of scandal to anyone, and that therefore "they were willing to bear witness that she was in word and deed and in all her habits of life a most honest creature."[26]

Unlike ecclesiastical proceedings against insects, secular punishments of animals were not usually imposed after a fair trial or any trial at all. The secular court rendered a plain statement of the facts, then sentenced the animal straightaway.[27]

These proceedings may appear amusing, primitive, and irrational, but they were carried out with the utmost solemnity. Brought neither to eliminate real physical threats nor to take vengeance, legal process, or at least its form, was enlisted to right what religious doctrine said were serious trespasses against the divine hierarchy. Boundaries had been breached, borders crossed. God had been affronted, justice insulted.[28] Repair was required. There was

> an unbridgeable gulf between mankind and the rest of creation, and there is beyond that an acute sensitivity towards boundary breaching between kinds within the world of living things . . . Animals that have killed persons were to be extirpated because the very fact of their having done so disturbed the cosmological environment in a way that could not be tolerated: the act appeared to negate the *hierarchically* differentiated order of creation by which man was granted sovereignty in the physical world. The visible evidence of the breach of this order had to be removed—and removed in solemn public procedure—in order that the cosmological equilibrium would be widely recognized as having been restored.[29]

The murder prosecution of an insane man before an ecclesiastical court in Bern, Switzerland, in 1666 emphasized the importance of hierarchy. Against the defendant's plea that he could not be held responsible for his actions, the prosecutor analogized to the law of the goring ox. It demanded the ox's death, though he was not responsible for his actions; so it should be for the insane killer. But the judges demurred: "(A)lthough God enacted a law for the ox, he did not enact any for the insane man."[30] The judges insisted that "the distinction between the goring ox and the maniac must be observed. An ox is created for man's sake, and can therefore be killed for his sake; and in doing this there is no question of right or wrong as regards the ox."[31] Even insane men were created by God for their own sakes.

THE ARCTIC NIGHT OF ANIMAL LAW

For animals, the river of injustice that flowed through the West was fed by streams of Hebrew, Greek, and Roman law, philosophy, and religion. Each emphasized the incommensurable difference and enormous gulf that separated humans from every other animal, and so it really didn't matter that each people weighted these traditions somewhat differently or interpreted each of them in its own way. The outcome for animals was always the same.

In England, killer animals and inanimate objects were not tried and punished as they were in continental Europe. The English responded to breaches of the divine hierarchy with the "deodand," by which offending things were "given to God." This unique institution filtered Roman and Saxon law through the law of the Old Testament.[32] The most famous and influential English legal commentator of the eighteenth century, William Blackstone, linked the deodand unmistakably to the goring ox.[33] But the community, instead of rubbing the offender from God's sight, gave it to him.[34] But God then, like the Internal Revenue Service now, needed a human agent to accept His gifts and the church was only too happy to act as His revenue agent on earth.[35]

Borrowing from Roman law, the ninth-century Laws of Alfred the Great required the "noxal surrender" of any slave, tree, and by implication, any animal, that killed a human being. The same was true for any cow who wounded a human. This "surrender" was to the injured person, or the person's relatives, if the victim no longer breathed.[36] After William the Conqueror and his successors wrested a large degree of spiritual control from the church, the religious belief in human transcendence began to have larger secular consequences.[37] If any thing, living or not, "cause(d) the untimely death of any reasonable creature by mischance" the offending thing was solemnly "given to God," that is, to the church, but was so given through the state, that is, the king, "God's Lieutenant on earth," according to Lord Coke, "to be distributed in works of charity for the appeasing of God's wrath."[38] That Blackstone inserted the deodand into his discussion of the king's revenue shows how the Crown actually used the proceeds not "in works of charity for the appeasing of God's wrath" but in the way a state uses sales tax. If one man killed another with a sword, the sword was deodand. If a man fell from a tree, the tree was deodand. If a cart killed an infant, if a man fell from a boat in fresh water and drowned, if a horse or ox killed a woman, all were deodand.[39] As late as 1845, the English Crown demanded £600 when the boiler of a derailed locomotive exploded and killed four people. The railroad balked at paying because the entire locomotive was only worth £500. The Court of Exchequer then allowed the company to limit its liability to just £500 "by giving up the engine."[40]

Bracton, the most famous English law commentator of the Middle Ages, thought that identical Roman law governed both animal and slave. As in Rome, domestic and wild animals and human slaves could be acquired by capture or birth.[41] Deer, peafowl, pigeons, and slaves who left their owners remained the owners' property as long as they intended to return. When slaves "cease(d) to have the habit of returning they begin to be fugitives, as in the case of domesticated animals."[42] Theft of both slaves and animals was a felony.[43] Three hundred years later, when Coke decided that certain swans were the king's, his reasoning was Bracton's and Justinian's.[44]

In the seventeenth century, Thomas Hobbes imagined an ancient mythical state in which all humans had the same rights to every thing and in which each waged perpetual war against every other. Nothing could be unjust, for there was no justice. Thus was "the life of man, solitary, poore, nasty, brutish, and short."[45] Humans purchased peace by entering into a permanent agreement and transferring nearly all of their natural rights to the sovereign.[46] The sovereign determined what was just, what was private property, and what was necessary for peace and security.[47] Because animals could neither understand speech nor the concept of rights nor contract, they could never purchase peace and were forced to remain in a state of nature.[48] Humans and animals could therefore kill each other without injustice.[49]

Hobbes' contemporary, John Locke, thought that fundamental rights pre-dated society. Nature was God's property to do with "in the way that clay is subject to the potter's will."[50] Every creature belonged to Him and He had given the animals to mankind.[51] Mankind had not just the right, but the duty, to use animals to God's end, which was mankind.[52] But God had given the animals to mankind in common.[53] How could an individual human have the right to possess a nonhuman animal? Locke answered that every human owned his body and the fruits of his labor.[54] Like a Roman judge explaining how one occupied a *res nullius*, Locke said that when a human mixed his labor with nature, whether by picking apples, hunting deer, fishing, or taming a bear, that something became his.[55]

Just before the American Revolution, Blackstone pronounced the legal thinghood of animals the child of Roman and Israelite law. Adding a pinch of Hobbes to two tablespoons of Locke, he located the source of humanity's claim of its right to own animals in Genesis.[56] Stirring in Justinian and Bracton, Blackstone found that humans owned any animal they occupied.[57] And so it remains in England. After the American Revolution, James Kent, former chancellor of New York, brushed the mantle of English common law to make it look homespun. Rejecting English custom, he dressed the new American common law in natural law and emphasized the importance of ancient Roman principles.[58] In the end, he produced a common law of animals that would have tempted Justinian, Bracton, and Blackstone to sue for plagiarism.[59]

A century ago, the United States Supreme Court could write that "the fundamental principles upon which the common property in game rests have undergone no change" since Roman times,[60] and Holmes could say, with respect to wild animals, that "we have adopted the Roman law."[61] Few would challenge either statement today. With almost no exceptions, the common law of wild animals in England and every American state still hews to Roman law either by (1) citing Justinian, (2) citing Bracton, Blackstone, or Kent, who incorporated the essentials of Roman law, (3) citing cases that adopted the essentials of Roman law, (4) calling Roman law common law, or (5) stating a common law rule that sounds like a Roman rule.[62] The common law of domesticated animals is even more overtly Roman. A leading modern American legal encyclopedia tersely states that "[g]enerally, all domestic animals are regarded as property, and an owner thereof has a property right therein as absolute as that in inanimate objects."[63]

THE COMMON LAW PUNISHES
NO ACT OF CRUELTY TO ANIMALS

The common law protected nonhuman animals no better under the criminal law than under the civil law. It has never been a crime to abuse them. One twentieth-century treatise on criminal law says that "[t]he right to take their life, and to make property of them, includes all other rights; so that the common law recognizes as indictable no wrong and punishes no act of cruelty, which they may suffer, however wanton or unnecessary."[64]

The Puritans of the Massachusetts Bay Colony, who did not hesitate to change the common law, enacted not just pathbreaking protections for women, children, and servants but the West's first animal protection laws.[65] Their 1641 "Body of Liberties" said that "[n]o man shall exercise any Tirranny or Crueltie towards any bruite Creature which are usuallie kept for man's use." But the Puritans were still trapped in the Great Chain of Being. Five years before the Body of Liberties was enacted, the legislature ordered the courts to follow the Old Testament unless it conflicted with a statute, no better sign for nonhuman animals then than it is for homosexuals today.[66] The Puritans never wavered from their Old Testament belief that animals were "made subject to man by the creation, from the largest and noblest to the smallest and most insignificant," or were "committed to their care and . . . were created for the beneficial use of mankind."[67] Their "Capitall Laws," also found in the Body of Liberties, were often linked to biblical provisions. One chillingly invoked the law of the goring ox—if any man or woman "shall lye with any beaste or bruite creature by Carnall Copulation" the human was to be put to death and the victim "slaine and buried and not eaten."[68] As a result, sixteen-year-old Thomas Granger was executed along with a mare, a cow, two goats, several sheep, two calves, and turkey he had sodomized in the Colony of Plymouth in 1642, while sixty-year-old Thomas Potter was hanged in New Haven with a cow, two heifers, three sheep, and two sows for the same offense.[69]

Nothing had changed in England or America when, in 1809, the House of Lords took up its first bill designed to protect animals.[70] The sponsor, Lord Erskine, was the chancellor of England. He was a famous animal lover, with a favorite goose, macaw, and two leeches (whom he named after surgeons). He brought his dogs to legal meetings.[71] It was said that once he protested to a carter who was beating his horse. When the man responded, "Can't I do what I like with my own?" Erskine then struck him with his stick and replied, "And so can I—this stick is my own."[72]

But even Erskine repeatedly referred in parliamentary debate to the Great Chain of Being and to the moral trust that the biblical grant of dominion over the animals implied.[73] Passed by the House of Lords, this bill stalled in the Commons.[74] Thirteen years later, Martin's Act, which prohibited any person from wantonly and cruelly beating or ill-treating any "horse, mare, gelding, mule, ass, ox, cow, heifer, steer, sheep or other cattle," finally succeeded, one year after Maine enacted the first anticruelty statute in the United States.[75]

Every American jurisdiction eventually passed anticruelty statutes. But judges often assumed that the statutes incorporated the biblical transcendence of human over nonhuman animals and that their purpose was to protect human morals, not animal bodies. A Mississippi Supreme Court judge eloquently summed up the common understanding in 1888.

Such statutes were not intended to interfere, and do not interfere, with the necessary discipline and government of such animals, or place any unreasonable restriction on their use or the enjoyment to be derived from their possession. The common law recognized no rights in such animals, and punished no cruelty to them, except in so far as it affected the rights of individuals to such property. Such statutes remedy this defect, and exhibit the spirit of that divine law which is so mindful of dumb brutes as to teach and command, not to muzzle the ox when he treadeth out the corn; not to plow with an ox and an ass together; not to take the bird that sitteth on its young or its eggs; and not to seethe a kid in its mother's milk . . . Cruelty to them manifests a vicious and degraded nature, and it tends inevitably to cruelty to men . . . The dominance of man over them, if not a moral trust, has a better significance than the development of malignant passions and cruel instincts. Often their beauty, gentleness, and fidelity suggest the reflection that it may have been one of the purposes of their creation and subordination to enlarge the sympathies and expand the better feelings of our race. But, however this may be, human beings should be kind and just to dumb brutes; if for no other reason than to learn how to be kind and just to each other.[76]

However, the anticruelty statutes assured that *completely* unjustified infliction of pain upon animals would no longer be tolerated. But because we can justify to ourselves nearly everything we do to animals, their effect has been minimal.[77] They give animals no legal rights. They are usually invoked only in rare situations in which humans harm animals merely "for the gratification of a malignant or vindictive temper," unnecessarily, and not in the pursuit of some legitimate benefit for which human beings had long been entitled to use them.[78] Typical is the view of the Indiana Court of Appeals in 1892. Indiana's anticruelty statute

> was evidently designed to inculcate a humane regard for the rights and feelings of the brute creation by reproving evil and indifferent tendencies in human nature in its intercourse with animals, but not to limit man's proper dominion "over the fish of the sea, and over the fowl of the air, and over every living thing that moveth upon the earth."[79]

And a New York Municipal Court concluded that

> [b]y biblical mandate man was given "dominion over the fish of the sea, over the fowls of the air and the beasts, and the whole earth and every creeping creature that moveth upon the earth." Man is superior to animals, and some of them he uses for food and is permitted to slaughter them. Many are the means he employs for such purpose, and in such cases the incidental pain and suffering is treated as necessary and justifiable.[80]

THE FOUNDATION COLLAPSES

In 1808, Lord Ellenborough, chief justice of the King's Bench in England, declared that "the death of a human being cannot be complained of as an injury."[81] Like the goring ox and the deodand, this rule testified to the extent that the law remained submerged in Hebrew, Greek, and Roman cosmologies even into the nineteenth century.[82] In 1868, the Michigan Supreme Court could say that

> [t]he reason for the rule is to be found in that natural and almost universal repugnance among enlightened nations to setting a price

upon human life, or any attempt to estimate its value by a pecuniary standard, a repugnance which seems to have been strong and prevalent among nations in proportion as they have been or become more enlightened and refined, and especially so where the Christian religion has exercised its most beneficent influence, and where human life has been held most sacred . . . [t]o the cultivated and enlightened mind, looking at human life in the light of the Christian religion as sacred, the idea of compensating its loss in money is revolting.[83]

By the midpoint of the nineteenth century, the thousand-year flow of the incommensurability of human life with anything else had begun to ebb. The deodand never really caught on in America, though legislatures occasionally passed statutes that demanded the forfeiture of things used in crimes, including animals, based on the fiction that the thing was the offender.[84] The Puritans, especially, had rejected what they saw as the deodand's blasphemous merger of political and sacred authority. By 1916, the Tennessee Supreme Court could call the deodand "superstition" and "repugnant to our ideas of justice."[85] Across the Atlantic, the English Parliament abolished the deodand on August 18, 1846. Eight days later, Parliament altered the common law bar to permit the recovery of a survivor's pecuniary, if not emotional distress, damages for wrongful death.[86] Soon every American jurisdiction followed suit.[87] The timing of these two statutes was no coincidence. They were the harbingers of a new world.

The law of the goring ox, the deodand, and the bar on recovery for the wrongful death of humans make absolutely no sense in a world that no longer accepted divinely placed borders and serious punishments for border crossings.[88] Nonhuman animals had no hope for legal rights in that world, for the rights of a being will not be recognized by a society that assumes that the Creator of the universe has designated it as inferior.[89] Only in the eighteenth century did the West begin to separate law from theology. Only in the nineteenth century did the scientific evidence that the universe was not designed at all, much less designed for humans, strengthen enough to convince those whose religious beliefs had not closed their minds to the possibility. The Great Chain finally snapped and a door cracked open to the *possibility* that at least some animals might *logically* transcend their legal thinghood.[90]

Surveys consistently reveal that about half of Americans reject any claim for biological evolution and embrace the hierarchical cosmology of Genesis.[91] But only the rare modern judge overtly rests her decisions upon these religious beliefs. As highly educated lawyers, most judges probably accept Darwinism, but in a form that has, in the words of Stephen Jay Gould, "been so spin doctored that we have managed to retain an interpretation of human importance scarcely different, in many crucial ways, from the exalted state we occupied as the supposed products of direct creation in God's image."[92] This will take some time to change, for logic alone is never enough. Scientific discoveries have powerfully supported Darwinian evolution and steadily and more truly revealed the natures of both human and nonhuman animals. Yet scientific facts that contradict old and cherished beliefs may take a long time to appear in the decisions of judges. The legal thinghood of nonhuman animals will not dramatically weaken—it cannot—until judges no longer *believe*, even unconsciously, in the disproven cosmologies upon which it depends. Only judges who almost intuitively understand how thoroughly modern evolutionary fact has upended the Great Chain of Being can truly be receptive to the idea that legal rights need not be restricted to human beings.

Because the common law values the past merely for having been, judges rely upon prior judicial decisions and the jurisprudential writings of those who lacked modern scientific understandings, because they wrote before Darwin or they misconstrue humanity's place in Darwin's world.[93] Despite the slow turning-out of the ancient cosmologies, twentieth-century judicial decisions have confirmed and reconfirmed the legal thinghood of animals. But none have ever tried to justify this anachronism because judges fail to realize that it requires justification. They mechanically cite earlier cases, which cite still earlier cases that inevitably reach back to Kent or Blackstone or further still to Locke and Hobbes, then to Coke, and Bracton, until we arrive at Justinian and the Old Testament. It is time that judges consider that as the ancient foundations have begun to rot away, so the law of animals that rests upon them should be changed.

The next two chapters will scrape this foundation to bedrock. Only then can we halt what the Greek scholar Moses Finley, described as the "final paradox" of the Greeks—the rise of both liberty and slavery. This is our paradox, too, as those animals for whom our most important principles of justice dictate fundamental liberties continue to be enslaved in a world in which the liberties of human beings are so often on the rise.[94]

5

What Are Legal Rights?

THREE CASES

Every human has the basic legal right to bodily integrity. We are all legally disabled from invading each other's bodies without consent. Every human has the basic right to bodily liberty as well, so that we're legally disabled from enslaving and kidnapping each other. (The reason I use the word "disabled" will soon be made clear.) But no nonhuman has these rights. Before I can argue that chimpanzees and bonobos are entitled to them, we must be clear about what legal rights are and who, or what, might be eligible for them. Three cases will set the stage. The first is the most famous English slavery case, the 1772 case of *Somerset v. Stewart*.[1] The second is the most famous American slavery case, the 1857 case of *Dred Scott v. Sandford*.[2] The third is a case I litigated in 1993 on behalf of a dolphin named Kama.[3]

"Can a Man Become a Dog for Another Man?"

James Somerset was captured in Africa, thrust onto a slaving ship, and shipped to Virginia in the year 1749. Purchased there by Charles Stewart, he was brought to Massachusetts, where Stewart worked as a customs officer. Twenty years after touching down in

America, Somerset sailed with Stewart to England. In October 1771, twenty-two years a slave, Somerset made a bid for freedom. But he enjoyed just a month at liberty before Stewart's men cornered him, dragged him aboard the *Ann and Mary*, and chained him in its bowels. Its captain, John Knowles, made ready to sail for Jamaica where Somerset was to be sold again. And sell him Knowles would have done, had Granville Sharp not learned of Somerset's plight.

Before the *Ann and Mary* could sail, Sharp and other English citizens begged a writ of habeas corpus from the great chief justice of the King's Bench, Lord Mansfield (a habeas corpus writ tests the legality of a loss of personal liberty; it literally orders a defendant to "produce the body"). These men and women had been testing the legality of slavery in English courts for nearly five years. Now they invoked the Great Writ. Early on, Mansfield urged settlement. But all refused. What compromise exists between freedom and slavery? At a preliminary hearing, Stewart's lawyer argued that slavery could exist in England because no law forbade it.[4] Somerset's lawyer demanded to know "upon what Principle is it—can a Man become a Dog for another Man[?]."[5] Very well, Mansfield replied. "If the parties will have judgment, *'fiat justicia, ruat coelumtet'* ('let justice be done though the heavens may fall')."

On June 22, 1772, the heavens fell.

> The state of slavery is of such a nature that it is incapable of being introduced on any reasons, moral or political, but only by positive law ... It is so odious, that nothing can be suffered to support it but positive law. Whatever inconveniences, therefore, may follow from the decision, I cannot say that this case is allowed or approved by the law of England: and therefore the black must be discharged.[6]

"No rights which the white man was bound to respect"

When Peter Blow died on June 23, 1832, in St. Louis, he may have owned Dred Scott. Or he may have sold him to Dr. John Emerson shortly before he died. We don't know. We also can't tell whether Emerson owned Scott at Blow's death or later paid $500 to the estate to buy him. But we do know that Scott accompanied Emerson when he took up his post as an assistant army surgeon at Fort

Armstrong, Illinois, in December 1830 and that Scott was then about thirty years of age and, some say, just five feet tall.

In 1836, Emerson was transferred to Fort Snelling in the Wisconsin Territory. There Scott met and married Harriet Robinson, owned by Major Lawrence Taliaferro. Of the four children their marriage produced, two boys died in infancy; they named the girls Eliza and Lizzie. The Scotts were brought to Fort Jesup in Louisiana in the spring of 1838, returned to Fort Snelling in the fall, then were hauled back to St. Louis the following spring. There they remained, even as their master set up a private medical practice in Davenport, Iowa Territory, in 1843. That Christmas, Emerson died. Where the Scotts resided for the next three years, and with whom, is hazy, but they probably returned with Mrs. Emerson's brother, Captain Bainbridge, to Fort Jesup and lived there until 1845, when they were taken to Corpus Christi, Texas. There they lived until the following February, when they were returned to St. Louis.

On April 6, the Scotts filed petitions in the circuit court in St. Louis requesting permission to bring suit against Mrs. Emerson to establish their freedom. Permission was granted. They were free, they argued, because they had lived in the Illinois and Wisconsin Territories, where slavery was illegal.[7] This *had* freed them. But had their slavehood reattached when they were returned to Missouri? The Missouri courts almost always ruled that it did not.

The Scotts, however, were not fated to share the happy destiny of James Somerset. Nearly six years later, the justices of the Missouri Supreme Court, growling at the growing aggressiveness of the Northern abolitionists, ruled against them.

> Times are not now as they were. . . . Since then not only individuals but States have been possessed with a dark and fell purpose with relation to slavery, whose gratification is sought in the pursuit of measures, whose inevitable consequences must be the overthrow and destruction of our government. Under such circumstances it does not behoove the State of Missouri to show the least countenance to any measure which might gratify this spirit.[8]

The Scotts were promptly sold to John Sanford, Mrs. Emerson's brother and a citizen of New York.

In 1853, the Scotts' lawyers seized upon this sale as reason for seeking their freedom once again, this time in a federal court. They contended that this court had jurisdiction because the Scotts' freedom was being contested by citizens of different states and the Congress had given the federal courts jurisdiction over suits in which a citizen of one state sued a citizen of another. Their new petition alleged that John Sanford had illegally assaulted Dred, Harriet, Eliza, and Lizzie Scott. But a second trial produced the same dismal result. Their appeal propelled the nation toward civil war.

Blacks, said Roger Taney, Chief Justice of the United States, could not be citizens of the United States and therefore could not sue in a federal court.[9] They

> had for more than a century before [the adoption of the U.S. Constitution] been regarded as beings of an inferior order, and altogether unfit to associate with the white race, either in political or social relations; and so far inferior, that they had no rights which the white man was bound to respect; and that the negro might justly and lawfully be reduced to slavery for his benefit. He was bought and sold, and treated as an ordinary article of merchandise and traffic, whenever a profit could be made of it.[10]

The historian Don Fehrenbacher concluded that "[t]he effect of Taney's statement was to place Negros of the 1780's—even free Negros—on the same level legally as domestic animals."[11]

Can a Dolphin Sue?

Kama was born into captivity at SeaWorld in San Diego on 1981. Eventually he was flown to the New England Aquarium in Boston. In 1987, he was transferred to the Naval Oceans System Center in Hawaii. Despite assurances from the navy, many were concerned about Kama's safety. To what point had he been trained for the 3,500 man hours the navy claimed and why had it spent $700,000 to train him?

Suit was filed in 1991 in the names of Kama and several animal protection organizations. The judge had to decide whether Kama could sue. Federal Rule of Civil Procedure 17(b) provides that the capacity of any individual to sue or be sued is to be determined "by the individual's domicile." Was Kama an "individual?" He was. "[T]here is no indication that [Rule 17(b)] does not apply to other non-human entities or forms of life," the judge wrote.[12] Did

Kama then have a "domicile"? He did. According to the judge, it was either in Hawaii or Massachusetts. Did Kama then have the capacity to sue? He did not, as neither state permitted dolphins to sue.

THE MEANING OF LEGAL RIGHTS

Potter Stewart, a late-twentieth-century justice of the U.S. Supreme Court, once observed about pornography, "I know it when I see it."[13] We have an intuitive "feel" for what legal rights are, the way we know pornography when we see it, even if, as Stewart confessed, we can't quite define it. At least we think we know what rights do. Legal rights act as

> "trump cards" that individuals can play against appeals to the society; armed with a right, the individual becomes a "small-scale sovereign." Rights are "side-constraints" or "limits or vetoes." They have a "preemptory or conclusory sound." And a right that does not stick in the spokes of someone's wheel is no right at all.[14]

We will need to get more specific. For most of the last century, legal scholars, judges, and lawyers have often classified legal rights as proposed by Wesley Hohfeld, a professor at the Yale Law School during World War I.[15] Hohfeld argued that a legal right was any theoretical advantage conferred by recognized legal rules.[16] We will use his system of classification because it is logical and helpful and will ignore the fact that the first year that Hohfeld taught at Yale, most of his students signed a petition asking the president of Yale to terminate his teaching appointment.

Hohfeld broke legal rights into their lowest common denominators, using terms that judges commonly employ.[17] He never formally defined them. Instead, he spelled out how they relate to each other. According to Hohfeld, legal relationships can exist only between two legal persons and one thing. One of the two persons always has a legal advantage (that's the right) over the other. The other person has the corresponding legal disadvantage. Just as a man can't be a husband without a wife and a woman can't be a wife without a husband, neither a legal advantage nor a disadvantage can exist all by itself.[18] A man is a husband because he has a wife, and one person has a legal advantage because another has the corresponding disadvantage. That's why the philosopher Samuel Pufendorf argued three hundred years ago that before Eve was created, Adam could have had no legal rights, for no other person was about the Garden of Eden to assume the corresponding disadvantage.[19]

Hohfeld's table of legal correlatives lies at the end of this paragraph. The four kinds of legal advantages (the rights) are on the top row. Each of the four correlating legal detriments is on the bottom row. For example, a "claim" is a legal advantage (or right) that pairs up (or correlates) with the legal disadvantage called the "duty." I'll explain what they mean as we go along. When we finish, you'll know more about the nature of legal rights than many lawyers and judges.

| liberty (no-duty) | claim | immunity | power |
| no-right | duty | disability (no-power) | liability |

Liberty and No-Right

The basic building block of legal rights is the "liberty." It allows one person to do exactly as she pleases with no duty to do otherwise. That's why a liberty is sometimes called a "no-duty." But here's the important thing about a liberty: No one is required to respect it.[20] It is merely "a permission without a protection."[21] Everyone has billions of liberties—the liberty to look left, the liberty to look right, to raise one finger, to raise two fingers, to raise three fingers—you get the idea. Some philosophers argue that every entity in the universe has liberties against everything else—say the star Alpha Centauri against the pope.[22] But remember that legal rights, at least as Hohfeld defined them, pertain only to the legal relationship between two legal persons. Legal things, whether they are the slaves James Somerset (before Lord Mansfield set him free) and Dred Scott, Kama the dolphin, Jerom the chimpanzee, a rock in New Mexico, or Alpha Centauri, have no legal rights.

The "no-right" correlates with the liberty.[23] This is the hardest correlative to grasp, because it is not what we normally understand to be a disadvantage. If we call a no-right a "no-claim," it becomes easier to understand. A person with the "no-right" simply has no claim that a person with a liberty has to do something or not do something. Professor H.L.A. Hart of Oxford University neatly illustrated the relationship between a liberty and a "no-right."

> The fact that a man has a liberty to look at his neighbor over the garden fence does not entail that the neighbor has a correlative obligation to let himself be looked at or not to interfere with the exercise of this specific liberty-right. So he could, for example, erect a screen on his side of the fence to block the view.[24]

Out of the more than two hundred recorded senses of "liberty," the British philosopher Sir Isaiah Berlin identified two of the most important: "negative liberty" and "positive liberty."[25] Berlin contrasted them succinctly: "Freedom for the pike is death for the minnow."[26] In the last year of the American Civil War, Abraham Lincoln explained them in a way that every Northerner understood:

> We all declare for liberty; but in using the same word we do not all mean the same thing. With some the word liberty may mean for each man to do as he pleases with the product of his labor; while with others the same may mean for some men to do as they please with other men, and the product of other mens' labor. Here are two, not only different, but incompatible things, called by the same name—liberty . . . The shepherd drives the wolf from the sheep's throat, for which the sheep thanks the shepherd as a liberator, while the wolf denounces him for the same act as the destroyer of liberty, especially as the sheep is a black one. Plainly the sheep and the wolf are not agreed upon a definition of the word liberty; and precisely the same difference prevails today among us human creatures, even in the North, and all professing to love liberty. Hence we behold the processes by which thousands are daily passing from under the yoke of bondage, hailed by some as the advance of liberty, and bewailed by others as the destruction of liberty.[27]

You can see why negative liberty is sometimes said to mean "freedom from."[28] It is the liberty of the sheep and minnow. Since World War II, the most fundamental negative liberties, bodily integrity and bodily freedom, have formed the core of what have become known as "human rights." We simply must be free from physical assault and battery and slavery if we are not to live in constant fear and have any chance to flourish. That is why the constitutions of nearly every country enshrine negative liberties. When absent or ignored for human beings, the horrors of the slaveholding American South, Nazi Germany, Rwanda, and Kosovo can ensue. When absent for chimpanzees and bonobos, the cruelties inflicted upon Jerom, the Yerkes chimpanzee, can occur.

"Positive liberty" is "freedom to." It is the freedom of the wolf to eat the sheep and the pike to swallow the minnow, of Charles Stewart and John Sandford to enslave James Somerset and Dred Scott, and of the Yerkes biomedical researchers to inject Jerom with a virus that they pray will kill him. Positive liberty allows the strong to impose their wills upon the weak with impunity.[29] Advocates of positive liberty see freedom as something more than mere lack of restraint; they emphasize the power of the strongest entity of which we know, the state.[30] But this is not always a bad thing. Isaiah Berlin thought that positive liberty has "animate[d] the most powerful and morally just public movements of our time."[31] While it led to Fascism and Nazism, it also freed the slaves and mandated equal rights.[32]

Claims and Duties

A "claim" entitles one person to limit the liberty of another, who then has a "duty" either to act or not to act in certain ways toward the claimant.[33] When we think about legal rights, we usually think about claims. Unlike liberties, claims demand respect. If I sign a contract to sell my house to my brother, Bob, he has the claim to buy it from me. I have the duty to sell it to him.

Claims come in two flavors. *In personam* claims exist against a small number of definite persons.[34] Brother Bob has a claim against me, and nobody else, for my house. We won't concern ourselves with *in personam* claims, because it is relatively unlikely that animals could ever have them. Instead, we're interested in *in rem* claims. These can exist against every person in the world.[35] Remember, I have the negative liberty not to be enslaved, as James Somerset was, or forced into biomedical research, as was Jerom. But no one has a duty to respect these rights when they exist all alone. Here is where we see the power of the claim. The common law, the U.S. Constitution, the Massachusetts Constitution, and numerous state and federal statutes, for example, all give me *in rem* claims against being kidnapped and forced into biomedical research. Every person then has the correlative duty not to do those things to me.

Must a person be able to physically *make* a claim in order to *have* one? The answer depends upon whether one emphasizes the "claim" part or the "duty" part of the claim-duty pairing. One school of legal scholars (we'll call them the Benefit/Interest School) emphasizes "duty." Any being with interests—an adult woman, a profoundly retarded man, an infant, a chimpanzee, or a dolphin—could, if allowed to be a legal person, have a claim that correlates to another person's duty. The opposing school (the Control/Choice School) accents "claims." These scholars argue that a person must actually have the mental wherewithal to be able to choose to make a claim and to control how it is made.[36] Profoundly retarded men and infants, who lack these mental abilities, cannot then have claims. An even stricter branch of the Control/Choice School—we'll call it the "Strict Control/Choice" School—says that claims and duties can only exist between members of a "moral community." Unless one has the capacity not just to choose but to act morally, one can have no claims.

If required to meet the more stringent requirements of the Strict Control/Choicers, none but the most extraordinary nonhuman animal could ever have a claim. But here's the rub: Millions of human beings would also be ineligible—and not just the profoundly retarded, but the insane, the permanently vegetative, and the very young. Many more human adults and older children, and perhaps even apes, whales, and parrots, might have claims if the Control/Choice School prevailed. But vast numbers of human beings would still be ineligible, as would most other animals. However, if the Benefit/Interest School triumphs, aside from the permanently vegetative, virtually every human being would be entitled to claims; but so would a large number of other animals.

Immunities and Disabilities

An "immunity" *disables* one person from interfering with the liberty of another, just as a thrown rod *disables* a car from traveling another mile.[37] That is why the "disability," which correlates with the immunity, is sometimes called a "no-power," just as the correlative of the liberty-right is the "no-right."[38] Because immunities are often nested like Russian dolls within more obvious claims, they are often overlooked and sometimes confused with claims.[39] But a vital difference separates them. Claims tell us what we *should not* do. Immunities tell us what we *cannot* do.[40] It is not that the First Amendment to the U.S. Constitution instructs Congress that it *should* not abridge the freedom of speech. Congress legally *can* not. It is not that a woman *should* not marry two husbands at the same time. She legally *cannot*. It is not that one man *should* not enslave another. He *cannot*.[41]

Two real cases illustrate the difference between an immunity and a claim. In the first, the U.S. Supreme Court agreed that a woman has an immunity to choose to have an abortion free from government intrusion.[42] The state is disabled from interfering with her choice. If it tries to interfere, whatever it does has no legal consequence. However, the judges still upheld the government's refusal to pay for a woman's abortion, even if her life is in danger, because her immunity-right to an abortion does not automatically bring with it a governmental duty to provide one.[43] A government has a duty to provide a woman with an abortion only if she has a claim to one, and the justices said she did not have a claim. In the second, the Texas Supreme Court refused to allow a teacher to sue the State of Texas for money damages after it violated a teacher's immunities guaranteed by the Texas Constitution.[44] The justices said that the Constitution's "framers intended that a law contrary to a constitutional provision is void. There is a difference between voiding a law and seeking damages as a remedy for an act. A law that is declared void has no legal effect."[45] The justices decided to give the teacher the power to sue to stop the state from continuing to violate his immunity.[46] But since Texas had no duty not to violate the teacher's immunities, he couldn't sue for money when it did.

Immunities often shield "freedoms and benefits now regarded as essentials of human well-being" and vital "for the maintenance of the life, the security, the development, and the dignity of the individual."[47] Blackstone referred to the Bill of Rights that emerged from England's Glorious Revolution of 1688 as declaring three immunities—personal security, personal liberty, and private property.[48] In an oft-quoted statement, the U.S. Supreme Court said that:

> [n]o right is held more sacred, or is more carefully guarded, by the
> common law, than the right of every individual to the possession

and control of his own person, free from all restraint or interference of others, unless by clear and unquestionable authority of law. As well said by Judge Cooley, "The right to one's person may be said to be a right of *complete immunity: to be let alone.*"[49]

Immunities insulate a person from the scholars' struggle over mental capacity that may plague claims by human infants or mental defectives. Immunities don't depend upon whether a person can choose, just upon whether she has freedoms, benefits, a life, security, and dignity that need to be protected. As we will see, legal personhood is the obstacle that prevents animals from enjoying basic immunities,

A person with just an *immunity* has neither a *claim* that another has a *duty* to respect her immunity nor the *power* to sue to stop a violation of that immunity. Of what use then is an immunity? Here are four. First, as James Somerset learned, an immunity can accomplish great things all by itself. Because of the unusual nature of the writ of habeas corpus, Somerset needed no power to bring suit himself. Once concerned English citizens sought the writ of habeas corpus on his behalf, Lord Mansfield could order him set to liberty. Second, most citizens will respect the important rights of others for no other reason than that these rights exist, irrespective of whether another can sue them.[50] Third, even if an immunity can't be enforced, it has value as "an organizing principle," a "great symbolic force," and "an accepted means to challenge the traditional legal order and to develop alternative principles" that may one day cause it to be enforceable.[51] Fourth, as the Texas Supreme Court showed, judges may give the holder of an immunity the power to sue to stop its violation.[52]

Powers and Liabilities

A "power" is an ability that the law gives a person to affect her own legal rights or the rights of someone else.[53] Its correlate, the "liability," carries "the sense of exposure to having one's legal status changed."[54] The power to sue for the violation of a claim or an immunity is so valuable that the U.S. Supreme Court characterized it as "the right conservative of all other rights."[55] American state courts have often declared it a fundamental principle or maxim of both the common law and equity.[56] However, the Permanent Court of International Justice, an arm of the League of Nations, reminded us that "the capacity to possess civil rights does not necessarily imply the capacity to exercise those rights oneself."[57] But rights without remedies are very rare and nearly unheard of when a fundamental immunity, such as bodily liberty or bodily integrity, is at stake. The judges on the European Court of Human Rights could "scarcely conceive of the rule of law without . . . access to the courts [which] . . . ranks as one of the universally recognized fundamental principles of law."[58] This may explain why the occasional court or commentator mistakenly believes that the power to sue correlates to a liberty, immunity, or claim.[59] As you can see from Hohfeld's table on page 54, it does not.

This power to sue for violations of human rights appeared dramatically on the international stage after World War II. Before the nineteenth century, both individuals and states were subject to the "law of nations," later known as international law.[60] But in 1789, the English philosopher Jeremy Bentham successfully argued that only states should be subject to international law, and that is how it remained until 1945.[61] Individuals had virtually no international remedy against abuse by governments; states were not even required to pay any money they received in reparations for injuries to its citizens to the injured themselves.[62] In short, states held the place in international law that humans today occupy in domestic law, while humans held the place in international law that animals have in domestic law today.[63]

After World War II, effective and adequate remedies sprouted in one major international human rights instrument after another.[64] The Nuremburg War Tribunals, and later the United Nations, imposed duties and liabilities upon individuals as well as states.[65] As we will see in Chapter 6, citizens were sometimes given the power to sue any state, including their own, that violated their rights.[66] Occasionally, this guaranteed access to an international tribunal.[67] More often, it has given individuals the power to sue in the courts of their own countries.[68]

How can a legal thing sue to challenge its thinghood? Here we encounter a legal paradox, for among the rights that a thing lacks is the power to sue for anything. The possibility, however, that a free man could be wrongly enslaved once motivated lawmakers to find ways to solve this paradox.

As long as there were any citizens willing to seek a writ of habeas corpus on his behalf, James Somerset needed no power to challenge the legality of his enslavement. The Scotts tried a second way in the Missouri courts. Slaveholding states sometimes refused to allow habeas corpus to challenge the legality of enslavement. But they still might want to provide a judicial forum in which the legality of an enslavement could be challenged. Their answer was the Freedom Suit Act. This was similar to a habeas corpus action but could be brought by slaves or their "guardians." Freedom Suit Acts assumed that slaves were things. But these things could challenge the legality of their enslavement. Judges simply *pretended* that slaves weren't things for the limited purpose of testing whether they were.[69] Dred Scott and Kama used a third way. Scott, having lost his Freedom Suit Act case in the Missouri state courts, filed suit in federal court and alleged that he was a citizen of Missouri and that John Sandford, a citizen of New York, had illegally assaulted him and his family. But the Supreme Court said that Scott was not a citizen of the United States and ordered his case dismissed. Inherent within this power to sue is the legal capacity and legal personhood of the rights-holder.[70] It turned out that Scott had no legal capacity, and therefore no power, to sue in a federal court. One hundred and forty years later, another federal judge said that Kama had no legal capicity either.

We can see that the might of these four basic kinds of legal rights is most concentrated when they cluster.[71] Negative *liberties* and *immunities* jump in value when infringements can be *claimed* by victims with the *power* to sue. But when and how should they cluster? And when should a legal thing be promoted to legal personhood? I will argue in Chapters 6 and 10 that the same principles that determine human entitlement to basic legal rights should apply to nonhuman animals.

6

Liberty and Equality

"Irrevocable and final limits to
the enactment of men"

Two brothers lie on a battlefield, each killed by his brother's
sword. The victorious king has decreed that the brother who died
defending his city, his nephew, is to be buried with highest honors.
The other brother must lie exposed to dogs and carrion birds. Anyone who tries to bury him will be put to death. Undeterred, the
brothers' sister openly buries her disgraced brother and is confronted by her enraged uncle, who asks how she dared ignore his
command. "For me," she replied,

> it was not Zeus who made that order.
> Nor did that Justice who lives with the gods below
> mark out such laws to hold among mankind.
> Nor did I think your orders were so strong
> That you, a mortal man, could overrule
> the gods' unwritten and unfailing laws.[1]

The battlefield was Thebes, the king, Creon, the sister and niece,
the heroine of Sophocles' play of 441 B.C., *Antigone*.

Twenty-three hundred years later, the descendants of Antigone and King Creon went to war in America. The founding fathers of the Confederate States of America immediately had to confront a major legal and political problem; how to form a Constitution that guaranteed the liberty of whites and the slavery of blacks without actually saying so? They decided to envision the new Confederacy not as a nation of individuals, but of states.[2] The only legal rights that Confederates possessed were those granted to them by the states. Abraham Lincoln responded for the Union at Gettysburg: "Four score and seven years ago our fathers brought forth upon this continent a new nation, conceived in liberty and dedicated to the proposition that all men are created equal."

In the century following, another decree set doctors to identifying the incurably ill as well as children and adults who suffered from schizophrenia, senility, insanity, idiocy, chronic diseases of the nervous system, epilepsy, imbecility, paralysis, and various kinds of serious hereditary diseases. Once identified, they were killed.[3] According to a 1943 opinion by a leading expert on Nazi constitutional law, Adolph Hitler's decrees constituted "the absolute center of the present legal order."[4] As Antigone had been, Hans and Sophie Scholl, who formed the core of the White Rose, a tiny student movement, were executed for distributing pamphlets that denounced Nazi crimes. The Scholls came to represent those in Nazi Germany "who were forced to pay a price for believing that human rights were more important than obedience to arbitrary laws."[5] Unlike Antigone and the Scholls, German doctors, nurses, and others obeyed.

In a 1947 Frankfurt judgment handed down against medical workers at Hadamar, one of the euthanasia centers, a German judge said:

It may be acknowledged that as a rule formal legal force is enough to lend the law validity and oblige all citizens to obey it. It is not therefore usually open to the jurists and moralists of a country to inquire into the validity of such a law. It is an urgent duty in the interests of the preservation of the uniformity and stability of law to recognize this circumstance explicitly. For otherwise legal instability, arbitrary action and eventually revolution would become permanent conditions and any communal life based on law and order would be rendered impossible. It is, however, equally essential to insist that there are certain limits to the positive character of law which must not be transgressed. The boundaries exist because the State is never the sole source of all law and can never arbitrarily determine what is right or wrong. There is one law superior to all formal legislation, one ultimate standard for assessment of the latter. This is the law of nature, which sets irrevocable and final limits to the enactment of men. There are certain legal maxims so deeply rooted in nature that every legal and moral obligation must in the end be adjusted to conform with the law of nature that stands superior to it. Such legal maxims exert compulsory force because they are independent of the vicissitudes of time and human beliefs millennium after millennium, remaining constant and valid for every era . . . The idea of a necessary equation between law and justice is basically acceptable, but only if the single limitation referred to is implied. If any legislation contravenes in any way the eternal standards of natural law, the content of such legislation will prevent its being equated with justice. It not only loses its obligatory force for the citizen but is actually invalid in law and must not be obeyed by him.[6]

Michael Ignatieff, who teaches the history of human rights at the London School of Economics, has written that "[t]he Holocaust laid bare what the world looked like when natural law was abrogated, when pure tyranny could accomplish its unbridled will."[7] The Nuremberg Charter authorized the prosecution of Nazi crimes "whether or not in violation of the domestic law of the country where perpetrated."[8] To Robert Jackson, a U.S. Supreme Court Justice and lead prosecutor at Nuremberg, the "real complaining party" was "Civilization."[9] The tribunal would later assert that "[h]umanity is the sovereignty which has been offended."[10] Adolph Eichmann, efficient organizer of much of the Holocaust, on trial for his life in Jerusalem in 1961, calmly insisted that "whatever he did he did, as far as he could see, as a law-abiding citizen. He did his *duty*, as he told the police and the court over and over again; he not only obeyed *orders*, he obeyed the *law*."[11]

Does the state have the final say about what is right and just? Or does it lie with some other entity, such as nature or God? Are there such things as natural law, natural rights, and natural justice? Or are law, rights, and justice what judges and legislators tell us they are and nothing more?[12] Antigone and Creon, Lincoln and Jefferson Davis, Hitler and the Scholls, the killing doctors of Hadamar and their German judge, and Adolph Eichmann and his Jewish judge all disagreed. The core question is this: Are things or beings or ideas valuable because we value them or because they are inherently valuable? If nonhuman animals, or humans, are valuable only because we value them, then they must lack value when we don't, and we must face the fact that Adolph Eichmann, Adolph Hitler, and the killing doctors of Hadamar, who did not value many kinds of human beings, were correct. It would then follow that the Final Solution, legal in Nazi Germany, was neither illegal nor unjust. Instead, it was the judges at Nuremberg and Frankfurt and Jerusalem who acted illegally and unjustly.

On the other hand, if humans, or nonhuman animals, are inherently valuable, then we ignore their value at the dreadful price of acting toward them with monumental injustice. In the first century B.C., Cicero wrote that "there will not be different laws at Rome and at Athens, or different law now and in the future, but one eternal and unchangeable law will be valid for all nations and all times."[13] "Wicked and unjust statutes," he claimed, "were anything but law."[14] Natty Bummpo, the hero of James Fenimore Cooper's *Leatherstocking Tales* of life on the eighteenth-century American frontier, earthily said, "When the colony's laws, or the king's laws, run a'gin the laws of God, they got to be onlawful, and ought not to be obeyed."[15] While the Frankfurt judge almost certainly had never heard of Natty Bummpo, he may have had Sophocles or Cicero on his mind, perhaps on his desk, when he wrote in judgment of the killing doctors of Hadamar. If they were right, then slavery was wrong even if everyone thought it was right. Then Hans and Sophie Scholl and the judges of Nuremberg and Frankfurt and Jerusalem were right and the Nazis were wrong. The Confederates were wrong and Mr. Lincoln was right.

"When I say that a thing is true, I mean that I cannot help believing it"

The question of whether humans should have fundamental legal rights must be answered, said Isaiah Berlin, by invoking the myriad, multi-faceted, and complex ways in which we "determine good and evil, that is to say, on our moral, religious, intellectual, economic, and aesthetic values." What rights we should have is "bound up with our conception of man, and of the basic demands of his nature."[16] The question of whether animals should have fundamental legal rights should be answered in a similar way, similarly bound up with our conceptions of who they are, the demands of their natures, and how we determine good and evil.

In the rest of this chapter we will seek to understand why humans are entitled to fundamental legal rights, for if we do not understand that, we cannot understand why nonhuman animals should have them. Western legal tradition determines human eligibility for fundamental legal rights by using values to channel "reasoned judgment."[17] We often hear of values. Family values. Judeo-Christian values. Community values. Values are the building blocks of our moral lives. They dwell within our souls. They mark the outposts of our irreducible beliefs.[18] Values have both a subjective, or *feeling*, aspect to them (we'll call them "beliefs") and an objective, or *thinking*, aspect. Decisions about fundamental legal rights pivot around them.

Beliefs are personal. We sometimes believe almost by instinct. Our beliefs need not be logical; they may even be irrational. They are often unconscious. Yet we credit them as true and may place them beyond challenge, though we cannot prove them. Anyone who tries to reshape them has her work cut out. Sophisticated judges understand that they decide cases under their influence. "When I say that a thing is true," declared Oliver Wendell Holmes, "I mean that I cannot help believing it."[19] As with us all, judges are occasionally swept upon the current of Holmes's "cannot helps." As did Holmes, U.S. Supreme Court Justice William Brennan understood that when judges decide cases, "emotional and intuitive responses . . . often speed into our consciousness far ahead of the lumbering syllogism of reason."[20] The neurologist Oliver Sacks recounts the story of a judge who suffered brain damage that stripped him of all emotion.

It might be thought that the absence of emotion, and of the biases that go with it, would have rendered him more impartial—indeed, uniquely qualified—as a judge. But he himself, with great insight, resigned from the bench, saying that he could no longer enter sympathetically into the motives of anyone concerned, and that since justice involved feeling, and not merely thinking, he felt that his injury totally disqualified him.[21]

A recent experiment provides scientific support for what Holmes, Brennan, and Sacks's brain-damaged judge intuitively understood. Card players were shown to start gambling advantageously before they could consciously articulate what strategy they were using. They displayed physical signs of stress whenever they pondered a move that was, in fact, risky, though they were not consciously aware of that risk. The experimenters concluded that a decisionmaker's prior emotional experiences may trigger "covert biases" that precede conscious reasoning and can influence decisions without the decisionmaker ever knowing it.[22]

Beliefs may change with time, place, religion, culture, history, and individual psychology. Values can shift within a single individual as feelings and attitudes change. The apostle Paul, who was led blind and helpless into the city of Damascus, had little in common with the arrogant Pharisee Saul, who had sallied forth from Jerusalem just a few days before, eager to find and bind the followers of Jesus. Indeed, Paul's conversion "on the road to Damascus" has come to symbolize how very rapidly beliefs can change. The beliefs of entire societies can alter. Today, with the exception of slavers in Sudan, nearly everyone on earth believes that human slavery is wrong. But three hundred years ago that belief was rarely encountered. Three thousand years ago it did not exist.

When *evidence* causes a judge to believe that a fact that does not affect his core beliefs is true, contradictory evidence can relatively easily convince him that the same fact is false. He may believe the oral testimony of a witness that she has never been to the house at 2221 North Dayton Street. But when he learns that the witness's fingerprints were lifted from a vase in its dining room, from a photograph hanging on the wall of its living room, and from the refrigerator standing in its kitchen, he may abandon his original belief without much of a mental struggle. But as we will see, when facts and opinions trespass upon core beliefs, a judge may not be flexible. He may believe that a fact is true, not because of the evidence but for personal or social reasons. If this belief is an emotional keystone in the structure of his world—one of Holmes's "cannot helps"—even starkly contradictory facts will often fail to convince him that his core belief is wrong.[23]

THE ADVANCE OF KNOWLEDGE
FUNERAL BY FUNERAL

We stubbornly fight to preserve our core beliefs, consciously and unconsciously, fair and foul, any way we can. We may ignore anomalous data or flat-out reject or exclude them from consideration. We may decide to hold them in abeyance, intending to deal with them "later." We may reinterpret and refashion them so that they no longer contradict our beliefs.[24] We may use what knowledge we have of arguments that oppose the threat to construct a barrier to arguments that threaten our beliefs.[25] If none of these mental tricks works and contradiction begins to penetrate our defenses, we may consent to shave just the periphery of a belief, thereby allowing its core to survive.[26] In the unlikely instance in which we find ourselves forced to alter a core, we may depend upon its resilience and ability slowly to resume its original shape. In short, beliefs survive unless we are strongly motivated to examine contradictory data with as unbiased a mind as we can muster and are both able and willing to think deeply about it.[27]

In 1610, Galileo turned his primitive telescope past the dome of St. Anthony's and into the Paduan skies. What he saw gave the Catholic theologians and supporters of Ptolemy and Aristotle a spectacular opportunity to demonstrate how belief preservation works. St. Thomas Aquinas's recasting of Aristotle's universe—ordained by God as earth centered and human centered—had long hardened into dogma. Celestial bodies were perfect spheres, mirrorlike, unchanging, and fixed into an equally perfect and unchanging heaven. Galileo, however, spotted four small bodies orbiting Jupiter in the way that the moon orbited the earth. When seen through the telescope, the phases of Venus so strongly resembled the phases of the moon that they could have resulted only from solar orbiting. As his eyes swept the sky, Galileo realized that the earth was not the stationary center of a universe around which everything else turned. This was going to be a problem.

Celestial bodies were believed to be composed not from the four elements from which everything earthly and changeable was made but from another unearthly and unchangeable element. But as Padua slept, Galileo watched shadows slowly build and fall from mountains that soared four miles above a lunar surface pitted with valleys and craters. This was going to be another problem. Almost as upsetting, during the day, Galileo watched ugly black spots roam the reflected solar surface.

How would the Aristotelians and Catholic theologians react to these observations? Caesare Cremonini, a prominent Aristotelian from Padua, scorned to peer through the telescope. He said it made him dizzy. Some refused to look on the grounds that if God had meant for humans to acquire knowledge in such a way, He would have equipped us with telescopic vision. Others muttered that they just hadn't gotten around to looking; perhaps they would later. Magini, who had beaten Galileo for a mathematics professorship twenty years before, assembled a group of observers in Bologna. Galileo stood beside his tube as the men peered through it. We can imagine what he thought as each man straightened and swore that he could see no satellite near Jupiter. Others were more creative in their dismissals of what Galileo saw. Father Claviius, the most respected mathematician in the world, thought that the lunar mountains were optical illusions. He suggested that perhaps Galileo had built fake sunspots into the telescope and that he, Claviius, could manufacture a similar instrument. Others proposed that some of the new objects were merely imperfections of the observer's own glasses. Or his eyes. Or they were merely crystals floating inside the instrument. Or that since limbs and muscles allowed animals to move and the earth had no limbs or muscles, it could not move.

Ludovico della Colombe, a Florentine philosopher, conceded there could be mountains on the moon. But if there were, these peaks were deeply buried inside the necessarily perfect and transparent crystal sphere. Because Galileo thought they were on the lunar surface, he erred. Colombe, whose name means "doves" in Italian, became Galileo's most persistent and irrational critic. Galileo twisted Colombe's name into "pigeons" and referred to the detractors who were irrationally fighting to protect their core Aristotelian beliefs as the "Pigeon League."

Then there was the problem of the Bible. The Word of God said that the sun rose and set. God Himself had stopped the sun to give Joshua and the victorious Israelites more daylight in which to slay the Amorites. One Psalm spoke of the sun circuiting the earth, another of a world that "cannot be moved." But if the earth was just another planet orbiting the sun, there must be people living on the others as well, for God made nothing in vain. But how could they have descended from Adam? And what about the Flood? Ominously, Cardinal Robert Bellarmine, next to Pope Paul V the most powerful man in Rome, warned that Scripture was always correct and could be reinterpreted only when a scientific truth was incontrovertible. Scientific truths, of course, never are. And what man could argue with a straight face that the earth did anything but stand stock-still?[28]

Galileo and the Pigeons were in thrall to conflicting core beliefs. The historian of science Thomas Kuhn observed that "[m]en who believed that their terrestrial home was only a planet circulating blindly about one of an infinity of stars evaluated their place in the cosmic scheme quite differently than had their predecessors, who saw the earth as the unique and focal center of God's creation."[29] In 1610, Galileo hoped that the objects and phenomena that his telescope revealed alone would convince the Pigeons to change their minds. By 1629, he understood that changing the minds of Pigeons was not merely a matter of matching beliefs to fact. In the book that had him hauled before the Inquisition, which threatened him with torture and then placed him under house arrest for the rest of his life—*Dialogue Concerning the Two Chief World Systems*—the wise man Sagredo springs to the defense of the fool, Simplicio. "[I]f you wish to persuade one to abandon an opinion which he has imbibed with his mother's milk, you will need powerful reasons." This stubbornness of the Pigeons has been explained as coming from "nothing more than a mental block, but it was so deep-seated, so ingrained, that only a few very independent spirits would or could make the break."[30] But it went deeper than that.

In a book called "the most influential work of philosophy in the latter half of the twentieth century," Thomas Kuhn explained.[31] Even powerful reasons might be insufficient to make a break. Because "a nonrational incommensurability" exists between competing paradigms, roughly defined as a common vision produced by shared beliefs, a switch can never be compelled by logic or experiment alone. There is only recourse to competing values.[32] "Just *because* it is a transition between incommensurables," Kuhn wrote, "the transition between competing paradigms cannot be made a step at a time, forced by logic and neutral experience. Like the gestalt shift, it must occur all at once (though not necessarily in an instant), or not at all."[33] The flashes of intuition that mark a

Pauline-like gestalt shift are quantum and never in transition. They are here. Then they are—there. Kuhn wrote,

...the continued opposition to the results of telescopic observation [wa]s symptomatic of the deeper-seated and longer-lasting opposition to Copernicanism during the seventeenth century. Both derived from the same source, a subconscious reluctance to assent in the destruction of a cosmology that for centuries had been the basis of everyday practical and spiritual life.[34]

It wasn't just that the Pigeons could not burst their coop; at some level, they didn't want to. It was all too frightening.

Many philosophers and historians of science and law alike agree that even anomalous data may not compel their colleagues to change their beliefs about what is true.[35] Copernicus's claim that the earth was not the center of the universe was rejected by most for more than a century, by some for several centuries. The physicist Max Planck complained that "a new scientific truth does not triumph by convincing its opponents and making them see the light, but rather because its opponents eventually die, and a new generation grows up that is familiar with it." Darwin despaired of convincing even his colleagues of the truth of evolution by natural selection.[36] In the face of attacks upon core beliefs, knowledge tends to advance, in the words of the economist Paul Samuelson, "funeral by funeral."[37]

During the 1999 NATO-Serbian War, many attacked NATO's decision to conduct the war entirely from the air, as no war had ever been won that way. The criticism that most convinced me came from John Keegan, an English military historian whose writing I admire. When events proved Keegan wrong, he admitted his mistake: "I turned the thought around for a while and looked at it from several directions, rather as a Creationist Christian might have done on being shown his first dinosaur bones. I didn't want to change my beliefs, but there was too much evidence accumulating to stick to the article of faith."[38] Keegan's exception helps prove the rule. He does not even realize that he is blessed with an unusual ability to change his beliefs, for it is the singular Creationist Christian indeed who fails to stick to his Creationist guns even after examining his first dinosaur bones.

BELIEVING IS SEEING

Incommensurabilities exist in the vacuum of a common scale of value.[39] Parents don't normally sell their children for any sum, because they value them on a different scale than money.[40] For Holmes, such "[d]eep-seated preferences" were akin to "liking a glass of beer." Two people who hold incommensurable values can create common ground when one or both change their values, empathize, or agree to disagree and work toward a common goal. Einstein thought it "a hopeless undertaking to debate about fundamental value judgments. For instance, if someone approves, as a goal, the extirpation of the human race from the earth, one cannot refute such a viewpoint on rational grounds."[41] Beliefs "can not be argued about," Holmes said, "and therefore, when differences are sufficiently far-reaching, we try to kill the other man rather than let him have his way."[42] As the seventeenth century turned, inquisitors pushed an iron stake through Giordano Bruno's tongue as they carted him to a fiery death in Rome's Plaza of Flowers, so that he could not pollute the minds of the jeering onlookers with his heresies one last time. Galileo was made to kneel before the Inquisitors-General of Pope Urban VIII and "abjure, curse, and detest" his belief that the earth moved. Appomattox decided that the positive liberty of the southern slaveholder would not trump the negative liberty of the southern slave.

Beliefs are incommensurable when differences of religion, tradition, or culture cause people to value something so divergently that no rational way exists to choose or because particular kinds of comparisons are considered inappropriate, immoral, or unjust in that they degrade or depreciate important beliefs. Perhaps one man "cannot help" but believe that the value of an individual of another race, sex, religion, nationality, or species is merely instrumental. This was how many Southern slaveholders viewed both their slaves and domesticated animals. One of them, Senator William Harper, told the South Carolina Society for the Advancement of Learning that "[i]t is as much in the order of nature, that men should enslave each other, as that other animals should prey upon each other."[43] It should come as no surprise that Harper asked rhetorically in the same address, "Who but a driveling fanatic, has thought of the necessity of protecting domestic animals from the cruelties of their owners?"[44] But another person "cannot help" believing that the value of that individual is not instrumental at all but is intrinsic, important because of the individual's nature. This was the belief of many an Abolitionist.

Reigning scientific paradigms dictate "not only what sorts of entities the universe does contain, but . . . those that it does not."[45] Paradigms blind believers to entities that are not supposed to exist. If they do exist, the paradigm cannot. In other words, what scientists see often depends upon what they believe *can* be seen.[46] But paradigm blindness is not confined to science. Even "one's ethical . . . framework is determined by what entities one is prepared to notice or take seriously."[47] And so is one's legal framework. Animals are invisible to law today because paradigm-blinded lawyers and judges long ago stopped entertaining the thought that nonhumans could possibly be legal persons. The most deeply embedded beliefs are the hardest to dredge from the unconscious level of "feeling right" to consciousness, where they stand some chance of reasoned examination.[48] That is why the ancient belief that "all law was established for men's sake" presents such a daunting obstacle to judges' being able to "see" the legal personhood of any nonhuman animal.

But incursions upon the traditional incommensurability between human and nonhuman animals have begun. In April 1999, the primatologists Roger and Deborah Fouts presented the results of some of their decades-long work with Washoe, Loulis, and the other chimpanzees at the Chimpanzee and Human Communication Institute at Central Washington University to thirty Washington State appellate judges. One judge, who obviously believed that "all law was established for men's sake" commented that it "[c]ontributed nothing to my role as a judge." But the thinking of a second judge, Supreme Court Justice Faith Ireland, provides a glimpse into the legal world of the coming millennium in Kuhnian terms. Justice Ireland wrote a fellow judge that the Foutses' talk "was a paradigm-shifting experience for me and challenged some of my basic assumptions and presumptions on the subject . . . The ethical challenges . . . have many parallels in our historic experience of judging, such as slavery, women's rights, and desegregation."[49] When I wrote to Justice Ireland to inquire into the process by which a judge such as she can change from thinking that the interests of nonhuman animals are irrelevant, or even a waste of time, to thinking that they may be worthy of consideration, she responded,

"What opened my mind was to see the dramatic portrayal in so many ways of how fine the line is between man and chimpanzee."[50]

A common scale of value is under construction from both sides. Nonhuman life is being infused with intrinsic value. The Preamble to the United Nations World Charter for Nature states that human beings are "a part of nature," that humanity's "[c]ivilization is rooted in nature," and that "[e]very form of life is unique, warranting respect regardless of its worth to man, and to accord other organisms such recognition, man must be guided by a moral code of action."[51] Respected international law commentators have argued that the legal right of whales to life is, after the better part of a century of development, becoming a part of binding customary international law.[52] In 1996, the British government banned the use of great apes in biomedical research. "This," the government said, "is a matter of morality. The cognitive and behavioural characteristics and qualities of these animals mean that it is unethical to treat them as expendable in research."[53] In October 1999, the New Zealand Parliament passed a statute that forbid the use of a nonhuman hominid in research testing or teaching, unless the director-general determines that the use of the non-human hominid is in her best interests or in the interests of her species, and that the benefits to her species are not outweighed by the likely harm to her.[54] The U.S. Endangered Species Act forbids harm to individuals of endangered or threatened nonhuman species or their habitats, except under unusual circumstances.[55] In some respects, at the species level it has even reversed the traditional incommensurability between human and nonhuman interests, treating endangered species as bearing "incalculable value" and affording them "the highest of priorities," even over the economic priorities, at least, of human beings.[56]

At the same time, the intrinsic value of human life is becoming more commensurable with other values. For more than a millennium after St. Augustine's fifth-century condemnation, suicide was forbidden by the Catholic Church under all circumstances.[57] In Kant's eighteenth-century view, suicide treated human "value as that of a beast . . . it degrades human nature below the level of animal nature and so destroys it."[58] As early as the thirteenth century, English common law made crimes of both suicide and assisting suicide.[59] But by 1798, six of the new American states had abolished penalties for suicide.[60] In 1900, fewer than half the states punished suicide; today none do.[61] A majority of the U.S. Supreme

Court recently left open the door to the possibility that some circumstances might support a constitutional right to physician-assisted suicide.[62] Meanwhile, the legislature of every American state has replaced a common law dictated by the Hebrew incommensurability of human life with Mesopotamian-like wrongful death statutes that enact a more commensurable and utilitarian view that makes human life at least partially compensable by money.[63]

The Hebrew incommensurability between human and nonhuman animals might have been appropriate in a society in which law was formally based upon a religion that demanded it or even when secular knowledge supported it. But modern Western law is no longer theologically based. The Great Chain of Being snapped long ago. Because even the bedrock values of an individual and a society can shift, we must choose among ultimate values to determine what is right.[64] However, the traditional responsibility of Western judges to reach *reasoned* decisions forbids their arbitrarily picking among ultimate values the way they might choose the color of their next car. We will see that there is not just an irreducible feeling component to valuing but an irreducible thinking component as well.[65] All values are not necessarily equal, and the choice of any one is not necessarily arbitrary.

Justice Brennan believed that a judge's "internal dialogue of reason and passion does not taint the judicial process, but is in fact central to its vitality."[66] It promotes justice by tempering reason with conscience and conscience with reason. But there must be a dialogue. When a core belief reaches a certain strength, dialogue stops and a monologue begins. I have encountered many courtroom Colombes, Claviiuses, Bellarmines, Maginis, and Cremoninis in the course of twenty years of fighting for the interests of nonhuman primates, dogs, bald eagles, dolphins, cats, goats, sheep, parrots, deer, and other species in courts across the United States. Some have been only too anxious to sweep their insignificant lives and deaths and sufferings out of their courtrooms. In 1993, one Chicago federal judge became so enraged when I filed a lawsuit on behalf of a State Department employee whose two beloved African Gray parrots had vanished from an Africa-bound flight that he ordered me to appear before him on twenty-four-hours notice to explain why I was clogging his court with such nonsense. Dissatisfied with my response that my client was simply seeking justice for stolen family members, hoping even to recover them, he ordered me to appear in his courtroom the next business morning, and every morning thereafter, until the case vanished from his docket. I was living and practicing law in Boston.

Well-educated, intelligent, and articulate, these judges live in an intellectual world that Galileo and Darwin have not yet penetrated. Perhaps religious texts have reminded them from childhood that they belong to the sacred species for whom the universe was made. Legal texts say that all law was made for them, too. Like Cremonini, some disdain to educate themselves further. Just thinking about the problem makes them dizzy. Others never get around to it. Like Magini, still others look at modern science but cannot see it. They are blinded by a belief that all and only human beings have some unprovable quality, such as an incorporeal soul, that justifies their legal rights. These judges are deaf to arguments in favor of extending legal personhood beyond human beings.

Are nonhuman animals thereby condemned to perpetual legal thinghood? Not at all. Judges are beginning to arrive who ripened after World War II, when the principles of equality and liberty were sweeping the world in response to the horrors of Fascism and Nazism and as the modern environmental movement was beginning to sprout. This decade, judges who matured alongside the newer animal-rights movement have begun to take their places. Many judges from these slowly filling pools will have watched Jane Goodall respectfully document the lives of the chimpanzees of Gombe on *National Geographic*, will understand and believe basic principles of Darwinian evolution and ecology, will reject the hierarchy of the Great Chain of Being, and will not be personally invested in the wholesale exploitation of animals or saturated with religious and other arguments against their legal personhood. They will be better equipped to examine the objective data and hear—not just listen to—the supporting arguments. They will begin to rattle the cage.

ABSOLUTE BARRIERS

The "thinking" part of values is no less important than are beliefs. Critics of Thomas Kuhn (he of the paradigms) have argued that he strayed too far in the direction of incommensurability in describing the processes of scientific change.[67] And despite the claims of a small number of scientific illiterates that even the mathematical constant pi and Einstein's famous $E=mc^2$ equation are merely the products of a specific culture and not a description of the physical world, scientific truth is not just a matter of either taste or fashion.[68] Some theories really can be tested and found to be correct or wanting. But Kuhn had latched onto a truth about how we humans believe. The physicist Steven Weinberg said that twenty years ago, when arguing for the truth of a certain theory:

[Because] any other way of interpreting the data was ugly and artificial, some physicists answered that science has nothing to do with aesthetic judgments, a response that would have amused Kuhn. As he said, "The act of judgment that leads scientists to reject the previously accepted theory is always based upon more than a comparison of that theory with the world." Any set of data can be fit by many different theories. In deciding among these theories we have to judge which ones have the kind of elegance and consistency and universality that make them worth taking seriously.[69]

Of course, the more strongly a theory threatens core beliefs, the bloodier will be the battle for acceptance. Whether Holmes's beer tastes good may be lustily debated for hours without either the beer lovers or the beer haters able to offer a single principled argument. But scientific judgments turn more on objective facts than does beer tasting.

No magic proportion of thinking to believing always results in a "reasoned judgment." Judging is neither like baking a cake nor conducting a chemistry experiment. But just as inflexible core beliefs can produce a judge who makes arbitrary decisions, excessive thinking can spawn judges contemptuous of imagination, instinct, and intuition. Because they have "no room for compassion in the cold calculus of judging," they issue chilly opinions that deny or disregard the realities of the lives they affect.[70] Unable to empathize, "to see one thing as another, to see one thing in another," oblivious to their biases, they cannot hope to do justice, but only to decide cases.[71] Worse, this disdain may lead them to undervalue those whom they believe to be less rational than they—women, children, foreigners, and animals.[72]

Three centuries ago, advocates of then-emerging liberal democratic values began to insist that the violation of fundamental human interests was wrong and that they must be protected by a

sturdy barrier of near-absolute legal rights. In 1864, the year between the Emancipation Proclamation in which he freed the slaves
in the rebellious Southern states and signed the constitutional
amendment that abolished slavery in the whole United States,
Abraham Lincoln wrote that "if slavery is not wrong, nothing is
wrong."[73] In May 1999, the writer Susan Sontag examined the objection to the NATO bombing of Kosovo on the ground that it was
an illegal intervention in the affairs of a sovereign state. "Imagine,"
she wrote, "that Nazi Germany had had no expansionist ambitions
but had simply made it a policy . . . to slaughter all the German
Jews. Do we think a government has the right to do whatever it
wants on its own territory? Maybe the governments of Europe
would have said that sixty years ago."[74] No longer. These liberal
democratic values were solidified, and their acceptance acceler-
ated in the aftermath of World War II, so that almost universal
agreement now exists that fundamental rights exist in some objective sense.[75] That punishment without trial, torture, and murder,
"even if they are made legal by the sovereign," wrote Isaiah Berlin,
"cause horror even in these days . . . springs from the recognition
of the moral validity—irrespective of the laws—of some absolute
barriers to the imposition of one man's will upon another."[76] "Dignity-rights" rather than "human rights," the term that has been
used most often since World War II, more accurately describes
these fundamental rights. This is because "dignity-rights" emphasizes that fundamental human rights derive not merely from being
a member of the species *Homo sapiens* but from the dignity that is
associated with qualities alleged or assumed to be shared by human beings everywhere.

The thinking component of Western law has led to wide agreement on core values and principles. Liberty and equality are preferred over unfairness, invidious discrimination, and arbitrary
restraints. The dignity-rights of nonhuman animals should be derived from these same principles and values and in the same ways.
This is not just because it is the strongest ground for securing the
dignity-rights of nonhuman animals, though it is. It is not just because both reason and fairness demand it, though they do. Arbitrary refusals to extend dignity-rights to those who would be
entitled to them, if only they were human, looses a virulent injustice upon beings entitled to justice. Their awful irrationality, arbi-

trariness, and invidiousness undercut the foundations of our own fundamental rights.

LIBERTY

Today liberty "stands unchallenged as the supreme value of the Western world."[77] It was not always so. One student of liberty has argued that it, and not slavery, is the "peculiar institution," because for hundreds of years it was valued only by Westerners and those influenced by Western values, and then only from the rise of the Greeks in the sixth century B.C.[78] But once liberty appeared, it seized the Western imagination.[79] Today, liberty rights emphasize the importance of certain fundamental human interests.[80] Chief among them is bodily integrity, for no one can long endure a life of beatings, torture, and enslavement and flourish. This is why the Supreme Judicial Court of Massachusetts recently labeled "the right to live in physical security" as "the most basic human right."[81]

Few siding with Antigone over Creon, with Lincoln versus the Confederacy, or with the Frankfurt and Jewish judges against the killing doctors of Hadamar and Adolph Eichmann believe that opposition to slavery or genocide derives from some indefinable absolute and leave it at that. There must be some kind of a thinking yardstick that a judge can use to measure right and wrong, justice and injustice. Actually, there might be two.

At Nuremberg, Frankfurt, and Jerusalem, the judges borrowed Cicero's changeless measure. The values they used were inherent, immutable, and self-evident, no less laws of nature than are speed of light and the pull of gravity. They exist apart from human experience. The most famous example rests in a bomb-proof vault in Washington, D.C. The Declaration of Independence says: "We hold these truths to be self-evident, that all men are created equal, that they are endowed by their Creator with certain unalienable Rights, that among these are Life, Liberty, and the pursuit of Happiness." Humans may strive mightily to alter or extinguish these standards, as the Nazis did when they set the Final Solution into motion and began to execute the infirm and incurably ill. But every attempt will flounder.

There is a second yardstick, one that we create ourselves. Its standards are not immutable but change as new facts are discovered and society's values change. Reason plays a substantial role in any decision made with this yardstick, and any final decisions must be consistent with the facts as we can best determine them.

These thinking yardsticks are used round the world in support of dignity-rights. A phalanx of post–World War II international treaties, agreements, declarations, and resolutions declare the "inherent dignity," "fundamental freedom," and "inalienable rights" of human beings. Human liberty is not a gift from the state; it derives from human nature and society. Bodily integrity is *always* preferred to torture and genocide. Bodily liberty is *always* preferred to slavery. These dignity-rights cannot be waived. Their violation can never be excused. Here are just a few examples.

The Preamble to the Convention Against Torture and Other Cruel, Inhuman or Degrading Treatment or Punishment provides that "recognition of the . . . inalienable rights of all members of the human family is the foundation of freedom, justice and peace in the world" and declares "that those rights derive from the inherent dignity of the human person."[82] The International Court of Justice has said that the object of the Convention on the Prevention and Punishment of the Crime of Genocide "on one hand is to safeguard the very existence of certain human groups and on the other to confirm and endorse the most elementary principles of morality."[83] The Stockholm Declaration on the Human Environment acknowledges the human "fundamental right to freedom."[84] The most powerful international laws are *jus cogens*, which is Latin for "compelling law." With roots plunging deep into the soil of a natural law that Antigone would instinctively embrace, these treaties prohibit torture, slavery, and genocide and function as an unwritten international constitution, a sort of "natural law of nations."[85] Their principles can never be abolished, never be annulled, never be waived.[86]

Not just international law but the domestic law of nearly every country recognizes dignity-rights. The brand-new South African Constitution provides that "[e]veryone has inherent dignity and the right to have their dignity respected and protected."[87] The Constitution of Bolivia assures its people that "[t]he dignity and freedom of the person are inviolable."[88] The Japanese Constitution guarantees "fundamental human rights" that are "eternal and inviolate."[89] Both "human rights" and "the dignity of man" are so "inviolable" under the German Basic Law that they can never be altered, even by constitutional amendment, a state of affairs of which Cicero would have approved.[90]

Such common law rights as physical security, privacy, and due process are anchored in natural law notions of justice, fundamental fairness, and reasonableness.[91] No American government may violate fundamental notions of liberty, justice, decency, or fairness or act in ways that shock the conscience of its judges. One of them, U.S. Supreme Court Justice Stevens, has written that "law . . . is not the source of liberty, and surely not the only source. I had thought it self-evident that all men were endowed by their Creator with liberty as one of the cardinal unalienable rights. It is that basic freedom which the [Constitution] protects."[92] Often inspired by the Declaration of Independence, each of the constitutions of the fifty states declares that every person possesses certain fundamental, often explicitly natural, rights.[93] Even though often attacked by the heirs of Creon, these fundamental principles have proven so hardy that international and national courts and legislatures commonly deploy them to shield human beings against all manner of threatened trespasses.[94] We would be shocked and outraged if they did not.

EQUALITY

Aristotle taught Alexander the Great what Plato had taught him; equality means that likes should be treated alike.[95] So it remains today.[96] Unlike the right to liberty, determined by independently examining a being's nature, the right to equality demands that situations or beings be compared. If alikes are treated differently or if unalikes are treated the same way for no good and sufficient reason, equality is violated. But likes and unalikes are not as easy to determine as one might first think.

In a recent article in *Nature*, linguists demonstrated that how humans categorize even such a basic biologically determined perception as color affects how like or unalike are the ways we classify different colors.[97] English-speaking peoples routinely distinguish between green and blue, while the Berinmo tribe of hunter-gatherers in Papua New Guinea do not. What English speakers call yellow, Berinmos divide between "wor" and "nol," while English speakers have the same problems classifying their previously uncategorized yellows as "wor" or "nol" that Berinmos have in classifying blues and greens. One investigator concluded that "categories affect the way we perceive the world."

Victorian England's scientists wrote their racist and sexist beliefs into their biological taxonomy.[98] Dr. Samuel Morton, a nineteenth-century American advocate of dividing humans into separate species, spiced up his racist rhetoric with allegedly objective skull measurements that he said proved that Caucasians had much larger skulls than blacks or Mongolians. But when Stephen Jay Gould remeasured these very same crania almost one hundred and fifty years later, he found that Morton had imagined the disparities. Morton was no fraud, Gould emphasized; his errors had been made unconsciously.[99]

Early in the nineteenth century, that astute observer of everything American, Alexis de Tocqueville, wrote that "democratic communities have a natural taste for freedom . . . But for equality their passion is ardent, insatiable, incessant, invincible; they call for equality in freedom; and if they cannot obtain that, they still call for equality in slavery."[100] He was exaggerating, of course, though not much, for even as he wrote, 2 million Americans were enslaved and the hottest of proslavery firebrands, John C. Calhoun, was vice president of the United States and figuring to move up. But equality *is* deeply rooted in Western imagination, culture, and law. It played an important role in toppling the natural hierarchy of the Great Chain of Being and was a linchpin in the American ("We hold these truths to be self-evident, that all men are created *equal*") and the French ("*Equality*. Liberty. Fraternity.") Revolutions.

Equality is violated when some unfairly enjoy benefits that others who are like them do not enjoy or escape burdens that others who are like them must bear.[101] But the promise of equality always collides with the reality that lines must be drawn somewhere. The question is where. Any two beings and any two situations are infinitely different and infinitely alike. By nature, everyone tends either to "lump," that is, they emphasize similarities, or to "split," they emphasize differences. Judges' beliefs may act as "differences and similarities sieves."[102] In the same way, lumper biologists are often drawn to similarities between creatures and may group those with a small number of similarities into the same species, even though a large number of differences may exist. Lumper physicists argue that we understand the fundamental laws of nature the way blind men understand elephants. They believe that someday all the seemingly separate physical laws of the universe will be lumped into a single "beautiful" and universal law.

Splitters may seize upon tiny differences to establish the unalikeness that justifies for separate categories. Engineers and trial lawyers are notorious splitters. They tend to see the world as messy and riddled with exceptions. The nineteenth-century Harvard biology professor Louis Agassiz was a splitter supreme. He not only claimed that each of the human races was a different species, but he once set up three separate genera (the taxonomic classification above the level of species) of fossil fishes from what were eventually determined to be the teeth of a single fossil fish.[103] This was as if the famed paleontologist Louis Leakey had stumbled upon a few teeth belonging to a single *Homo erectus* individual while digging at Olduvai Gorge and then announced that he had found the molars not just of a single *Homo erectus* but of an *Australopithecus* and a chimpanzee to boot.

Those who struggle to extend legal personhood to nonhuman animals may find themselves charged with "unreasonable lumping," that is, accused of emphasizing overly general criteria for legal personhood and erroneously thinking that one or more of the essential elements is irrelevant. Whenever reformers have agitated to transform such "legal things" as slaves, women, children, and fetuses into "legal persons," their ideas are, in Professor Christopher Stone's words, "bound to sound odd or frightening or laughable. This is partly because until the rightless thing receives its rights, we cannot see it as anything but a thing for the use of 'us'— those who are holding rights at the time."[104] I will inevitably be charged with unreasonable lumping for demanding legal personhood for chimpanzees and bonobos and ignoring allegedly relevant differences between human and nonhuman animals. This book is one long argument against that charge.

Legal splitters invariably want to limit legal personhood to those who have it. But their arguments are often based upon overly specific criteria, and they end up erroneously believing that one or more of the nonessential elements of legal personhood is essential.[105] I will argue in Chapter 11 that being human is an overly specific criterion for legal rights and not an essential element of legal personhood. "Unreasonable splitting" is the charge I levy throughout this book against those who refuse to extend legal personhood, for no adequate reason, to chimpanzees and bonobos.

All this means that inequality can be fair or it can be unfair. Inequality is fair when it turns on a relevant and objectively ascertainable difference.[106] If one class of beings poses a lesser harm or deserves a greater benefit, then discrimination in favor of that class is probably rational and not arbitrary.[107] If two classes pose the same harm or deserve the same benefit, then discrimination in favor of one class against the other is probably irrational and arbitrary.[108] And discrimination in favor of a class that poses a greater harm or deserves a lesser benefit than another class is not just arbitrary and irrational; it is perverse.[109] Imagine that two high-school seniors, who tied for class valedictorian, and a third, who was "socially graduated," apply to the University of Wisconsin. So as not to muddy the waters, we'll assume they are identical triplets. If the university accepts the two valedictorians and rejects the social graduate, it acts rationally and nonarbitrarily. If the university accepts one valedictorian and rejects the other but accepts the social graduate, it acts irrationally and arbitrarily. If the university accepts just the social graduate and rejects the two valedictorians, it acts not just arbitrarily and irrationally but perversely.

Unequal treatment must be not just rational; it must be morally acceptable.[110] Even rational classifications can infringe fundamental rights or be born of naked prejudice and fail to protect the powerless or despised, especially when based upon immutable characteristics or status.[111] The U.S. Supreme Court consistently strikes down statutes that violate the constitutional equality for just these reasons. In 1886, it erased an ordinance "fair on its face and impartial in appearance" that arose from a "hostility to" Chinese.[112] A Florida statute that punished the habitual cohabitation of men and women of different races was clearly rational in light of its purpose to deter interracial sexual relations. But the Court eliminated it in 1964 because this purpose was abhorrent.[113] In 1996, the Court voided a Colorado constitutional amendment that disabled state government from enacting or enforcing laws giving preferences to homosexuals because it "identifies persons by a single trait and then denies them protection across the board."[114] As ethical standards evolve, what once appeared reasonable may no longer seem that way. Judges once approved racially discriminatory systems of education because to their nineteenth-century eyes, these were not unreasonable.[115] But by the 1950s, these "separate but equal" systems were widely seen as inherently unequal.

Not just the American Constitution but the common law has criticized "unfair inequality" as "manifestly contrary to the first principles of civil liberty and natural justice"; it requires classifications be reasonable.[116] Judges insist that classifications not "unnecessarily produce incongruous and indefensible results" nor generally result in "a wrong inflicted for which there is no remedy."[117] That is why the Wisconsin Supreme Court allowed suits for injuries suffered by a viable fetus. To have done otherwise would have been arbitrary.[118] Many judges have interpreted state wrongful-death statutes to allow the estates of both viable and nonviable fetuses killed in utero to sue as "persons," because to do otherwise would be arbitrary, inequitable, unreasonable, and unjust.[119]

Proportionality rights involve another kind of equality that, as we will see in Chapter 11, carries great potential for the legal rights of nonhuman animals. The ubiquitous and very busy Aristotle explained these rights as well. Just as likes should be treated alike, sometimes unalikes should be treated not just differently but *proportionately* to their unalikeness.[120] Yale Law School professor Alan Gewirth has proposed an important principle of proportionality that attributes legal rights in proportion to the possession of qualities to less-than-normal degrees. It goes like this: Anyone who possesses a quality that justifies a legal right should possess that right. However, the degree to which a being may *approach* having that quality can make that being eligible for some part of that right.[121] As Gewirth says, "[A]lthough children, mentally deficient persons, and animals do not have [fundamental] rights in the full-fledged way normal human adults have them, members of these groups *approach* having [fundamental] rights in varying degrees, depending on the degree to which they have the requisite abilities."[122]

Washington University professor Carl Wellman has taken this an important step further by postulating three dimensions along which legal rights might vary as one approaches having the necessary quality.[123] First, one might possess *fewer* legal rights. A severely mentally limited human adult or a child might not be able to participate in the political process but should still have the right freely to move about. Second, one might possess *narrower* legal rights. A severely mentally limited human adult or a child might not have the right to move about freely in the world at large but should have the right to move within the confines of a home.

Third, one might possess only *partial elements* of a complex right. A profoundly retarded human might possess a claim-right to bodily integrity but lack the power-right to waive it and thus be unable to consent to a risky medical procedure or the withdrawal of life-saving medical treatment.[124]

Those who demand the recognition of the legal personhood of such nonhuman animals as chimpanzees and bonobos and those who refuse these demands are heir to the disputes between Antigone and Creon, Lincoln and the Confederacy, Hitler and Hans and Sophie Scholl, the killing doctors of Hadamar and their German judge, and Adolph Eichmann and his Jewish judge. There is no doubt where Frederick King, longtime director of the Yerkes Regional Primate Research Center—where Jerom suffered and died–and one of the most prominent critics of the animal-rights movement throughout the 1980s and into the 1990s stands: "I don't believe that rights fall like manna from heaven. Rights are not magical or absolute. I know our Constitution talks about inalienable rights. But I haven't seen any evidence of these. Rights are given by one group to another."[125] In Chapter 11, I will show how liberty and equality determine that chimpanzees and bonobos should have legal rights under the common law. But first we must understand how the common law works and what chimpanzees and bonobos are. These are the purposes of the next three chapters.

7

The Common Law

There was no common law on Sunday morning, October 15, 1066. But a Sussex wayfarer, chancing upon Senlac Bottom at the edge of Telham Hill near Hastings, would instantly have realized that great changes were afoot. Dead Saxons lay twisted and bloody across a long row of oaken shields, swords, and broad-axes, many with arrows sticking from their bodies. A smaller number of more Nordic Normans were sprawled among them, some fallen alongside dray horses, others with a death grip on bows, swords, or crossbows. Near the center of this carnage our wayfarer might have spied an imposing-looking man, obviously used to command, barking orders in French. Beneath the flapping banner of the Pope, he was outlining the foundation for a great Abbey to be built on that field to atone for his slaughter. Its high altar he positioned directly over the spot where lay the rigid corpse of a man, disemboweled by a spear, his right leg and chest sliced open, his face grotesquely swollen where his right eye had been. Had our wayfarer dared inch closer, he might have recognized the corpse of his king, Harold II, though probably not, for Harold had been king for just ten months and dead for half a day. He certainly would not have recognized his new king, Duke William of Normandy.

The Conqueror was laying the foundation not just for his abbey, but for the common law. Over the next hundred years, William and his successors struggled to control the land of England and the intricate mesh of feudal relationships that were tied to it.[1] Centralized royal courts slowly displaced the welter of local courts that had idiosyncratically dispensed Saxon justice for hundreds of years.[2] The "Norman French" and later the French-English hybrid, "law French," spoken in these royal courts over the next 700 years injected legal terms into English law that continue to confound anyone enmeshed in a lawsuit, beginning with "plaintiff" and "defendant"[3] The first treatise on the infant common law soon appeared bearing the name of Glanvill, Chief Justiciar to King Henry II, great-grandson of William the Conqueror.[4] Glanvill simply ignored the welter of local customs and discussed just a single common custom, the law of the King's Court.[5]

Today the common law breathes wherever English is spoken. Yet, after spending much of his life studying and writing about it, one historian confessed that "(t)he greater a man's knowledge of the law, the more hesitant he will be in answering the question: What is common law?"[6] But where historians can hesitate, judges must tread. The 19th century Chief Justice of the Supreme Judicial Court of Massachusetts, Lemuel Shaw, provided a working definition as good as any and better than most. He said that common law "consists of a few broad and comprehensive principles, founded on reason, natural justice, and enlightened public policy, modified and adapted to all the circumstances of all the particular cases that fall within it."[7]

The decisions of common law judges are quite unmistakable. For a century, every aspiring common lawyer has flipped the cover of her casebook on the first day of law school and realized, as did our Sussex wayfarer following the Battle of Hastings, that great changes were afoot. Armed with a bachelor's degree in chemistry, I opened my Property Law casebook in the first minute of my first morning of law school and saw that it was filled with the decisions of judges.

The first decision I read was Shelley's Case. I found it nearly impossible to understand, but much of what I read in those first months of law school I found confusing. It was two decades before I read the story of the young attorney who sent a request to the editors of the *American Law Review* in 1892 begging "a plain, com-

mon-sense, easy-to-be understood definition of the rule in Shelley's Case." The editors promptly refused, "not having," they said,
"the capacity to understand the rule in Shelley's Case, or to acquire
an understanding of it within the limits of a lifetime."[8] I failed that
day to appreciate the legal profession's sense of humor in demanding that I comprehend the incomprehensible . . . as my *first* assign-
ment. But one thing snagged my attention. *Shelley's Case* had been
decided when Shakespeare was a teenager![9] That was my first notice that common law judges tend to value the past simply for having been.[10] But they are not interested in the past the way an
historian or a sociologist or an anthropologist is. They care about
that narrow slice that earlier judges have written. That is quite
large enough for, over 900 years, millions of decisions have accumulated. So when old law starts to wheeze, judges can easily reach
back to replace an old principle here, install a new policy there,
and roll out shiny new law, if they are so inclined.

While some common law judges salaam to the past and others
merely nod, most new common law is forged from old and judges
often reason by analogy to earlier cases. Here's how they do it:
first, they compare the problem before them to others they believe
are similar. Then they try to determine what rule of law those cases
stand for. Finally, they apply this rule to the case before them.[11] The
third step is, thankfully, uncontroversial. Some think that the first
and second steps are really one, because no judge can see any two
cases as similar unless she is simultaneously determining what
rule of law the earlier cases stand for.[12] But whether it's one step or
two, the way that judges do it leads to fights.

Few common law judges have been formally trained in logic.
But, like Moliere's student in "The Misanthrope," who is amazed
to discover he had been speaking prose for forty years, every time
a judge reasons by analogy, she uses a formal logic that resembles
an "induction" followed by a "deduction."[13] A deduction involves
reasoning from a broader proposition to a narrower one. Aristotle's Axiom, which we met in Chapter 2, is an example. If the broad
proposition that all humans have legal rights is true, and you are
human, then the narrower proposition you have legal rights must
be true. You can see how deductive reasoning always leads to
truth. Not so with inductive reasoning. An induction involves reasoning from a narrower proposition to a broader one. Here's the
battleground. If we now start with the proposition that all humans

have legal rights, that doesn't necessarily mean that *only* humans have rights. That may be true. But it may not be. The statement "all humans have legal rights" is true not only if all and *only* humans have rights, but also if all apes have legal rights, if all primates have rights, if all animals have rights, or even if every living creature has legal rights. The probability that inductive reasoning will lead to truth will depend upon how well judges choose from among the infinite number of broader propositions that might possibly exist.

In the end, common law mapmakers rely on just three signposts, precedent, principle, and policy. The skilled common lawyer labors to convince judges that they needn't choose one against the others for, amazingly, all three have converged to compel a decision in his favor. Occasionally this is even true. But much more often they point in different directions and judges must decide which to follow. How are they to decide? It's actually not that hard, for values, temperament, education, experience, even religious training and esthetic sense causes most judges to arrive at many decisions with a predilection to follow one road or another.[14] Some judges will be most receptive to arguments from *precedent*. These boil down to demands that we should keep doing things the way we always have because that's how we've always done them. But no judge can step twice into the same river. Every case she decides will be infinitely similar to and infinitely different from every other case that has ever been or ever will be decided.[15] She will somehow have to fashion order from this chaos by deciding which of the thousands of possible cases that have been decided are "similar" to the one sitting on her desk.

Other judges will tend to favor arguments from *policy*. Policy involves judges predicting what will happen, good and bad, if they rule one way or another. These judges think that law should be used to achieve ends that they think are important, and that they think that society thinks are important too, such as the promotion of economic growth or national unity or democracy or the health or welfare of a community. Still other judges will be drawn toward arguments from *principle*, that is whether their rulings are morally right or wrong.[16]

Judges do not agree, have never agreed, and never will agree on what the ideal proportions of precedent, policy, and principle should be in law. And so their decisions must finally rest upon a "reasoned judgment" channeled through each judge's vision of what law should be.[17] How they choose and how their choices affect the ability to win fundamental common law rights for nonhuman animals will occupy the rest of this chapter.

WHY JUDGES DISAGREE

Imagine that I visit my friend, Sarah, in her spacious London condominium. She offers me a cold beer in a clear bottle that she purchased fresh that morning from the microbrewer owned by a lovely man just around the corner. I quaff the beer, but fail to notice that the rear end of a mouse is bobbing about on the bottom of the bottle until it soggily knocks against my lips. Though I am a friend of live mice, soggy mouse rears from that close up make me very upset. I manage to avoid becoming physically sick. The lovely man around the corner apologizes. Not enough. I sue the microbrewer anyway.

The microbrewer resents being sued and demands that the judge throw my case out. Judge Athena takes the case. As any common law judge would, she cracks open her law books to see how her predecessors handled similar problems. Soon she encounters a famous case, *Donoghue v. Stevenson*, decided by the House of Lords (the highest English Court) in 1932. It seems to resemble mine. Ms. Donoghue was sitting in a cafe with a friend who purchased a bottle of ginger beer for her. The bottle was opaque. After downing most of the ginger beer, she noticed a decomposed snail resting on the bottom. After suffering a nasty bout of gastroenteritis, she sued the manufacturer and won, despite the fact that the law had, until then, limited the liability of manufacturers of defective products to buyers.[18]

Judge Athena decides that *Donoghue v. Stevenson* is similar to my case. But is it? Remember that the probability that her inductive reasoning will lead to truth and justice depends upon how well she chooses from among a large number of possible broader propositions. *Donoghue v. Stevenson* could be characterized at very different "levels of generality." It could mean that an injured plaintiff can sue only if she suffers severe shock and gastroenteritis, any kind of physical and emotional injuries, a physical injury, an emotional injury, or any old kind of injury. The defendant must be a nationwide distributor of ginger beer, a local distributor of ginger beer, a company that manufactures any kind of beer, a company that manufactures any liquids, or a company that makes anything. The plaintiff has to be an animal, a primate, a human being, or a woman. The injury has to have been caused by a snail in an opaque bottle of ginger beer, any decomposing animal in an opaque bottle of beer, any nasty thing in any opaque bottle of anything, anything in any bottle, or anything in any container from which people eat or drink. I haven't come close to exhausting the possibilities and the combinations are endless, but you get the point.

Our judge has no way of knowing what the judges who decided *Donoghue v. Stevenson* sixty-eight years ago would have ruled in my case.[19] But Judge Athena has to decide today whether I should win or lose. If she decides that any company is liable to any human who consumes any tainted drink meant for humans and suffers any kind of injury, I win! But if she decides that only a nationwide company that manufactures ginger beer in opaque bottles with snails that cause serious physical harm to a woman, I lose.

She will have a lot of leeway in deciding. Other judges might fairly and reasonably make different choices. The interesting thing is that most judges believe that a "right" choice exists and that they choose it. But they can't *logically* explain why. The thoughtful Chief Judge of the Seventh Circuit Court of Appeals, Richard Posner, has written that "(a) set of cases can compose a pattern. But when lawyers and judges differ on what pattern it composes, their disagreement cannot be resolved either by appeal to an intuitive sense of pattern or by the methods of scientific induction."[20] Each judge will find the "right" answer for her by filtering the past through the vision of law that she has both consciously and unconsciously been constructing from childhood.

FORMAL JUDGES

Judges with a "formal" vision of law submerge their decisions in existing legal rules. Formal Judges, more or less faithfully, lash themselves to the past in accord with the common law maxim, *stare decisis set non quieta movere*, which roughly means "to adhere to decided cases and not unsettle established things." Lawyers usually call it just plain "*stare decisis*."[21] In other words, judges follow "precedent," rulings set down by earlier judges. They try to treat like cases alike because they think that is the fair thing to do.[22] Better that the law be settled than it be settled right![23] Settled law encourages both the parties to any one case and citizens in general to accept the legitimacy of judicial decisions as coming from something other than the whims of judges.[24]

The rallying cry of the Formal Judge was succinctly penned in the joint opinion of three Justices of the United States Supreme Court in the 1992 abortion case of *Planned Parenthood v. Casey*: "Liberty finds no refuge in a jurisprudence of doubt."[25] Thus, while it was common knowledge that two, and perhaps all three, of these Justices personally disagreed with the reasoning and outcome of *Roe v. Wade*,[26] the 1973 decision that established the constitutional right to an abortion in the United States, they refused to overturn it. Formal Judges rule the way earlier judges ruled, not because those judges ruled correctly, but because that was how they ruled. They will rule that way even if they think that the earlier judges made a mistake or that circumstances or public values or morality have shifted.[27] They want law to be stable, certain, something that people can rely upon, a system of coherent rules that they can look up in law books, then more-or-less mechanically apply. They think that value judgments can be squeezed from the law the way orange juice can be squeezed from an orange, leaving dry pulpy facts that can be compared to other facts.

In one way, the most formal of the Formal Judges bridge the common law system and the civil law system that dominates the legal systems of Continental Europe. While it is not the only value in the civil law, certainty is often thought of as its supreme value.[28] Civil law judges are often said not to make law; legislatures do, and they darn well expect judges to apply, and not interpret, the law they make. Viewing judges suspiciously as potential sources of undesired ambiguity, civil law legislators ideally try to "judge-proof" the law by enacting numerous and detailed statutes that form such a detailed roadmap that a judge can follow it without recourse to her own values and vision of law.[29] Even in the United States, the California lawyer, practicing in a jurisdiction influenced by the Spanish civil law, has to buy more bookcases to hold his statute books than does the lawyer practicing in Massachusetts, which was settled by the English.

Three kinds of Formal Judges exist. Obsessed with the quest for legal certainty and stability, one kind believes that precedents set out very narrow rules for them to follow the way a cookbook might tell them how to make bread. They won't change the recipe. We'll call them Precedent (Rules) Judges. Kathleen Sullivan, dean of the Stanford Law School, has written that

> Rules aim to confine the decision-maker to facts, leaving irreducibly arbitrary and subjective value judgments to be worked out elsewhere. A rule captures the background principle or policy in a form that from then on operates independently. A rule necessarily captures the background principle or policy incompletely and so produces errors of over- or under-inclusiveness. But the rule's force as a rule is that decisionmakers follow it even when direct application of the background principle or policy to the facts would produce a different result.[30]

These "cookbook judges," who bridge the civil and common law systems, want to know as precisely as possible what specific rules earlier judges thought they were announcing and what facts they considered important in the cases in which they announced them. They think they can skip the messy inductive reasoning step and immediately reason deductively to a "true" and "just" answer in any case they decide.[31] In 1955, the highest court of Massachusetts considered the case of a plaintiff who had been injured by a defendant driving negligently. The defendant had automatically won the case in the lower court–it never went to the jury–solely because in the early days of automobile driving, the high court had ruled that a plaintiff whose car was not properly registered could not sue for injuries suffered by a negligent driver. The judges now conceded that

> (t)he doctrine has been called unique . . . It has been very generally criticized .. As an original proposition, it could hardly find favor with us today. The rule, however, has stood for more than forty-six years without repeal by the legislature. Some of us would prefer to overrule the . . . case, but the majority of the court think that its termination should be at legislative, rather than judicial, hands.[32]

In their text on legal process, Hart and Sacks heaped scorn upon these judges for saying, in essence: "Fellow citizens, forgive us, while we shed these great crocodile tears. Because the legislature has not seen fit to correct this one among many other mistakes we have made, our hands are tied; and we are bound to pass the buck of responsibility to that body."[33] But Lord Farwell had already retired the Ahab Award for Pursuit of Stability by announcing for the English House of Lords that it was "impossible for (the judges) to create any new doctrine of common law."[34]

There are also Precedent (Principles) Judges. They believe that precedents lay down, not narrow rules, but broad principles, and that it is these broad principles, and not the narrow rules, that they should follow and apply directly to the facts of a case. They don't feel confined by the specific ways in which earlier judges may have applied the principles.[35] They may even use them to radically reconstruct the law when they think that it's time to bring old legal rules into line with modern thinking. One of the classic cases in which this occurred was decided on 1916 by the great common law judge, Benjamin Cardozo, for the New York Court of Appeals. The owner of a new Buick was injured when a wheel that the Buick Motor Company had purchased from another manufacturer collapsed. This collapse had been caused by a defect in the wheel that the Buick Motor Company should have detected. However, a long-standing legal rule said that the manufacturer of a defective product was liable only to the immediate buyer unless the product was inherently dangerous, as with a jar of poison or a painter's scaffold. Cardozo said that "(p)recedents drawn from the days of travel by stage coach do not fit the conditions of travel today. The principle . . . does not change, but the things subject to the principle do change. They are whatever the needs of life in a developing civilization require them to be."[36]

Finally, there are Precedent (Policy) Judges. They resemble Precedent (Principles) Judges in their belief that precedents are not just recipes that, if followed faithfully, will produce just results. But they differ from Precedent (Principles) Judges in that they will follow the policies, not the principles, that their predecessors followed, even if they do not adhere to the exact same rules.

SUBSTANTIVE JUDGES

Other judges look not to the past, but to the present or the future. Their vision of law is "substantive." Substantive Judges don't care as much as Precedent Judges about what earlier judges have said. Sometimes they may not care at all. Instead they reason from the touchstone of such social considerations as morality, economics, and politics, which can be roughly allocated into principle or policy.[37] They think that law should express a community's sense of justice.[38] Principles and policies live and breathe, and just as they can live and breathe, so they can die away. When they die they should be given a burial with the respect to which an old soldier who fought the good fight in a long-ago war is entitled. Then they should be forgotten. Precedents, they think, embalm a past that may not only lack the answers to the troubling questions of today, but may not even have asked the questions. Worse, the answers they give may actually mislead. Their rallying cry is that of Oliver Wendell Holmes: "It is revolting to have no better reason for a rule of law than that so it was laid down in the time of Henry IV."[39] When they choose to follow a precedent, it may be the newest, and not the oldest one, because they think that old precedents embody anachronistic principles and policies and that newer precedents are far more likely to embody modern values.[40]

Substantive Judges want to know *why* judges once decided a certain way and whether their reasons make sense in today's world. They want issues settled right and they will take it upon themselves to say what it was that earlier judges meant. In another sense, these judges also bridge the common and civil law systems in that they care about what the law is and not what other judges have said it is and because overarching principles are sometimes believed to predominate even over rules that a Legislature has enacted.[41]

Professor Leonard Levy reveals Chief Justice Shaw, who defined the common law for us a few pages back, as a paradigm of a common law Substantive Judge.

The infrequent use of citation–often its total absence–was habitual with Shaw. Although he was duly respectful of the value of old formulas, he was more often compelled to make them serviceable. Consequently, his opinions give the impression of an imperious scorn for precedent. But it was his conception of the law as a growing science that made him impatient with mere authority for its own sake. He could not content himself with precedent, well though he could conscript it to his service when he wished. His inventive spirit would not permit him to be the prisoner of someone else's opinions. For him the vigor of the law depended upon its keeping abreast of the changes wrought by human endeavor. Therefore he constantly searched for ways to adapt the old to the new, reconcile conflicting doctrines, and so restate the law as to make it practical and plastic.[42]

Substantive Judges wouldn't continue to treat blacks as slaves after public morality or policy had shifted against it. They wouldn't treat Chinese as mentally and morally inferior to whites, as the California Supreme Court did in 1854, or women as naturally unqualified to practice law, as the Wisconsin Supreme Court did in 1875, after the facts showed otherwise.[43]

This is a good place to emphasize that Substantive Judges are not necessarily liberal judges and Formal Judges need not be conservatives. Certainly judges who wanted to keep blacks enslaved, Chinese at the margins of society, and women in the home were nineteenth century conservatives. But conservatives may be Substantive Judges. Some modern American Substantive Judges rebel against what they see as the unjust and wrongly decided decision of *Roe v. Wade*. When the abortion issue comes before the United States Supreme Court, Chief Justice Rehnquist and Justice Scalia care nothing for *Roe*. These modern Substantive Judges think that courts should alter constitutional law when they believe it to be wrong.[44] Substantive Judges believe that courts, no less than legislatures, should change the common law to promote justice and keep law current with public values and morality, new discoveries, and modern understandings of history, psychology, and biology. When the reason for a legal rule ceases, Substantive Judges change the rule.

We'll call the Substantive Judges who try to peer into the future "Policy Judges." They value policy or "goal reasons" for deciding cases. Policy Judges are usually indistinguishable from Precedent (Policy) Judges. Only when Policy Judges adopt policies that have not been used before do differences emerge. "Principle Judges" value principles or "rightness reasons" when deciding cases.[45] Remember that Precedent (Principles) Judges rely on principle, too. As Precedent (Policy) Judges differ from Policy Judges in that they rely on policies that have been relied upon in the past, so Precedent (Principle) Judges differ from Principle Judges in that they borrow their principles from the past as well. Principle Judges look to the present and don't see principles as arising just from precedent, but from religion, ethics, economics, politics, almost anywhere, and ranging from the values of representative democracy or the maximization of wealth to the two most important with respect to dignity-rights, liberty and equality.[46]

Tens of thousands of common law cases testify to the often acrimonious struggle between Formal Judges and Substantive Judges. One of the purest Substantive Judges of the twentieth century was England's Lord Denning.[47] On March 6, 1999, he died. In summing his career, one Substantive lawyer said admiringly that Denning had "steered the law towards the administration of justice rather than the administration of the letter of the law." But one Formal Lord Chancellor groused that, "The trouble with Tom Denning is that he's always remaking the law, and we never know where we are."[48] This enduring conflict marked two areas of law highly relevant to the dignity-rights of nonhuman animals, the legal donnybrooks over human slavery in England and the United States, and the struggle to establish the common law personhood of human fetuses in the United States. To them we will now turn.

SLAVERY BOWS TO ENGLISH PRINCIPLE

In the 1569 Cartwright's Case, a judge declared that "England had too pure an Air for Slaves to breathe," even for the Russian slave involved.[49] But a hundred years later, the judges of the highest English court, the Kings Bench, allowed a plaintiff to collect damages for the loss of his black slaves in *Butts v. Penny*. He could own blacks because they were "usually bought and sold . . . as merchandise" and because they were infidels.[50] Seventeen years later another court allowed another plaintiff to collect damages for the loss of his black slave because the boy was a "heathen."[51]

Soon, in *Chamberline v. Harvey*, the Kings Bench judges faced the question of whether a slave who (smart man!) had gotten himself baptized was still a slave.[52] The slave-master's lawyer began by taking aim at any Precedent (Rules) Judge who might be sitting in judgment. In light of the *precedent* of *Butts v. Penny*, "it cannot be denied" that the man was a slave.[53] But hadn't the *Butts* slave been a "heathen"? And hadn't this slave been baptized? Back-up arguments were clearly called for and the lawyer switched to an argument calculated to net any Policy Judges who might be sitting. Allowing a slave to free himself by seeking baptism was like

handing inmates the key to the jail. And who would run the plantations if all a slave had to do to gain freedom was to get himself
baptized?[54]

The opposing lawyer, a Mr. Dee, argued *principle* for the benefit
of any Precedent (Principles) Judge or Principle Judge primed to
listen. Slavery couldn't exist in England because it violated the law
of nature, the common law had a presumption in favor of liberty,
and this slave was now a Christian and Christianity was inconsistent with slavery.[55] As Chief Justice Holt began to speak for the
court, Mr. Dee must have known that he had struck the bullseye.[56]
Though a black man might be a "slavish servant," Holt said, he
could not be a chattel; but more like an apprenticed laborer.

Holt was finished with *Chamberline*, but not with slavery. As the
eighteenth century rang in, a Mr. Smith sold a slave to the firm of
Brown and Cooper in the Parish of the Blessed Mary of the Arches
in the Ward of Cheap, named because its buyers never paid up.
When this firm didn't pay, Smith sued.[57] Holt got straight to his
point. "As soon as a negro comes into England, he becomes free."[58]
Then Holt noticed that the slave hadn't actually *been* in the Ward of
Cheap when he was sold. He had been laboring in Virginia at the
time, where slavery was perfectly legal, and to which the laws of
England did not extend.[59] Man of principle that he was, Holt's
principles stopped at the English shoreline.

Six years later, a London jury awarded damages to a Mr. Smith
(who may have been the same Smith, or may not have been) in an
action "for a singing Ethiopian negro" and "other goods" against a
Mr. Gould.[60] Gould's lawyer knew–by now every lawyer should
have known–that Holt was a Precedent (Principles) Judge on slavery, and so he argued the *principle* that no man could own another.
But, amazingly, Smith's lawyer, Mr. Salkeld, mistook Holt for a
Precedent (Rules) Judge. Hadn't *Butts v. Penny* said they were
slaves? Negroes were "merchandize," he said, like "monkeys."
Not satisfied with invoking a *precedent* from the Kings Bench,
Salkeld invoked even higher authority. Exodus 20:21 said that a
slave was a chattel and the same as his master's money. Levitticus
gave a master the power even to kill his slave. Unimpressed, Holt
"denied the opinion in the case of *Butts and Penny*" and said that
"the common law takes no notice of negroes being different from
other man . . . there is no such thing as a slave by the law of England."[61]

The next "decision" was pronounced, oddly enough, in the refectory of Lincoln's Inn. In 1729, a delegation of planters and merchants interested in maintaining slavery in the West Indies invited the Attorney-General of England, Philip Yorke, and the Solicitor-General, Charles Talbot, to dinner. Over after-dinner wine, Yorke and Talbot merrily agreed that a slave coming to England was not freed when he touched English soil and that a baptized slave remained just that, a baptized slave. Remarkably, this joint opinion became the near-equivalent of a judicial decision overruling those of Chief Justice Holt.⁶²

Philip Yorke became Lord Hardwicke and "one of the hardest men ever to become lord chancellor of England."⁶³ He used the case of *Pearne v. Lisle* to turn his after-dinner opinion into law.⁶⁴ Pearne claimed that his agent had rented fourteen baptized negroes in Antigua to a Mr. Lisle, who had refused to return them. Harwicke showed himself to be either a Precedent (Rules) Judge (if you think he was being honest) or just a lawless judge. He had "no doubt" that a negro slave was "as much property as any other thing" and there was no precedent gainsaying it. Holt, he announced, had decided *Smith v. Brown & Cooper* on a mere technicality. Confusing the two *Smith* cases, he said that Smith had sought damages "for a singing Ethiopian". But Holt had never said that the man was a *slave*. He could have been a free servant! Any doubt that baptism did not free a slave had been laid to rest in his after-dinner opinion at Lincoln's Inn. But he refused to order Lisle to deliver up the fourteen slaves. They were not unique things, like a finely engraved cherry-stone or an extraordinary wrought piece of plate. One slave was "as good as another." And negroes "wear out with labour, as cattle or other things." Hardwicke told Pearne to sail to Antigua and sue for their value.⁶⁵

We arrive at the most famous case of all, *Somerset v. Stewart*. We saw in Chapter 5 that James Somerset had run away from his American master in England, only to be re-captured and clapped into chains on board the *Ann and Mary*, to be sold in Jamaica. A group of citizens, led by Granvill Sharp, induced Lord Mansfield to stop the departure of the ship by issuing a writ of *habeas corpus*.

Mansfield's history hinted that he was a Precedent (Policy) or Policy Judge, who might let the *Ann and Mary* sail. And he was known to admire that "hard man," Lord Hardwicke. The year before, he had judged the case of Thomas Lewis, a former slave who had been abducted by his master, Robert Stapylton, then thrown into a ship to be sold in Jamica. The Crown prosecuted Stapylton for assault and false imprisonment. Mansfield said that whether one could "have this kind of property or not in England never has been solemnly determined," and urged the suit be dropped. As to slavery, he advised Lewis' attorney that "perhaps it is much better" that it "never be finally discussed or settled."[66] Mansfield was also famous for founding the modern commercial law of England and injecting it into the common law. He based much of it on the customs of English merchants. Everyone involved in *Somerset* knew that it was the custom of English merchants from time out of mind to buy and sell slaves.

But Mansfield also admired the common law's unceasing tendency "to work itself pure."[67] Ten years before, he had written that "(t)he reason and spirit of cases make law; not the letter of particular precedents." Two years later he was to write that "(t)he law of England would be a strange science indeed, if it were decided on precedents only. Precedents serve to illustrate principles, and to give them a fixed certainty. But the (common) law of England . . . depends upon principles."[68] And even the customs of merchants had to "square with the standards of morality, of honest dealing, and of good faith."[69]

Somerset's lawyers took no chances. They argued *precedent*: England had never permitted slavery, except in the special case of villeins. Slavery conflicted with the common law.[70] Slaves "could not be imported into England without fundamentally undermining the common law doctrine in favor of personal rights and freedom of all Englishmen."[71] They argued *principle*: freedom was "the grand object" of English law."[72] Slavery contradicted "natural justice."[73] England's air was "too pure for slaves to breathe in it." And they argued *policy*: though slavery had long existed, it had begun to decline and this decline evidenced an evolving disapproval.[74] Slavery not only corrupted the morals of the master, by freeing him from the usual restraints on how one human normally dealt with another, but actually endangered the state because it corrupted some of its most valuable citizens.[75] Here they tiptoed around one final delicate policy issue. "Tis said let slaves know they are all free as soon as they arrive here, they will flock over in vast numbers, over-run this country, and desolate the plantations."[76] Not to worry; slaves hardly had money enough to book passage to England in large numbers.[77]

Stewart's lawyers hastened to meet every argument. *Precedent* was with them. In *Cartwright's Case*, the judge had only frowned on Russian slavery, because that species was so virulent. Somerset was in no danger of being either killed or "eaten." He merely was required to perform certain services for his master.[78] *Butts v. Penny* had said that a master could own a slave. And Chief Justice Holt, in *Chamberline*, had not said that negroes were free who came to England, had he, just that they couldn't be sold in England. Lining up behind Lord Hardwicke, they dismissed Holt's contrary opinion as "mere dictum, a decision unsupported by precedent."[79] And hadn't Hardwicke "pronounced a slave not free by coming into England"?[80]

Here Mansfield interrupted. Hadn't Hardwicke's opinion on slavery been offered "after dinner?" It had. Mansfield, who had lifted many a glass at the Inns of Court, doubted that it therefore should "be taken with much accuracy."[81] Stewart's lawyers shrugged and pressed on. Then *policy* was with them. Freeing the slaves would lead to an influx of blacks into England, as many really had money enough to entice sailors to sail them to England.[82] (Where they had gotten it the lawyers didn't say). *Principle* was with them, too. Though freedom was claimed as a natural right, "there is perhaps no branch of this right, but in some at all times, and in all places at different times, has been restrained." English air had not been too pure for villeins.[83]

Mansfield set five hearings in as many months. At one he implied that, as feared by Somerset's lawyers, he thought that slavery should be governed by policy, for "the setting 14,000 or 15,000 men at once free loose by a solemn opinion, is much disagreeable in the effect it threatens." He tried to get the parties to settle, then suggested that "(a)n application to Parliament, if the merchants think the question of great commercial concern, is the best, and perhaps the only method of settling the point for the future." But neither side would budge. Finally Mansfield threw up his hands. "If the parties will have judgment, *'fiat justitia ruat coelum'* ('let justice be done though the Heavens may fall')."[84]

It turned out that Mansfield thought policy irrelevant to slavery. Long and cherished principles of liberty controlled and they demanded Somerset's freedom.[85] When James Somerset had departed and refused to serve . . . he was kept, to be sold abroad. So high an act of dominion must be recognized by the law of the country where it is used. The power of a master over his slave has been extremely different, in different countries. The state of slavery is of such a nature, that is it incapable of being introduced on any reasons, moral or political; but only (by) positive law . . . It's so odious, that nothing can be suffered to support it but positive law. Whatever inconveniences, therefore, may follow from a decision, I cannot say that this case is allowed or approved by the law of England; and therefore the black must be discharged."[86]

SLAVERY IS SUPPORTED BY AMERICAN POLICY

The United States Constitution implicitly recognized slavery's legitimacy in several clauses, the most important being the Fugitive Slave Clause. Anyone "held to Service or Labour in one State" who escaped to another state had to be "delivered up on Claim of the Party to whom such Service or Labour may be due."[87] American legal struggle for half a century centered about around how to interpret these clauses. Judges who reasoned, as Mansfield had, from honored principles of natural law, were not uncommon in the early American republic. By the 1820's, however, they had begun to give way to policy. Like locomotives, Policy Judges were determined to pull the young Republic down the track of economic growth, facilitation of commerce for its westward, and national unity, all of which they feared would be irretrievably fractured if they were to strike at slavery.[88] The American anti-slavery movement, rising on the principled foundations of the higher law of God and inherent human rights, placed themselves squarely before that train, determined to derail anyone who tolerated slavery in the name of public policy.[89]

United States Supreme Court Justice Joseph Story and Chief Justice Shaw were two of the leading Policy Judges in the decades before the Civil War. Both personally found slavery repugnant, both embraced national unity as the supreme policy value.[90] From his thirty year perch on the Supreme Court, Story helped forge a country in which the powers of the states were subordinated to those of the federal government, especially when commerce was involved. In *Prigg v. Pennsylvania*, Story upheld the constitutionality of the Fugitive Slave Act of 1793, in which Congress had deprived runaways of such a fundamental bulwark against arbitrariness and oppression as the right to a jury trial, and overrode the ability of a state to remedy the defects by layering protections atop the federal law. The Fugitive Slave Clause, he said, was "so vital to the preservation of (the) . . . domestic interests and institutions (of the Southern colonies) that it cannot be doubted that it constituted a fundamental article, without the adoption of which the Union could not have been formed."[91]

Shaw's moment of truth arrived on April 7, 1851. Four days before, Boston officials had detained a fugitive slave, Thomas Sims, claiming he was owned by a Georgia planter. To forestall any repeat of the rescue by a Boston mob just six weeks before of Shadrach, another fugitive slave, from a federal Courtroom, the state courthouse was draped with heavy chains. The sight of the aged and shaggy-haired Shaw stooping low beneath the chains to enter his own court was of immense significance to the abolitionists. Sims' lawyers now begged a writ of *habeas corpus* from Shaw, then furiously attacked the constitutionality of the new Fugitive Slave Act of 1850. But Shaw pointed out that Mansfield himself had said that, odious as slavery might be, it could be sanctioned by positive law.[92] And all the positive law he needed was written plainly in the Constitution.

Shaw's disdain for precedent marked him as anything but a Formal Judge. When national unity was threatened, he was a Policy Judge to the core, certain that the Fugitive Slave Clause formed "the essential element" of the compromise that had led to the Constitution's enactment and that it "was necessary to the peace, happiness and highest prosperity of all the states."[93] And so he swept principle aside, upholding Congress' refusal to allow Sims to testify, to benefit from trial by jury, to appeal, and the outrageous incentive Congress had dangled before magistrates of paying them double if they found for the master. Shaw ordered Sims returned to Georgia, where he was horse-whipped in the public square and thrust back into bondage.

Professor William Nelson has noted that the conflict in the courts between slavery and antislavery was a conflict "between men possessing different views about the proper role of law and government."[94] Principle Judges "believed that courts should decide cases consistently with standards of what, in some ultimate sense, was right and wrong."[95] These were the Northern citizens and lawyers of the North, if not their judges. Policy Judges, like Story and Shaw, "believed that courts should explicitly pursue expedient policy goals."[96] These were the judges, who almost unanimously upheld the rights of slaveholders in the policy interests of national commerce, national unity, and economic progress and denied the liberties of slaves for nearly three quarters of a century.[97]

The line of pro-slavery victories in the American courts was broken only by the Civil War. In the decades following the war, the public's respect for Policy Judges plummeted, as people associated the emphasis on policy with an immoral justification of slavery. As policy's grip on law loosened, Principle Judges began to reassert natural rights and higher law.[98] But Precedent (Rules) Judges also began to rise. Higher and higher they soared, past all the other judges, determined to shape law once and for all into a science in which all the pieces fit into a neat logical whole. Then a tiny hole pricked by Oliver Wendell Holmes, Jr. in 1880 began to take them down. "The life of the law," Holmes said, "has not been logic: it has been experience. The felt necessities of the time, the prevalent moral and political theories, intuitions of public policy, avowed or unconscious, even the prejudices which judges share with their fellow-men, have had a good deal more to do than the syllogism in determining the rules by which men should be governed."[99]

PRECEDENT (RULES) AND POLICY SAY
NO PERSONHOOD FOR FETUSES

In 1883, Mrs. Dietrich, then four or five months pregnant, fell on a defective road in Northampton, Massachusetts and miscarried. When her fetus lived only ten or fifteen minutes, a suit was filed on its behalf against the Town of Northampton for not keeping the road in good repair. As far as anyone could tell, this was the first suit in the English-speaking world to claim damages for injury to a fetus. In a brief decision for the Massachusetts Supreme Judicial Court, Holmes both created a precedent that Precedent (Rules) Judges could follow and fired the first shot in what became a furious cannonade of precedent, principle, and policy. The fetus, he said, had been "a part of the mother at the time of the injury" and not a separate being.[100] Over the next six decades, Precedent (Rules) Judges, often clasping hands with Policy Judges, steadfastly boxed the fetus into legal thinghood.

The next case rose in Ireland in 1890. A train operated by the Irish Great Northern Railway Company tossed Annie Walker, quick with her daughter, Mabel, about so badly that the baby was delivered crippled and deformed. The Irish Court of Appeals rejected Mabel's claim for damages. One judge declared that "such a thing was never heard of before" and hadn't Holmes, in far-off Massachusetts, dismissed a similar claim just seven years before?[101] Another announced that Mabel "was not a person, or a passenger, or a human being. Her age and her existence are reckoned from her birth, and no precedent has been found for this action." A third noticed that Mabel had not purchased a ticket and "had no existence apart from her mother, who was the only person whom the defendants contracted to carry on their line." A fourth, Justice O'Brien, worried aloud about policy. There was an "inherent and inevitable difficulty or impossibility" in proving why any fetus might be born deformed.[102] But his lecture on the nature of common law change stands as an anthem for Precedent (Rules) Judges everywhere.

> The law is, in some respects, a stream that gathers accretions, with time, from new relations and conditions. But it is also a landmark that forbids advancement on defined rights and engagements; if these are to be altered–if new rights and engagements are to be created–that is the province of legislation and not decision.[103]

Six weeks into the new century the second American case involving fetuses was decided. Ada Allaire, nine months pregnant with Thomas Edwin, was told by employees of Chicago's St. Luke's Hospital to take an elevator to the obstetrics department. But the hospital had poorly maintained its elevator. It malfunctioned, broke Ada's body, and crippled Thomas Edwin.[104] The Precedent (Rules) Judges of the Illinois Supreme Court briskly pointed to Holmes's decision in *Dietrich* and the Irish *Walker* case and said "(w)e concur in the foregoing views," and left Thomas Edwin to fend for himself for the rest of his life.

One lonely Precedent (Principles) Judge–or was he a Principle Judge?–dissented. Hadn't Lord Mansfield, Justice Boggs asked, said that "the law of England would be an absurd science were it founded on precedents only?"[105] The principles of the common law provided "redress for personal injuries inflicted by the wrong or neglect of another."[106] Blackstone had declared that one's right of personal security was absolute.[107] The boy's claim should not "be denied on a mere theory–known to be false–that the injury was not to his person, but to the person of his mother."[108] Ada could have died in that elevator, yet Thomas Edwin could have lived. Baby Dietrich's case was irrelevant because he or she had not been viable. Baby Walker's was irrelevant, too, because St. Luke's Hospital, unlike the Irish Great Northern Railway Company, knew that Thomas Edwin would be riding up inside his mother. Principles of "natural justice" demanded that he be able to sue.[109] Boggs' was to remain a lonely voice for much of the next half-century.

Joseph Drobner was born prematurely after his mother, Sarah, fell into a coalhole that August Peters had negligently permitted to remain open on a New York public sidewalk. Just after World War I, the New York Court of Appeals denied baby Joseph any remedy for his injuries. Allowing him a remedy "against precedent and practice may be a tempting task, to which sympathy and natural justice point the way; but I cannot bring myself to the conclusion that plaintiff has a cause of action at common law."[110] During the War, the Policy Judges of the Wisconsin Supreme Court hadn't even cared that science had shown that a fetus and his mother were separate. It was of no "aid in determining its legal rights. The law cannot always be scientifically or technically correct. It must often content itself with merely being practical."[111]

During the Depression, a Texas Coca Cola truck rammed a car driven by Mrs. H.P. Jordan, pregnant with twins, and crushed her abdomen. One twin was born bruised and lived for just nineteen days.[112] Policy Judges again denied justice to the dead fetus.

> In many cases it would be impossible to establish that the death or condition of the child was proximately caused by the injury. But, far worse, than the indulgence of such speculation and conjecture and the insurmountable difficulty of satisfactorily proving viability, there would follow in the wake of this character of litigation many fictional claims, with false testimony in their support, which the defendants would always find difficult and often impossible to refute. These considerations, we think, outweigh the denial of justice in the abstract to the meritorious case."[113]

Principle grabbed its first toehold in the Supreme Court of Canada in 1933. Mrs. Laveille, seven months pregnant, had been thrown from a tram negligently operated by a conductor of the Montreal Tramways. Jeanine was born two months later with a club foot.[114] It was "natural justice that a child, if born alive and viable, should be allowed to maintain an action in the Courts for injuries wrongfully committed upon its person while in the womb of its mother."[115] But a fetus' personhood would be contingent. If born alive, she was a legal person able to sue for any injuries sustained. But if she died before birth, she died a nonperson.

American courts began to slouch towards common law fetal personhood only in 1946. Bette Gay Bonbrest was injured at birth through malpractice. Judge McGuire, a Principle Judge commanding a federal court in the District of Columbia, refused to follow *Dietrich* for the same reason Justice Boggs had given. Baby Dietrich had immediately died, while Bette Gay Bonbrest was alive, if not well.[116] Precedent had only a limited role to play. "The common law is not an arid and sterile thing, and is anything but static and inert." It was clear that a fetus and her mother were different and the common law "is presumed to keep pace with the sciences and medical science certainly has made progress since 1884."[117] Mere policy was not to obstruct fundamental rights. "That a right of action in cases of this character would lead to others brought in bad faith and might present insuperable difficulties of proof–a premise with which I do not agree–is no argument."[118] As a matter of principle, Bette Gay Bonbrest would get the chance to prove her injuries, for "what right is more inherent, and more sacrosanct, than that of the individual in his possession and enjoyment of life, his limbs and his body?"[119] So began "a rather spectacular reversal."[120]

PRINCIPLES PASSING THROUGH GHOSTS

On April 4, 1941, Mina Margaret Williams was riding in the womb of her mother, Ruth, on a bus operated by the Marion Rapid Transit, Inc. Ruth tumbled from the bus steps and died of her injuries. Mina Margaret was born permanently crippled and with heart trouble, anemia, and epilepsy. The Ohio Constitution guaranteed every injured "person" the right to "have remedy by due course of law, and shall have justice administered without denial or delay."[121] Mina Margaret was clearly a person when she filed suit. But was she when her mother had fallen to her death?

Precedent (Principles) Judges on the Ohio Supreme Court scorned the previous rejections of fetal personhood. They had been "based mainly upon precedent and a high regard for stare decisis."[122] The law must "keep pace with science."[123] To pretend that Mina Margaret had been a part of her mother would "appl(y) . . . a time-worn fiction not founded on fact and within common knowledge untrue and unjustified."[124] "It is elementary that if a wrong has been committed there should be a remedy." Fundamental principles were at stake.

And so began the ascent of fetal personhood that continues today in the common law. Here I take a brief side-track Some Americans might say, "What about *Roe v. Wade*? Fetuses aren't persons under our federal constitution. That's why a woman has the right to an abortion." They would be correct. But because a fetus is not a person under the federal constitution doesn't mean that a fetus cannot be a person under the law of a state. Any time a woman's constitutional right to an abortion is not triggered, a fetus will be able to assert his or her rights. That is precisely what has occurred in many states.[125]

Rita Verkennes died with her mother, Beatrice, from medical malpractice in the delivery room of Maternity Hospital in Minneapolis. The Minnesota Supreme Court, promising to "adhere to the principles" set out by Justice Boggs, the Canadian Supreme Court, and Judge McGuire, decided that Rita's estate could sue for her wrongful death. Rita was not just a person from birth, but from the moment of viability; the absence of Minnesota precedent afforded no refuge to an invader of an individual's rights; medical science had shown that a viable fetus lived independently of her mother, and the law was presumed to keep pace with science.[126]

Less than a month from a full-term birth, Robert C. Woods was maimed in New York. According to the New York Court of Appeals, "(t)he precise question . . . is this: shall we follow Drobner v. Peters, or shall we bring the common law of this state, on this question, into accord with justice? I think, as New York State's court of last resort, we should make the law conform with right."[127] Following precedent was the last thing on the minds of these five judges. Borrowing the words of a British judge, they said that "(w)hen the ghosts of the past stand in the path of justice clanking their mediaeval chains the proper course for the judge is to pass through them undeterred."[128] Undeterred they were. They believed they had "not only the right but the duty to re-examine a question where justice demands it" and were to "bring the law into accordance with present day standards of wisdom and justice rather than 'with some outworn and antiquated rule of the past.'"[129] Judges, they said, acted "in the finest common law tradition" when they adapted and altered prior cases "to produce common-sense justice." To do anything else would be "harsh" and "do reverence to an outmoded timeworn fiction."[130] *Drobner* was dispatched. The leading American professors of the legal process, Hart and Sacks, acidly rebuked the two dissenters. "Are not (the dissenters) in effect saying to the injured plaintiff: 'There can be no justice for you under the institutions of this state, you poor child, wrongfully injured though you have been?'" [131]

But *Allaire* and *Dietrich* still lived. The Illinois case fell first.[132] The common law was not a straitjacket, said its Supreme Court, but "a system of law whose outstanding characteristic is its adaptability and capacity for growth." It was built from elementary rules and general principles "which are constantly expanding with the progress of society, adapting themselves to the gradual changes of trade, commerce, arts, inventions, and the exigencies and usages of the country."[133] Everything else, lack of precedent, difficulty in determining whether a causal relationship existed between the injury and death or condition of the fetus, and identity of the fetus and mother must give way to science and the principle that a remedy should be provided when wrongs are inflicted.[134]

The impregnable claim that a fetus was part of her mother would barely survive to mid-century. "(N)o scientific or medical basis in fact," said the Wisconsin Supreme Court.[135] "No medical or scientific basis for such a proposition," said the Oregon Supreme Court.[136] "(I)t is clear that the medical authorities recognize that before birth an infant is a distinct entity," said the New Jersey Supreme Court. "The myth, perpetrated by Mr. Justice Holmes in 1884, that the unborn fetus is but a part of its mother has long since been put to rest in both the law and medicine," and the Rhode Island Supreme Court:[137] That "(t)he unborn child is a human being distinct from its mother . . . is an established and recognized fact by science . . . (and) the law should endeavor to keep abreast with the marvelous developments of science and the rapidly-changing conditions of the world," said the California Court of Appeals.[138] "If the common law has any vitality it has been argued that it should be elastic enough to adapt itself to current medical and scientific truths," said the Wisconsin Supreme Court.[139] What science tells us animals are, not what some want them to be, will be the bedrock upon which I build my argument in Chapter 10 that chimpanzees and bonobos are entitled to fundamental legal rights.

Precedent (Principles) Judges everywhere began to re-assert fundamental principles of justice. "Obviously courts heeded the teachings of Lord Mansfield and Justice Boggs that *stare decisis* does not require static doctrines, but instead permits law to evolve and to adjust to changing conditions and notions of justice," said the Supreme Court of Appeals of West Virginia.[140] Precedents are "valuable so long as they do not obstruct justice or destroy progress," said the California Court of Appeals.[141] "It would be contrary to every principle of right and justice, which are the very essence of law, to deny such rights to the injured child," said the Georgia Supreme Court.[142] "Justice requires that the principle be recognized that a child has a legal right to begin life with a sound mind and body," said the New Jersey Supreme Court, and "(t)he nature of the common law requires that each time a rule of law is applied it be carefully scrutinized to make sure that the conditions and needs of the times have not so changed as to make further application of it the instrument of injustice."[143]

Most courts abandoned the idea that viable fetuses were merely persons contingent upon birth.[144] The Supreme Court of Appeals of West Virginia even extended legal personhood to a *nonviable* fetus because the principle of "justice is denied when a tortfeasor is permitted to walk away with impunity because of the happenstance that the unborn child has not yet reached viability at the time of death. . . . Our concern reflects the fundamental value determination of our society that life–old, young, and prospective–should not be wrongfully taken away."[145]

What of *Dietrich*, the case that started it all? It lived the longest life and died the slowest death, fighting and scratching like a cat in a bath. In 1950, sixty-seven years after *Dietrich* and well after the counter-revolution was underway, a viable Massachusetts fetus was killed. The Precedent (Rules) Judges of the Supreme Judicial Court acknowledged the strength of the arguments for fetal personhood that had been developing.[146] But purely on the ground of *stare decisis*, they refused to overturn it.[147]

Two years later, the judges revisited the same troublesome issue. Marshall C. Cavanaugh was six months in the womb when his mother went shopping for turkey. On Christmas Day, 1945, it had made her so ill that Marshall was born prematurely. By now, half a dozen high courts had abandoned *Dietrich*..[148] But, being Precedent (Rules) Judges, they played by the rules, even if those rules were obsolete and unjust. They still were "not prepared to overrule our earlier decisions, which began nearly seventy years ago."[149]

Soon afterwards, Duncan Reed was injured in a Massachusetts automobile accident. Born prematurely, he died from his injuries.[150] Eight years after denying justice to Marshall Cavanaugh, the justices conceded that *Dietrich* had been decided mainly due to lack of precedent and that their later decisions had been " founded on the doctrine of stare decisis . . . Although this doctrine may be salutary it may be important in a given case that the court be right, in the light of later examination of authorities, wider and more thorough discussion and reflection upon the policy of the law, than it adhere to its previous decisions." Now the judges thought it "advisable" to adopt new law "in harmony with that of the large and growing proportion of the other States which have adopted in principle the rule proposed by Judge Boggs."

Never underestimate how much Precedent (Rules) Judges *hate* to overturn precedent and these were Precedent (Rules) Judges of the first order. Even now they saw "no need to reverse the Dietrich decision which doubtless was right when rendered but we recognize that in view of modern precedent its application should be limited to cases where the facts are essentially the same."[151] So *Dietrich* remains alive and ready to be applied to any fetus who is born prematurely and lives just ten or fifteen minutes when his four or five month pregnant mother falls on a defective road in Northampton, Massachusetts.

The Primacy of Principle in Determinations of Legal Personhood

Because Precedent (Rules) Judges fight for certainty and stability, they always bring up the rear of legal change. They will almost certainly reject the argument that a chimpanzee or bonobo is entitled to the legal right not be experimented upon in a biomedical research laboratory. While the common law honors such principles as "likes should be treated alike" and "fundamental interests should be protected" and "for every wrong there is a remedy," these principles have only been applied to human beings. That is all a Precedent (Rules) Judge needs to know.

Both Principle Judges and Precedent (Principles) Judges will want to know why these principles *should* or *should not* be applied to nonhuman animals. Precedent (Principles) Judges will invoke the great time-tested principles of liberty and equality, now deeply embedded into the common law, as standards against which to measure the justice and moral rightness of their rulings. They will consider themselves bound by them. But they will not consider themselves bound to apply them only in the old ways. Principle Judges invoke principles because they believe them to be morally right. Hard as it may be to imagine, these may or may not include liberty and equality. What principles might a Principle Judge who was a dedicated Nazi have applied to the Jews in Germany during the Second World War? But because they almost always believe in the moral rightness of liberty and equality, the decisions of modern Principle Judges will almost always be grounded in the same principles of liberty and equality as will be the decisions of a Precedent (Principles) Judge. Both kinds of judges will then use these bedrock principles as benchmarks against which to re-examine the justice of ancient legal rules, whether they produced the legal thinghood of human slaves or nonhuman animals, and to construct new and more just legal rules.[152]

Policy Judges and Precedent (Policy) Judges will want to know how society will be affected *if* these nonhuman animals attain legal personhood. Policy Judges will look to the policies of today. Precedent (Policy) Judges will follow the policies that earlier judges deemed important. Because yesterday's policies may have almost nothing to do with the problems of today, these kinds of judges are virtually nonexistent.

Each approach–precedent, principle, and policy–has its drawbacks. Excessive reliance on precedent can produce decisions soaked in arbitrariness and inequality that lead to stagnation and create the very instability that Precedent (Rules) Judges hope to avoid.[153] Excessive reliance upon principle **or** policy can also produce decisions laden with arbitrariness and inequality that equally lead to unpredictability and chaos. With such rare exceptions as Lords Farwell and Denning, every common law judge rules in every way (except as a Precedent (Policy) Judge). But nearly every judge will *tend* to rule in one consistent along one or another line, if not across the board, then in specific legal areas, depending upon her values, the vision of law that her values have produced, and how strongly that vision grips her. A judge may be a Precedent (Principles) Judge on fundamental rights, but a Precedent (Rules) Judge with regard to trusts or wills, and a Policy Judge on tort law.[154] A judge is most likely to act as a Precedent (Rules) Judge when she either believes that the law of a legal area has been decided correctly or when she highly values legal stability and certainty for their own sakes. A judge is least likely to act as a Precedent (Rules) Judge when she believes that the law of a legal area is wrong or when she dismisses the value of legal certainty and stability.

Today most common law judges are Precedent (Principles) Judges, in the mold of a Lord Mansfield, when the most fundamental human rights are at stake. I believe that this is how it should be when new candidates for legal personhood present themselves. The "reasoned judgment" of the common law judge on the momentous issue of legal personhood should not turn on the mechanical operation of a narrow legal rule. No being should ever be denied legal personhood just because others like him were denied it before or have always been denied it. This blindly perpetuates the most pernicious and invidious biases of which we humans are capable.

Turning legal personhood on policy is as bad as linking it to narrow rules of precedent. No being should ever be denied legal personhood and fundamental rights as a matter of policy. Fundamental rights are intended precisely to protect a rights-holder *from* others who might think that harming her is either good for them or good for society. For example, there are many reasons to support the argument that a woman should have the legal right to an abortion. But when judges of the New York Court of Appeals asserted in 1972 that the legal personhood of fetuses was a *policy* question, they were not just wrong, they were plain wrong, laughably wrong, disastrously wrong. As one dissenter complained, "(t)his argument was . . . made by Nazi lawyers and Judges at Nuremberg."[155] And so it was. Connecting fundamental rights to policy betrays the right, betrays the supplicant, and finally undermines the rights of the betrayer. The Shaws and Storys, who betrayed the fundamental rights of slaves to what they thought was a greater good, may have gone to their graves with honor. But they did not escape the judgment of history that, in the end, what they betrayed *was* the greater good.

A judge might conceivably hold that a human adult or infant or fetus or a chimpanzee or a worm should be a legal thing in the eyes of the common law. But such a momentous decision should be made only after a careful weighing of the highest *principles*, for never does what is *right* more clearly trump what is *good* or what has *been* than when legal personhood, from which every legal right flows, is itself at stake. A judge might decide that human fetuses should never be legal persons under the common law, because autonomy is necessary for legal personhood and fetuses merely have the *potential* to achieve this autonomy, not the autonomy itself. But this judge must be ready to apply this principle wherever it may reasonably lead him, and it may lead him to infants, children, the severely mentally-defective, and the comatose. A judge might also conceivably hold, and reasonably so, that human fetuses are legal persons under the common law, because the *potential* for autonomy is all that is necessary for legal personhood. But that judge too must be ready to apply that principle wherever it may lead him. As we will see, he might be casting a wider legal net than he suspects or wishes.

In arguing that using high principle is always the best way to determine legal personhood, I make no claim here to objective truth. The evidence that every judicial decision is saturated with a judge's own values is overwhelming and the more contentious the issue, the more influential those values will be. My argument–and I freely admit that it is based on a value judgment–is this: those judges who, like me, supremely value liberty, equality, and reasoned judgment, who despise slavery, genocide, and who may even be disturbed by abortion, should be prepared to analyze every claim for common law personhood, above all other claims, through those prisms of fundamental principles.

But doors can swing both ways. The danger in always using returning to fundamental principles to analyze claims for common law personhood is that, like everything else in the common law, no one's legal personhood will ever be as secure as it would be if those principles could be allowed to harden into legal rules that even a Precedent (Rules) Judge could mechanically apply. What would be wrong with constructing the legal rule that "All humans, chimpanzees, and human beings are legal persons?"

Flexibility is inherent in the common law and that is a good thing. None of us prefers to conduct our most important business in indelible ink. Judges who understand that changing morality or scientific findings might one day justify their unmaking the legal personhood for nonhuman animals that they made will be the more willing to experiment with it in the first place. At some point, as with human personhood, it will become clear that the experiment has succeeded. That will be the time to chisel the rights of nonhumans into the same constitutions in which human personhood has been chiseled.

8

Consciousness, Taxonomy, and Minds

THINKING ABOUT CONSCIOUSNESS

When I see my twelve-month-old twins invade a room, I know how the universe will end. The Second Law of Thermodynamics dictates that the force of entropy is inexorably pushing the universe into an increasing state of disorder. The mini-universes for Christopher and Siena switch as they totter from room to room, but each expires in the same way. I've watched them die. In our family room, brimming chests are emptied. Toys are minutely examined or hammered, pushed, pinched, fingered, then skittered across the floorboards. After everything is sufficiently scattered, they toddle into the kitchen. There they force and enter cupboards headfirst. Shifting their Biter Biscuits from hand to hand, they emerge with containers of cocoa to pop open and pour out. They reenter and reemerge with pans to bang, pots to sit in, spaghetti-sauce jars and plastic cups to twirl and spin, and flatware to be laughingly tossed at the dog. Finally, these exhausted twin Soldiers of Entropy are trundled upstairs for naps and the parental Canutes slouch back to battle the tides of pots, pans, toys, and other flotsam still floating on the linoleum.

Some claim that chimpanzees and bonobos aren't conscious, that no nonhuman animal is conscious, that my Soldiers of Entropy aren't conscious, and I can't prove them wrong.[1] Some claim that I'm not conscious, and I can't prove them wrong. And if I claim that you're not conscious, you can't prove me wrong, either. My twins certainly *seem* conscious to me. So do bonobos and chimpanzees. So do you. But I can't prove that any of you are. Try to think about consciousness. Define it. Describe it. Try to imagine someone else's consciousness without leaning on an analogy. Then imagine how anyone might be able to prove to your satisfaction that she was conscious. So is it any wonder that the philosopher Ted Honderich complains that "[t]rying to think about it can make you unhappy"?[2]

Analogy by Duck

A hundred years ago, the pioneering psychologist William James thought that consciousness and mind were the same thing; unconscious processes were physical, not mental. Few think that today.[3] We care about consciousness here because species with no capacity for it lack that quality of mind that matters for legal rights. They're not aware that they, or anyone else, exist; they can feel neither pain nor pleasure; they can't feel anything at all. Because, as philosophy professor Colin McGinn has pointed out, "to any sensible person consciousness is the essence of mind: to have a mind precisely is to endure or enjoy conscious states—inner subjective awareness" and because entitlement to legal rights rests upon the existence of conscious states, I will use the words "consciousness" and "mind" interchangeably throughout the rest of the book, though we understand that they are not really the same thing.[4] Each will stand for the totality of one's conscious awareness, thoughts, and feelings.[5]

The strongest available argument that you, chimpanzees, bonobos, and the Soldiers of Entropy are conscious is "by analogy." It goes like this:

1. I know I'm conscious.
2. We are all biologically very similar.
3. We all act very similarly.
4. We all share an evolutionary history.
5. Therefore, you, apes, and the twins are conscious.

In short, "if it walks like a duck and quacks like a duck . . . "

"Analogy by duck" has at least three problems. First, I might be deluded into thinking I'm conscious when I'm not. We can politely exclude this possibility. As the philosopher John Searle points out, "Where the existence of conscious states is concerned, you can't make the distinction between appearance and reality, because the existence of the appearance *is the reality* in question. If it consciously seems to me that I am conscious then I am conscious."[6] Second, I might be the only being who's conscious and therefore be King of the Universe. My wife tells me I can safely exclude this possibility. She has patiently explained to me that consciousness would bestow an extraordinary advantage upon me over every other human being. Since I don't appear to be in line for a Nobel Prize, a Pulitzer Prize, or an honorary knighthood, I reluctantly conclude that—as usual—she's right. Third, the ways in which you and the Soldiers of Entropy are similar to me may have nothing to do with consciousness. The ways in which we are *dissimilar* may hold the key. This is the toughest objection, though no tougher for chimpanzees and bonobos than it is for you and probably easier than for the Soldiers of Entropy. You might stand six feet tall, weigh 200 pounds, and have blond hair, blue eyes, and fair skin. I stand five feet ten inches tall, weigh 160 pounds, and have brown hair, brown eyes, and dark skin. I could go on. But I won't, because there is no evidence that any of these differences, or any of the thousands of other differences between us, have the slightest thing to do with consciousness.

If analogy by duck doesn't prove that we're both conscious, at least it throws the burden onto the skeptic. In this chapter, we will compare brain anatomies, shared genetic material and evolutionary histories, and behaviors, then add bits of seemingly objective evidence. But in evaluating the arguments for and against the minds of nonhuman animals, keep three rules in mind. Each is meant to neutralize the worst of Aristotle's Axiom, which is that everyone assigns the group to which he belongs to the top of any hierarchy of rights. Wise's Rule One: Only with the utmost effort can we ever hope to place ourselves fairly in nature. Wise's Rule Two: We must be at our most skeptical when we evaluate arguments that confirm the extremely high opinion that we have of ourselves. Wise's Rule Three: We must play fair and ignore special pleading when we assess mental abilities.

The idea that nonhuman animals are mindless was standard until about a hundred and fifty years ago. Darwin fired the first shots of the revolution. Any difference in mind, he said, was "certainly one of degree and not kind."[7] But, as revolutionaries often do, Darwin and his prominent supporter, George Romanes, overshot their target. They told a few too many stories about nonhuman animals in which they ascribed too wide a variety of humanlike cognitive and emotional abilities to too many species that were too evolutionarily distant from us. This cascade of anecdotes about cats and foxes and rats acting in overly human ways ignited a counter-revolution that fought fire with fire. Nonhuman animals didn't just lack human minds. No, the counterrevolutionaries were determined to push Darwin and Romanes into the sea. Animals had no minds at all!

This counter-argument was inevitable. Descartes well knew three hundred and fifty years ago that the benefits of this all-encompassing argument to humans were substantial. If nonhuman animals have no minds, humans are absolved "from the suspicion of crime when they eat or kill animals."[8] The philosopher Daniel Dennett, who can locate no animal minds today, admits it relieves "our hunters and farmers and experimenters of at least some of the burden of guilt that others would place on their shoulders." The claim is, in Dennett's word, "convenient."[9] But Dennett denounces anyone who would distinguish nonconscious humanlike zombies from real humans whose consciousness can't be observed. This kind of argument, he says,

> echoes the sort of utterly unmotivated prejudices that have denied full personhood to people on the basis of the color of their skin. It is time to recognize the idea of the possibility of zombies for what it is: not a serious philosophical idea but a preposterous and ignoble relic of ancient prejudices. Maybe women aren't really conscious! Maybe Jews![10]

Indeed!

Back to the counter revolution. The whitest flame was fanned by the English psychologist Conway Lloyd-Morgan, who modestly named it "Lloyd-Morgan's Canon."[11] It gets a regular workout even today. Lloyd-Morgan thought that Darwin and Romanes had been hornswoggled by a pervasive and powerful human tendency to believe that whenever nonhuman animals act like us, they think like us. So he created the opposite bias: Never believe animals think as we do unless you must. No one, the eponymous Canon instructed, should interpret the behavior of any nonhuman animal as caused by a "higher psychical faculty," if it can be explained in terms of a mental faculty that "stands lower in the psychological scale."[12] This fine example of Occam's Razor was based on sound evolutionary theory and told us that, all things being equal, one should prefer simpler explanations for behavior over more complicated ones. Sounds like a sensible, scientific, logical, easy-to-apply rule, doesn't it? Well, it's not, and the closer one examines it, the less sense it makes, the less scientific it is, the more illogical it appears, and the harder it is to apply.

Lloyd-Morgan probably wouldn't even recognize his Canon today, because he actually thought that all the "higher animals" were wonderfully conscious and he just wanted the Canon to mow down the more outrageous claims that nonhuman animals could engage in complicated reasoning, which he thought they could not.[13] Well, either Lloyd-Morgan actually thought that or he later changed his mind with respect to the minds of chimpanzees.[14] But it really doesn't matter what Lloyd-Morgan thought his Canon meant. What's important is what we think it means today. Generations of scientists have been taught that Lloyd-Morgan's Canon means that they should never . . . ever . . . think that a nonhuman animal has any mental capacity that can possibly be otherwise explained and so, by golly, that's what it means and that's what they think.

Janus-like, Lloyd-Morgan stares from one direction, his Canon clipping all who would believe that the mental abilities of nonhuman animals are the same as humans, while Darwin stares the other way, his principle of evolution by natural selection reminding us that human and nonhuman animals differ just in degree and not in kind.[15] But Lloyd-Morgan's Canon, with its assumption that as creatures evolved, their mental capacities marched from lower to higher, seems an antievolutionary throwback to the Great Chain of Being.[16] The philosopher Elliot Sober argues that the Canon has nothing to do with evolution at all and would work exactly the same way "even if organisms are separately created by God."[17] So what is this struggle about? Like Claude Rains's Vichy police inspector in the movie *Casablanca*, we are "Shocked! Shocked!" when the primatologist Frans de Waal gives us the answer—"humanity's place in nature."[18]

Most mammals and every primate act in ways that cause most reasonable people to think they have minds of some kind. This is certainly not because we are biased in favor of believing that—the more than two thousand years of intellectual history reviewed in Chapters 2, 3, and 4 demonstrate that we have quite the opposite bias—but because the evidence is convincing. It is circular thinking to dismiss this belief as mere anthropomorphism (which is the erroneous attribution of human characteristics to nonhumans), as some do. They begin by assuming that only humans are conscious, then label any contrary claim as anthropomorphic. Why? Because only humans are conscious.[19]

A clear-eyed analysis will require judgment and de Waal lends us a crisp rule of thumb to explain how scientists actually use "analogy by duck" to puzzle out the mental processes that underlie the behavior of nonhuman animals. I'll call it "de Waal's Rule of Thumb": "[S]trong arguments would have to be furnished before we would accept that similar behaviors in related species are differently motivated."[20]

By contrast, what de Waal's Rule of Thumb does not mean, and what scientists do not do, is what the primatologist Daniel Povinelli and some of his colleagues claim it does and they do. They argue that scientists use "analogy by duck" in the undiluted form proposed more than two hundred and fifty years ago by the English philosopher David Hume: "[W]hen . . . we see other creatures . . . perform like actions, and direct them to like ends, all our principles of reason and probability carry us with an invincible force to believe the existence of a like cause."[21] Povinelli writes that Hume's brand of "argument by analogy" means that "similarity in behavior *guarantees* similarity in mental processes."[22] Because this is untrue, Povinelli argues for "the end of the argument by analogy."[23] The problem for Povinelli is that no scientist actually believes that "similarity in behavior *guarantees* similarity in mental processes." But when animals who are closely related behave in a way that strongly suggests similar mental processes to ours, it seems reasonable and fair to shift the burden of proof to those who would argue that what we are seeing is not what we think we are seeing.

Even an animal's complex behavior, all by itself, is not necessarily enough to prove that a mind is driving it.[24] But the more parsimonious explanation might easily be that an animal is consciously thinking.[25] Standing by itself, Lloyd-Morgan's Canon may not even be an example of Occam's Razor.[26] It's part of a larger proposition, says the animal behaviorist Marian Stamp Dawkins:

Logic says two *other* things; first, that on the same grounds we would have to allow that other *people* may not be conscious either, and, second, that some rather special pleading is going to be needed to maintain that similarities in behavior co-exist with a lack of similarity in conscious awareness.[27]

Here Dawkins puts her finger on a deficiency in the Canon. Anyone's behaviors can be explained in terms of no mental processes at all, though not too convincingly. Here's the famous behaviorist, B. F. Skinner, calling a baseball game:

And the pitcher's sternocleidomastoideus and trapezius muscles are extended towards home plate. The catcher exhibits the dorsal surface of the second digit of the right hand. The pupils of the pitcher's eye are momentarily occluded by the movement of the palpebral portion of the orbicularis oculi.

I would call the game differently:

And the pitcher leans toward home plate expecting the catcher's sign. He's thinking that the batter can't hit the fastball. The catcher extends one finger. He's calling for the fastball. The pitcher's blinking in surprise. Hadn't he struck this batter out with curve balls in the last two at-bats?

The philosopher Stuart Shanker has pointed out that Skinner's call of the game "might well have considerable interest in the context of an anatomy class, but it does not provide a more rigorous analysis of a baseball game: of what the players are *really* doing."[28] Yet that is precisely how many scientists, most notably behaviorists, thought and still think actions must described. It often remains, in Dennett's, words, "the language of *Science* even if it is no longer exclusively the language of science,"[29] but only when nonhuman animals are concerned, a clear violation of Wise's Rule Three (play fair and ignore special pleading).

LEVELS OF CONSCIOUSNESS

Consciousness seems to be a process and not a thing.[30] It results entirely from natural, and not supernatural, causes.[31] It has an "aboutness" about it; it is always "about" things, "about" sights, "about" noises, "about" fear, "about" sex. It is selective, interested in some things but not in others.[32] Nobody is conscious about everything happening around him at once. We all have one-track minds, and our conscious minds pick and choose what they will be aware of at any moment. That is why two conscious beings in the same place may be aware of different things or different degrees of the same thing. If you disagree, interview five witnesses to a bank robbery or an automobile accident (as I have), then cross-check their stories. You'll come away hardly believing that all five could have witnessed the same event. Consciousness flows so continuously and is so ever-changing that we often think of it as a stream.[33] But we don't know if it is really a stream or just *feels* like one.[34] Consciousness is lonely. Mine can only experienced by me, yours by you.[35] No one can experience another's "qualia," those incessant experiences of sensation, emotion, and cognition that have a particular "feel" to them, like the "redness" of roses, their "rose-y" smells, their "velvety" touch, the "happiness" we feel when we look, smell, and feel them.[36] The intensely personal and subjective quality of consciousness explains why investigations into minds, and especially the minds of beings with no complex and sophisticated language, whether they are children or chimpanzees, might drive even a Pangloss to suicide.

The philosopher Peter Carruthers thinks that nonhumans aren't conscious: "[I]f consciousness is like the turning on of a light, then it may be that their lives are nothing but darkness."[37] But why should he think that? It seems unlikely that all other animals lack consciousness. The primatologist Alison Jolly has noted how strange it would be

if all animal species had the same kinds of consciousness. . . . The ultimate goal for a primatologist is to trace the evolution of our own mental complexity to its first beginnings in creatures very different from ourselves. In the same way we trace the perfection of the vertebrate eye back to the first light-sensitive pigment spot, while showing how each form of eye may have been adaptive for the animal which evolved it.[38]

The neuroscientist and evolutionary anthropologist Terrence Deacon probably speaks for most scientists when he says that "[a]nimals can have conscious minds without sharing all the attributes of human consciousness."[39] Dawkins says that "[d]ifferent animals might possess some or all of these attributes to different extents, so that it may not be possible to say that an animal is either conscious (possessing all elements) or not (possessing none)."[40] If one appreci-ates artificial illumination analogies, a better one might be the din-ing-room dimmer switch, for consciousness probably ranges cross a vast continuum.[41] Understanding that levels of animal conscious-ness can exist is one reason that the cognitive psychologist Bernard Baars has concluded "that the scientific community has now swung decisively in its favor. The basic facts have come home at last. We are not the only conscious creatures on earth."[42]

Animal consciousnesses probably differ. The dimmest, cloudiest bulbs are the less complicated animals and the very youngest of humans, who probably experience in dreamy and incoherent ways. The blazing chandeliers are the most deeply conscious of human adults, who can use recursive symbolic memory, especially language, to think incredibly complex thoughts and experience highly sophisticated feelings.

One Sunday afternoon in April 1999, I found myself eating a salad across the table from Donald R. Griffin. Now semiretired, he taught with distinction at Cornell, Harvard, and Rockefeller Uni-versities and not only discovered that bats use echolocation but founded the field of cognitive ethology, which is the study of the cognitive processes behind the behaviors of nonhuman animals. Griffin thinks that "[a]nimals are probably conscious of a different and more limited range of subject matter, appropriate for their ways of life rather than ours."[43] Having led science toward animal minds for almost thirty years, he has suffered uncounted barbs. Unperturbed, he has steadfastly maintained that a fair viewing of the evidence shows that many animals do have minds. Puzzled by the continuing onslaught, he suggests that the "antagonism of many scientists to suggestions that animals may have conscious experiences is so intense that it suggests a deeper, philosophical aversion that can reasonably be termed 'mentophobia.'"[44]

Griffin asked me to imagine what was happening in the mind of one of two female gazelles being chased by a lion. He has little doubt that she is conscious, though he cheerfully acknowledges that others might disagree. Chatting over lunch, I felt just the latest in a long line of participants in discussions about whether a non-human animal could not just perceive something but perceive that it *was* something. In the covered public walkway of the Lyceum on the outskirts of Athens 2,400 years ago, Aristotle had debated whether a lion could understand that an ox was near. The first Stoics argued in the Athenian Agora about what an ox understood when the lion was near. Around the turn of the last millennium, the Persian philosopher Avicenna talked about what sheep thought about wolves.[45] Now Griffin explained how our gazelle must realize that it was *she* who was being chased, as well as another gazelle who was not her, and how she must understand, however dimly, that dire consequences will flow for her if she, and not the other gazelle, is caught.[46]

Everybody working in this field has an opinion about what levels of consciousness exist. Most opinions distill to this: Some consciousness tends to allow one to experience the present, another to realize that one is experiencing the present and to anticipate, to some degree, the future and to think about the past. *The International Dictionary of Psychology* warns against tumbling into the trap of confusing consciousness with self-consciousness—"to be conscious it is only necessary to be aware of the external world."[47] Gerald Edelman, a Nobel laureate in medicine or physiology, thinks that animals can have a "primary consciousness" that allows them to see "the room the way a beam of light illuminates it. Only that which is in the beam is explicitly in the remembered present; all else is darkness. "Higher-order consciousness" allows them to be conscious that they are conscious and to be aware not just of the present but of the past and future.[48] Griffin contrasts "perceptual consciousness" with "reflective consciousness." Perceptual consciousness occurs when one is "mentally conscious or aware of anything." It may "entail memories, anticipations, or thinking about nonexistent objects or events as well as immediate sensory input." Reflective consciousness is "the recognition by the thinking subject of his own acts or affections."[49] Griffin has suggested an intermediate category: the awareness of self, though not a reflective awareness. Animals with this intermediate consciousness might be eating, be unable to reflect upon the fact that they are eating, yet be aware that it was, indeed, *they* who are eating.[50] Just as our gazelle might be unable to *reflect* upon the fact that she is being chased, she may be acutely aware that is indeed *she* who is being chased. Others agree. "From an evolutionary point of view, as soon as there is locomotion there is perceptual awareness and as soon as there is perceptual awareness there is self-awareness (meaning awareness of an animal's own body)."[51] The philosopher Owen Flanagan proposes a spectrum of self-consciousness. At the lower end lies the "self-consciousness involved in experiencing my experiences as mine," while the upper end involves "the sort of self-consciousness involved in thinking about one's model of one's self."[52]

Secondary, or reflective, consciousness is no unalloyed good. The neurologist Antonio Damasio, taking his lead from the seventeenth-century philosopher and mathematician, Pascal, observed that we almost never think of the present, and when we do, it is to weigh its effect on the future. "It is easy to see," Damasio says, "how perceptive (Pascal) was about the virtual nonexistence of the present, consumed as we are by using the past to plan what-comes-next, a moment away or in the distant future."[53] I know tourists who obsessively document their vacations on video so they can experience their trips on their return home. I say "experience" their trips, rather than "reexperience" them, because they never experienced them the first time. Their trip experiences become the experience of watching their videotapes. Beings with just a primary or perceptual consciousness can't plan for a future they don't know is coming. But they may experience the here and now with a fullness that most self-reflectors will never know. Some of us may try through meditation or religion or therapy.[54] Every meditation novice knows the struggle to shut off the spigot that spills thoughts of the past and the future into our minds and experience nothing but the present. Secondary consciousness also allows, and probably demands, that we dread our deaths for most of our lives and fear the future as well as anticipate it.

The Top Ten Theories of Consciousness, or Nobody Has a Clue

The dance over the *levels* of consciousness is a cotillion compared to the brawl over what consciousness actually *is*. The *International Dictionary of Psychology* tells us that consciousness "is a fascinating but elusive phenomenon: it is impossible to specify what it is, what it does, or why it evolved." Nothing worth reading has been written about it.[55] The philosopher Jerry Fodor concedes that "nobody has the slightest idea how anything material could be conscious."[56] The neuroscientist Susan Greenfield thinks that "[e]ven the vaguest and most speculative of answers is currently beyond us."[57] The cognitive psychologist Steven Pinker admits, "It beats the heck out of me."⁵⁸ The neuroscientist Michael Gazzaniga sums up our knowledge as this: "[N]obody has a clue."⁵⁹ There is no shortage of speculation. In just the eight weeks over which I wrote this chapter and the next, five new books on consciousness were published. If you're not unhappy thinking about consciousness now, you might be after you read the next paragraph. In it, I offer outrageously simplified versions of what Western scientists and philosophers claim to be the Top Ten Theories of Consciousness. Take three aspirin and read on.

(1) *"Identity theorists"* believe that mental states exist and that we are natural identity theorists because we believe that others believe, desire, and love. It has also been called "mindreading" and "folk psychology," because so many of us naturally believe it. We experience the brain states that make up consciousness the way we experience molecular motion as "temperature" through our sense of touch (the faster they move, the hotter they feel) or the way we see certain wavelengths of electromagnetic radiation as "light."⁶⁰ Consciousness is not *caused* by physical brain processes; it *is* them. (2) *"Eliminative materialists"* dispute this "folk psychology." Cognitive science, they assure us, will replace these outdated ideas with neuroscientific explanations, the way chemistry supplanted the alchemy of the Middle Ages and astronomy replaced astrology.⁶¹ (3) *"Logical behaviorists"* deny that consciousness even exists. Instead, what we normally think are mental states are just different kinds of behaviors.⁶² (4) *"Emergent materialists"* allege that consciousness is merely a property of the physical structures and events within our brain in the way that the liquidness of water is a property of what you get when two atoms of hydrogen hook up with an atom of oxygen ("water," for the chemistry-impaired).⁶³ (5) Closely associated with artificial intelligence, *"functionalists"* insist that mental states are functional states that drive our behavior the way computer software drives hardware. Identical functional states produce identical mental states.⁶⁴ Build a robot just like me and it will be conscious, just like me. (6) *"Biological naturalists"* declare that neurophysiological brain processes cause mental events and that consciousness is a physical property of the brain that cannot be reduced to any other physical property.⁶⁵ (7) *"Constructive naturalists"* say that consciousness is neither a thing nor a mental faculty but a set of processes generated by the brain; the mind is the brain processes experienced.⁶⁶ (8) *"Substantive dualists"* main-

tain that a brain is matter but that the mind is not of this world.[67] (9) *"New Mysterians"* claim that the physical properties of the brain can never be understood by our limited intelligence.[68] (10) *"Mind agnostics"* believe that though we don't now, we may one day understand consciousness and the mind and the mind's relation to the brain.[69]

Had enough? I have been waltzing with you to the different tunes of consciousness for one very good reason. Legal rights are connected to consciousness the way Christianity is connected to Jesus. Consciousness is not the *whole* of rights any more than Jesus is the *whole* of Christianity. But if you believe that Jesus and Peter Pan are equally real, Sunday services will just be a combination book lecture and group sing. If you think that nonhuman animals are never, ever conscious, then they are mindless, feelingless, thoughtless *brutes* that deserve no more legal rights than a toaster. Before the music stops, I hope to persuade you that this is not true.

THE GENETIC SIMILARITY OF CHIMPANZEES AND HUMANS

Modern skyscrapers aren't haphazardly thrown into the clouds. If builders casually threw a wall up here, a stairwell there, and an elevator round the corner, nothing substantial would ever get built. And if, by some miracle, a skyscraper began to rise, it would come crashing down with the first breeze. Builders have to follow a blueprint.

Every living being is fantastically more complicated than the most complicated building ever built. In the last two years, scientists have for the first time puzzled out the genetic blueprint for an animal, a tiny worm about the size of the tip of your baby fingernail, called the *Caenorhabditis elegans* (we'll come back to why it has two Latin names). Although it has just 959 cells, it would take 2,748 pages of the *New York Times* just to stamp out its blueprint. Its genetic plan can be found in the chromosomes that lie within the nuclei of its cells. The worm has six chromosomes. Just as a building is constructed from wood, cement, brick, and glass, chromosomes are fashioned from genes. The worm has more than 19,000 genes.[70] And as the wood, cement, brick, and glass are made from atoms constructed from protons, neutrons, and electrons, inside the genes are long ladderlike strips of twisted double strands of

DNA. Each half-rung is composed of one of four kinds of protein called a "base," or nucleic acid. The double strand of DNA hangs together because each half-rung of the DNA ladder is attracted to its opposite and complementary half-rung like a magnet to iron filings.

Our DNA and that of chimpanzees is more than 98.3 percent identical. That means that, on average, more than 983 out of every 1,000 base pairs along every double strand of DNA in both species lie in the same sequence. But we're actually more closely connected than that. Scientists have realized that chimpanzees and humans have a lot more DNA in their cells than they could possibly need. A lot of it doesn't do anything; it's "junk DNA." Of the DNA that actually does something, humans and chimpanzees share, on average, more than 995 of the same base sequences along every double strand of DNA, or more than 99.5 percent. Investigators now believe that humans and chimpanzees may differ by only several hundred genes out of approximately 100,000. A mere fifty genes may control differences in our cognition.[71] We apes (more on that as well) probably differ by only four or five base pairs for every thousand that populate the double strands of DNA and one out of every five hundred genes.[72]

The Similar Brains of Humans and Chimpanzees

We share almost identical DNA with chimpanzees and bonobos for an excellent reason: Not long ago in evolutionary time (forever in historical time, just a heartbeat in geological time), we were the same creature. Fourteen million years ago (give or take a couple million), there were neither chimpanzees nor humans but another kind of creature, our common ancestor. Orangutans then split away from the common ancestor of gorillas, chimpanzees, bonobos, and humans. Perhaps six or seven million years later, gorillas broke away, leaving the common ancestor of human beings, chimpanzees, and bonobos. Five or six million years ago, the common ancestor divided again, one branch evolving into chimpanzees and bonobos, the other into us.[73]

Because consciousness is almost certainly a natural physical process subject to normal biological evolution, it likely evolved because it bestowed environmental and social advantages.[74] That is why many scientists believe that the most primitive consciousness first flickered tens, perhaps hundreds, of millions of years ago. Bernard Baars points out that every mammal has the brain structures necessary for simple consciousness—a brain stem, a thalamus, and a perceptual cortex.[75] Gerald Edelman thinks that primitive consciousness is about 300 million years old and that most mammals, some birds, and possibly some reptiles, have it. Because he believes that at least a simple brain cortex, or its equivalent, is necessary for consciousness, he thinks that lobsters, who have neither, are probably not conscious.[76] Susan Greenfield cautions that Edelman may be "a bit hasty" even here.[77] Donald Griffin writes that

> the central nervous systems of multi-cellular animals all operate by means of the same basic processes regardless of the species or even the phylum in which they are found. Because we know that at least one species does indulge in conscious thinking, and take it for granted that conscious and unconscious thinking result from the activities of the central nervous system, we have no solid basis for excluding a priori the possibility that conscious thinking takes place in any animal with a reasonably well-organized central nervous system.[78]

Human and chimpanzee brains certainly appear similar. Human brains weigh perhaps three pounds. A chimpanzee's brain weighs about one pound. Our brains contain between 10 billion (10^{10}) and 100 trillion (10^{14}) neurons.[79] That humans have triple the number of neurons of chimpanzees almost certainly makes no difference when such vast numbers are involved. At least some of the physical structures believed to underlie consciousness in all mammals are found in the cerebrum and especially its outer layer, which is called the cerebral cortex. Many think the thalamus is also involved.[80] However, the complex behaviors of some birds, who lack a substantial cerebral cortex but possess highly developed striatal brain regions, suggest that complex cognition may not necessarily be dependent upon a cerebral cortex. Anatomical equivalents may also exist.[81]

Eighty percent of our brains and 75 percent of chimpanzee brains is cerebral cortex.[82] The area of our cerebral cortex is about 2,200 square centimeters, compared to about 500 square centimeters for the chimpanzee. Each square millimeter of the surfaces of the cerebral cortexes of both species contains about 146,000 neurons.[83] Both cortexes therefore probably hold on the order of 10^{10} or 10^{11} neurons, about the number of stars in the Milky Way. The map of our cortical layers is also similar.[84] The anthropologist Katerina Semendeferi, an expert on neuroanatomy, has written that the most forward section of the cortex, the frontal lobe, is often associated with the "most complex mental activities, such as language, creative thinking, planning, decision-making, artistic expression, some aspects of emotional behavior, and working memory."[85] Both human and chimpanzee frontal lobes have nearly identical relative volumes (the ratio of the frontal lobe to the rest of the hemispheric volume) and cortical surfaces. Humans average a 36.7 percent relative volume and 35.9 percent cortical surface compared to the chimpanzee's 35.9 percent relative volume and 38.1 percent cortical surface; each is about what would be expected of a primate with that size brain.[86] On average, each neuron in the cerebral cortex connects with the synapses of more than 1,000 other neurons and can potentially connect with tens of thousands more, so that the synaptic connections in the cerebral cortex number about 1 million billion (10^{15}) for both us and chimpanzees.[87] For both species, the combinations of neural connections is an estimated ten, followed by millions, perhaps trillions, of zeros. There are about 10^{87} elementary particles in the entire universe.[88]

The structures, numbers, and density of neurons, synaptic connections, and combinations of neural connections in both brains are of the same kind and order of magnitude, while the cortical layer maps are approximately the same. According to the psychologist Stephen Walker,

> Much work has been done since Huxley emphasized 100 years ago that "every principle gyrus and sulcus of a chimpanzee brain is clearly represented in that of a man," but there is nothing that contradicts his conclusion that the differences between the human and chimpanzee brains are remarkably minor by evolutionary standards.[89]

There must be differences, but we don't really know what they are.

THE GREAT CLASSIFICATION DEBATE

As we saw in Chapter 2, the history of law is chockablock with classifiers who obey Aristotle's Axiom while they busily fashion the world into a ladder upon which they occupy the top rung. Chief Justice Taney's statements in the *Dred Scott* case that blacks were seen as "beings of an inferior order" and "so far below [whites] in the scale of created beings" that "they had no rights which the white man was bound to respect" are probably the most famous examples. But instances abound in the history of science as well. The Swedish biologist Carolus Linnaeus classified African blacks as "ruled by caprice," and American Indians as "regulated by habit"; Asiatics were "governed by opinions," and white Europeans, like himself, were "governed by customs." Europeans were "gentle, acute, inventive"; the American Indian was "obstinate," the Asiatic, "severe, haughty, covetous," and the African, "crafty, indolent, negligent."[90]

Formal classification of the living world began just a few thousand years ago. Aristotle, the father of biological classification, tried to wring order from a tumultuous nature by dividing animals into those with blood, which he divided again into six groups, and those without blood, which he subdivided into four groups. His was no idle undertaking but rather a quest to answer what Stephen Jay Gould has described as the most important question in human intellectual history: "What is the role and status of our own species, *Homo sapiens*, in nature and the cosmos?"[91] Aristotle's system was to last two thousand years.[92]

We saw earlier that some of us tend to lump while some of us tend to split whenever we classify. Taxonomic lumpers submerge us within the natural world, whereas splitters place us above it. Some of the earliest taxonomic schemes classified plants and animals either alphabetically or with respect to their usefulness to humans. But that an animal that goes "moo" is called "cou" in Middle English, "kuo" in Old High German, and "bos" in Latin reveals one problem with an alphabetical system. Worse, the usefulness of plants and nonhuman animals is subjective and individual. Plainly, a more rational system was needed.

Just as the basic unit of French currency is the franc, the basic unit of classification is the now-familiar "species." To the disappointment of Frenchmen, only species can reproduce. Species was the brainchild of John Ray, Cambridge University class of 1648. He began by copying Aristotle. But soon he struck out and began sorting according to similarities. In the next century, Linnaeus dramatically improved on Ray's handiwork. Just as Moses was permitted to see the Promised Land, Linnaeus said he had glimpsed God's rational plan for the natural world.[93] "God created," the Swede's friends said. "Linnaeus arranged."

Linnaeus feared that revealing God's natural system would require him to compare every characteristic of every species in order to classify Creation correctly. Even if he lived to the age of the patriarchs, this was unattainable, even by Linnaeus, whose reputation as a science wonk was hard-earned. He would have to settle for classifying according to a small number of characteristics.

He was a creationist. God had created each species individually and divided it from every other species.[94] Because he believed in the Great Chain of Being, Linnaeus was ever-ready to classify as an intermediate species any creature that might link one species to another. That is why, in his great work, *Systema Naturae*, or *System of Nature*, he identified three members of the genus *Homo*. *Homo sapiens* ("wise man") was the name that Linnaeus unblushingly invented to describe himself. *Homo troglodytes* he had never seen. But he believed the sightings of wayfarers who said it was nocturnal and spoke in hisses, though he dithered over whether it was more closely related to the African pygmy or the orangutan. Last, there was *Homo caudatus*, a man with a tail who lived in the Antarctic. Linnaeus did not know "whether it belongs to the human or monkey genus."[95]

Plants and animals had often been named with one eye on classification and the other on description. Early on, Linnaeus named a plant of the morning glory family *Convolvulus foliis palmatis cordatis sericeis: lobis repandis, pedunculis bifloris*. He finally realized that a name might usefully label or it might usefully describe, but it could not do both well. So he hit upon the idea of using a "binomial nomenclature." A single genus name would be followed by a single species name, say, *"Homo sapiens."* Once a plant or animal was identified, other sources of information could be consulted. He also devised broader categories of classification than the genus, adding the empire, kingdom, class, and order. Others deleted empire and inserted the phylum after kingdom and the family after order, and the resulting basic seven ranks (kingdom, phylum, class, order, family, genus, species) have held up for over two hundred years.[96]

In his *Origin of Species*, Darwin observed that all naturalists classify according to a "Natural System." He thought the true "Natural System" was

founded on descent with modification; that the characteristics which naturalists consider as showing true affinity between any two or more species, are
those which have been inherited from a common parent, and, in so far [as]
all true classification is genealogical . . . the arrangement of the groups
within each class must be strictly genealogical in order to be natural.[97]

Species, then, actually existed in nature. Not so for genus, family,
order, and so forth. These flowed from the human imagination.
Stephen Jay Gould has explained that they

cannot be objectively defined, for they are collections of species and have no
separate existence in nature . . . [But they] are not arbitrary. They must not
be inconsistent with evolutionary genealogy [you cannot put people and
dolphins in one order and chimps in another]. But ranking is, in part, a matter of custom with no "correct" solution. Chimps are our closest relatives by
genealogy, but do we belong in the same genus or in different genera within
the same family?[98]

I propose to answer that question.

Taxonomy may not be arbitrary; but it's not like sorting chess
pieces, either. There is a lot of room for subjectivity. We have
Wise's Rule One ("Only with the utmost effort can we ever hope
to judge or place ourselves fairly in nature") and Rule Two ("We
must be at our most skeptical when we evaluate arguments that
confirm the extremely high opinion that we have of ourselves")
precisely to cabin that subjectivity. In a later edition of his *Systema Naturae*, Linnaeus complained: "I demand of you, and of the
whole world that you show me a generic characteristic . . . by
which to distinguish between man and ape. I myself most as-
suredly know of none. I wish somebody would indicate one to
me."[99] Although no one ever did, he still refused to place humans
and chimpanzees in the same genus. They made the same order
Anthromorpha, which he changed to *Primates*. But we were tagged
Homo sapiens, while chimpanzees were assigned first to the
genus, *Satyrus*, then to *Simia*.[100] As with many of Linnaeus's genera, this difference was later raised to the level of family, then superfamily.[101]

A fight has recently broken out among taxonomers over how to
classify not just chimpanzees, bonobos, and humans, but most animals. Each of the two main contenders (others exist) claims Dar-

win for its patron saint. They differ over whether taxonomy should tell us a creature's genealogy alone or whether it should tell us about its genealogy plus other things.[102] The school of "traditional taxonomy" uses the "genealogy-plus" method. It tries to cram as much information into a ranking as possible.[103] Harvard biologist Ernst Mayr says that "[t]he [traditional] school includes in its analysis all available attributes of these organisms, their correlations, ecological stations, and pattern of distributions and attempts to reflect both of the major evolutionary processes, branching and the subsequent diverging of the branches."[104] Traditional taxonomy then deteriorates into a battle of lumpers against splitters.[105]

The advocates of "genealogy alone" are called "cladists." They think that every rank, from kingdom to species, plant or animal, should represent all species that have descended from a common ancestor. Species that divided from a common ancestor should be placed at the same level and, if possible, be the same evolutionary age.[106] Cladists are often criticized because the only information they supply is about common ancestors. This is true, for they make no other judgments and attach no significance to traits that evolved after a plant or animal splits.

A cladist would draw my family tree showing straight-line descents from my mother and father, their descents from four grandparents, and all of our descents from eight great-grandparents. But a traditional taxonomist drawing my family tree would try to weigh all available attributes of everyone concerned. Great-grandma Bessie won the Nobel Prize in literature, though Uncle Elmo thought her writing stunk. Great-grandpa Arnie pulled six years in the Big House for embezzlement, though Great-grandma Joyce swore he was as honest as the day was long and was framed. Grandma Selma thought Grandpa Sidney looked like Rudolph Valentino, whereas his mother-in-law, Great-grandma Deborah, thought he looked more like Valentino's camel. Not only would my family tree look very different depending upon whether a cladist or traditional taxonomist drew it, but it might look different depending upon which traditional taxonomist drew it.

Traditional taxonomists argue that the human evolution of upright posture, toolmaking, intelligence, and speech since we split from the chimpanzees overwhelm all our similarities.[107] Every so often, one of them loses all self-restraint. The eminent nineteenth-century biologist Thomas Huxley demanded separate suborder status for humans. His equally eminent biologist grandson, Julian, sought to give us our own kingdom![108] But traditional taxonomists usually classify humans, chimpanzees, and bonobos in the same superfamily, *Hominoidea*. They constructed a family, *Hominidae*, just for us, then built another, *Pongidae*, for chimpanzees and bonobos, along with gorillas and orangutans, and sometimes gibbons. They forgot why Linnaeus concluded that a name could either label or describe but not do both.

The anthropologist Michael Ghiglieri says that if anthropologists followed the same criteria as mammalogists and ornithologists do with mammals and birds, chimpanzees and humans would inhabit the same genus.[109] Demands that chimpanzees and bonobos be grouped in one family and humans in another family clouds our common evolutionary relationship, for no other beings are so closely related yet placed so far apart. This troubled Stephen Jay Gould, a self-proclaimed "agnostic" about cladistics, for it contradicts evolutionary genealogy.[110] He writes that based on the latest evidence,

[h]umans arise within the space of the *Pongidae*, and cannot represent a separate family, lest we commit the genealogical absurdity of uniting two more-distant forms (chimps and gorillas) in the same family and excluding a third creature (humans) more closely related to one of the two united species. I surely cannot claim to be more closely related to my uncle than to my brother, but we make exactly such a statement when we argue that chimps are closer to gorillas than to humans.[111]

Wayne State University professor Morris Goodman is a prominent cladist. In the journal *Science*, the science writer Roger Lewin recounted that Goodman was "one of the first biologists in recent times to upset anthropologists by suggesting that family separation between humans and African apes should be torn down, an act of temerity he perpetrated in 1962." Goodman's idea, Lewin deadpanned, "was not well received."[112] Undeterred, Goodman has continued for almost forty years to criticize traditional taxonomy as a pre-Darwinian throwback to the Great Chain of Being, while arguing that cladistics represents real evolutionary groupings.[113]

But this is more than just a "family" squabble. The extremely close genetic relationship between humans and chimpanzees or bonobos is commonly found between such species of mammals within the same genus as the chimpanzee and bonobo, the fin whale and blue whale, and the horse and donkey.[114] We know that as recently as 5 or 6 million years ago, humans, chimpanzees, and bonobos were the same animal. That is why in recent years, an increasing number of biologists have begun to insist that humans, chimpanzees, and bonobos be placed in the same genus, *Homo*, and be separated only at the subgenus level; human beings should be *Homo (Homo) sapiens;* chimpanzees, *Homo (Pan) troglodytes;* and bonobos, *Homo (Pan) paniscus.*[115]

Because of the cladists' "just the facts, ma'am" attitude, David Pilbeam, a prominent Harvard anthropologist, wrote in 1996 that "we are almost all cladists now." Even the floor plan of the new fossil mammal halls in New York's famed American Museum of Natural History was cleverly constructed so that a visitor's walk through the evolution of mammals reproduces the cladist mind.[116] I cannot say that cladistics is the Holy Grail of taxonomy or even the best route through the Hall of Mammals. Frankly, I got lost, but that could have been because the twin Soldiers of Entropy were screaming in their stroller. But I can say that its reliance on DNA, its relative objectivity and lack of arbitrariness, and its greater resistance to lumping and splitting warfare has made cladistics absolutely indispensable for classifying *ourselves.* "Know thyself," Linnaeus said, after Socrates. But Freud showed us that we can't. We rarely expect to overcome our biases, and we rarely disappoint ourselves. We are all Muhammad Alis. We are the Greatest! That is why anthropologists warn that how we classify tells us more about ourselves than about what we're classifying.[117] That is why the law prohibits us from judging our own cases. It accounts for Wise's Rules One, Two, and Three.

There is not much running room in traditional taxonomy if we want to stay true to evolution. We are without doubt animals (kingdom). We have backbones (phylum). We are mammals (a class name derived from the female breasts; Linnaeus invented it just to underscore his argument that humans were of nature and not apart).[118] We are primates (order). But as soon as we could reasonably separate ourselves from the rest of nature, we did, and that was at the level of "family." And Chapters 2, 3, and 4 make it clear that seeing ourselves as a separate family justifies just about anything we do to the other animals. As though desperate to hide a family skeleton, traditional taxonomists muddy our apish lines. That is precisely what cladists want us to stop doing. As the goddess Circe warned Odysseus, the cladist admonishes the taxonomist to stopper his ears against the siren song of human arrogance.

IF WE ARE CONSCIOUS, SO ARE RABBITS

Some objective evidence does exist for the consciousness of nonhuman animals, if we are conscious. Here's some of the most intriguing evidence. Scientists think there are two forms of memory.[119] Some experiences can be remembered consciously. These are called explicit, or declarative, memories and are what we nonscientists normally think of as "memory." Our *explicit* (or declarative, or conscious) memories emerge in our consciousness as words or images.[120] But we can nonconsciously "remember" other experiences as well. We just don't *know* that we remember them. But they influence our behavior just the same. These memories are implicit or nondeclarative. The part of the brain known as the cerebellum, which all mammals and many other animals have, seems to be required for *implicit* memory.[121] Conscious awareness, however, is thought to require coordination between the medial temporal lobes of the brain, which includes the hippocampus and its supporting structure, and the cortex. The brains of all mammals have both.

Both kinds of memory can be tested. In a test of *implicit* memory, called "delay conditioning," a subject listens to a tone that is followed immediately by an air puff that causes a blink. Both the tone and the air puff stop at the same time. After a number of trials, the tone alone will cause a blink. A human will perform normally on this test as long as her cerebellum is intact.[122] The test does not require her to be consciously aware. Interestingly, the medial temporal lobes of some humans whose brain activity is being tested during delay conditioning begin working anyway, even though unneeded for delay conditioning.

The test for *explicit* memory, called "trace conditioning," looks just like the test for delay conditioning. But there's a difference. There is no overlap between the tone and the air puff. The tone stops and the air puff starts after a brief interval. Trace conditioning occurs only in humans with an intact medial temporal lobe and cortex, which is most of us. We are thought to be *consciously aware* of the relationship between the tone and the air puff. We know that damage to the medial temporal lobe destroys some conscious awareness, because human amnesiacs with damaged medial temporal lobes fail the trace conditioning test by not blinking.[123]

Rabbits have also been subjected to both delay- and trace-conditioning tests. As with humans, the medial temporal lobes of some rabbits whose brain activity is being tested during delay conditioning begin working, even though they are not needed for delay conditioning. Damage inflicted on their medial temporal lobes does not affect their delay-conditioning responses. Scientists believe that the reason that the medial temporal lobes of some humans and some rabbits are activated during delay conditioning is because both human and rabbit become *incidentally* aware of the relationship between the tone and the air puff, even though it was unnecessary.[124] Rabbits exhibit other so-called markers of declarative memory during delay-conditioning tests as well.[125]

There is even more impressive evidence. When either the medial temporal lobes or the prefrontal cortex of rabbits is damaged, so is their ability to perform on trace-conditioning tests.[126] Scientists have concluded that this means "that delay and trace conditioning could be used to study aspects of awareness in nonhuman animals" and that "by combining data from brain-injured patients, neuroimaging studies, and even nonhuman animals, it should be possible to gain an even greater insight into the neural processes that support memory, learning, and awareness."[127] The two University of California neuroscientists who conducted the latest air-puff tests concluded that

> an implication of the present findings is that learning and memory tasks, including trace conditioning, which are failed by animals with hippocampal lesion, are tasks about which intact animals must acquire declarative knowledge. Characteristics that have been helpful in extending the concept of declarative memory to nonhuman animals include its flexibility and the ability to use it in novel situations.[128]

In their opinion, "a large body of literature involving both humans *and experimental animals* can now be understood by recognizing that memory tasks requiring *declarative memory* depend on the integrity of the hippocampal formation and related structures."[129] In short, if we are conscious, so are rabbits. No bookie is going to take odds that rabbits are conscious but chimpanzees and bonobos are not.

BLINDSIGHT

Sometimes philosophers who claim that animals are not conscious point to an interesting phenomenon called "blindsight."[130] Here is how blindsight works. The major way in which our eyes connect to our brain is at the occipital lobe in a place called the striate cortex, or V1. But they connect in nine more minor ways as well. Humans whose occipital lobes have been damaged may *say* they are blind in that part of the visual field that corresponds to V1. Blind they think they are, but they remain able to identify accurately the positions of lights and lines in the visual field in which they say they are blind when they are required to guess and display frank amazement when shown videotapes of their actions.[131] Perception and awareness must somehow be able to be disconnected so that humans can "see" without being aware they can see. Humans, it seems, can perceive both consciously and unconsciously. If people can do that, then conscious and unconscious perception can be disconnected. Maybe animals perceive *unconsciously* all the time!

Scientists think that the anatomy of V1 is the same in human beings and monkeys.[132] Monkeys who have been unconscionably mutilated by having their striate cortexes removed behave in blindsight experiments just as humans who suffer the natural disaster of striate cortex destruction do. Of course, monkeys can't use language to communicate, so how can we know? In one experiment, it was initially determined that, as with humans, monkeys could detect a light that was flashed onto their "blind" V1 visual fields. They were then taught to respond to the position of a light shone onto their intact visual field by pressing a lit panel. In about half of these trials, the monkeys were rewarded for pressing an unlit panel when no light came on. That way the monkeys learned that when they saw a light, they should press the lit panel, and when they did not see a light, they should press the unlit panel. Then came the crucial part of the experiment. Lights were shone onto their blind V1 fields. Would the monkeys consciously see these lights and hit the lit panel or would they not see them (even though the researchers know that they could detect such a light) and hit the unlit panel? Just as blindsighted humans do, they hit the unlit panel.[133] The problem for the doubting philosophers, it seems, is that monkeys, like humans, perceive *both* consciously and unconsciously.

Are chimpanzees and bonobos conscious then, even a little? To the unhappy reader eagerly glancing at her watch to see if its time to leave for that root canal, I say this: Adult humans and infants, chimpanzees, bonobos, and a vast number of other nonhuman animals are almost certainly conscious—and not just "a little." If you agree, you may postpone all dental work and go directly to Chapter 9. The more skeptical may read on.

Rings of Consciousness

Human children have been studied far more often than the young of any other species of animal. No developmental expert thinks that human consciousness arrives full-blown, like Athena from the head of Zeus. The old biological idea that the way individuals develop strictly "recapitulates" the path along which its ancestors evolved has proven to be untrue. But Stephen Jay Gould has observed that "some relationship cannot be denied," and recapitulation is often "the dominant mode of evolution."[134] After thousands of experiments, reported in hundreds of articles and summarized in dozens of books with titles like *The Child's Discovery of the Mind*, *The Child's Theory of Mind*, and *The Growth of the Mind*, there seems little doubt of this: As human infants become children and children become young adults, increasingly complex layers of mind develop.[135]

The development of the human mind can be compared roughly to the growth of annual tree rings. But developmental psychologists disagree over the manner in which a child's mind grows. Some visualize young minds seamlessly expanding the way a balloon expands if inflated in one big breath. Others believe minds expand in stages, as a balloon does when it is inflated with a series of short breaths.[136] Much wider agreement exists about the kinds of attributes that emerge in human children and in what order they appear than when these attributes emerge and how gradually. Near-chaos reigns on how to correlate what is actually going on inside children's heads with their behaviors. Unsurprisingly, we will find similar grumblings and near-chaos when we explore the cognitive capabilities of chimpanzees and bonobos in Chapter 10. Here are just two examples—pretend-play and theory of mind— that will have important implications for chimpanzees, bonobos, and children. But first, a little background on the ability to mentally represent.

I had a zero-order ability to mentally represent anything when I was very young. I later developed a first-order ability to mentally represent ("John *intended* to insult me"). Then I could mentally represent at the second-order level ("John *intended* to insult me so that Sue would *believe* that he was her friend"). These are often called "metarepresentations." Eventually, I could mentally represent at the level of third-order ("John *intended* to insult me so that Sue would *believe* that John *wanted* her to be his friend") and fourth-order ("John *intended* to insult me so that Sue would *believe* that John *wanted* her to *think* that he would be her friend"). You get the idea. Daniel Dennett argues that "in principle" we might understand infinite levels of mental representations, "but in fact I suspect that you wonder whether I realize how hard it is to be sure that I understand whether I mean to be saying that you can recognize that I can believe you to want me to explain that most of us can keep track of only about five or six orders under the best of circumstances."[137]

Children begin to pretend-play around the age of eighteen months. My older daughter, Roma, used to stage tea parties at which her dolls, Mr. Bear, and other stuffed animals, were placed around the table with Roma, quaffed imaginary tea that Roma poured, and chatted about topics Roma chose. Jean Piaget, the influential child developmental psychologist, thought that pretend-play demonstrated Roma's ability to symbolically represent one thing for another. Professor Alan Leslie, a later child developmental psychologist, noticed that when children start to pretend-play, they seemed already to understand the pretend-play of other children. He thought this was the first indication that they are on their way to understanding the mental states of others. Roma, he suspects, harbored beliefs about beliefs and was thinking "I am pretending that Mr. Bear is drinking tea."

Professors Josef Perner and Angeline S. Lillard, also child developmental psychologists, disagree. They think that understanding that one thing can stand for another doesn't necessarily mean that it symbolically represents something else. Roma, they think, was just imagining a number of possible situations, then acting "as if" the world turned in the way she imagined. Two other child developmental psychologists, P. R. Harris and L. D. Kavanaugh, disagree with everyone and argue that Roma was not thinking about any mental states at all. Instead, she was viewing her tea party as an activity in which she "flagged" her dolls and animals and tea, then read the flags whenever she needed to keep the party-action going.[138]

Scientists disagree even more about theory of mind than they do about pretend-play. The term "theory of mind" was coined in 1978 to describe what two researchers, David Premack and Gary Woodruff, thought might be going on in the mind of Sarah, a chimpanzee. An "explicit" theory of mind exists when someone can understand and predict the behavior of another by attributing mental states to her.[139] This requires mental representation of the second-order. But some scientists argue that such animals as young human infants, who are limited to the ability to mentally represent at the first-order, might still engage in "intersubjectivity." An infant without the ability to mentally represent at the second-order might "feel" some subjective mental states of his mother by experiencing her mental state as inseparably linked to her behaviors in just the way he might perceive the color red not as an internal state of a red fire hydrant but as a property of it.[140] Professor Juan Carlos Gómez of the University of St. Andrews argues that this leads to an "implicit" theory of mind by which someone, an infant or an ape, with just the first-order ability to mentally represent could have a theory of mind that provides much the same information as the theory of mind of someone who can mentally represent at the second-order.[141] As scientists do, when I refer simply to "theory of mind," I will mean "explicit theory of mind," that is, the ability to function at the second-order and understand and predict the behavior of others by attributing mental states to them.

Not only can those with a theory of mind have mental representations about mental representations (second-order representations, or metarepresentations), but they can understand that other minds may know things that they don't.[142] In Western culture, theory of mind is essentially the "identity theory" (a.k.a. "belief/desire psychology," "mind reading," or "folk psychology") with which I led off my Top Ten Theories of Consciousness.[143] It snagged the number-one spot because we Westerners seem automatically to impute mental states right and left every day of our lives (I *think* he *wants* to go out with me. I *want* to go out with him. Or, my boss *wants* me to do that? I *think* he's nuts!). But are they true? The child psychologist Janet Astington doesn't care: "Whether our folk psychological theories and our commonsense understanding of mind are *true* or not, in any scientific way, is not really relevant here. This commonsense understanding is what at least children acquire, regardless of its standing in the philosophical debate."[144]

Western researchers studying the minds of children happily seized on Premack and Woodruff's idea of theory of mind to help explain what they were finding. Since 1980, they have generated a mountain of research about its development. There is nearly universal agreement that it emerges gradually, but not much agreement on anything else.[145] University of Virginia professor Angeline Lillard has shown that folk psychologies can vary dramatically from culture to culture; certainly compared to the flaming mind reading of most Westerners, some cultures appear to have very little theory of mind (the Baining and Kaluli of Papua New Guinea), while still other cultures believe that minds cannot be known and so are irrelevant (Samoans; Bimin Kuskusmin and Kaqchikel, both of Papua New Guinea).[146] But very little research has been done on theory of mind in either non-Western cultures or on Westerners over the age of five.[147] University of Michigan psychology professor Henry Wellman has diagrammed no less than thirteen warring claims about *when* it begins to develop, claims that range from the moment of birth, perhaps even before, to the child's fourth year.[148] Three major, and other minor, schools spar over *how* theory of mind originates. "Nativist" theorists claim that humans are born with an "innate processing device" that tells us how to understand the mental states of others. "Theory-theorists" argue that children develop something akin to an abstract scientific theory, or a series of theories, about how other minds work. Then they appeal to this abstract theory, or theories, to explain and predict how others behave. "Simulation theorists" believe that children instead "slip on the mental shoes" of others and imagine how they would feel in the same situation. Then they attribute their feelings to that other person.[149]

Professors Vittorio Gallese and Alvin Goldman have illustrated the differences between the latter two schools with the following question posed by other researchers: Mr. A and Mr. B, whose planes are scheduled to depart at the same time, share a limousine that arrives at the airport thirty minutes late. Mr. A learns that his flight took off on time, while Mr. B is told that his flight left just five minutes before. Which traveler will be more upset? Only 4 percent of the research subjects chose Mr. A, while 96 percent chose Mr. B. Theory-theorists would argue that the research subjects arrived at their answers by applying an abstract psychological law that generally determined which traveler would be more upset. Simulation theorists will argue that the research subjects placed themselves in the "mental shoes" of each traveler and imagined how they would have felt.[150]

The point of airing psychology's dirty laundry is not to taunt them for their cluelessness about theory of mind and pretend-play, for they aren't clueless. It is to demonstrate the difficulty that they have in understanding what goes on inside the minds even of Western adults who can tell you about it. It should therefore come as no surprise in Chapters 9 and 10 that researchers trying to puzzle out the minds of apes often clash. But on to the first child "consciousness ring."

The First Ring

I'll use that Soldier of Entropy, Siena, as a prototype of the Western mind in its first year. In the first months of life, she explored her own body, reached for and grasped her toy dog, and understood that her toy cat couldn't be in two places at one time, pass through her dog, or do other physically impossible things. She imitated me when I opened my mouth or stuck out my tongue, detected the relationship between her own movements and her corresponding movements being played live on video, and identified a new toy by sight that she had only heard or touched. She recognized that, unlike her toy dog and cat, I was an animate being and could interact directly with objects, humans, and other animals.[151] The early eye contact that she made with me may have signaled the beginning of her ability to engage in "intersubjectivity."[152]

The psychologist Karen Wynn's series of brilliant experiments has made it clear that Siena knew that if one jar of the World's Best

Baby Food was added to another, the sum was not one, not three, but precisely two, and that when one jar was subtracted from two, precisely one jar remained.[153] Wynn took advantage of the fact that infants, like adults, stare longer at events that interest them than at events that don't. And when the seemingly impossible occurs, we're all interested even in our first year of life.

Wynn set up a little puppet theater in front of four-month-olds. A disembodied hand placed a Mickey Mouse doll on the stage. Then Wynn lowered a screen and blocked the infant's view of the doll. Now the hand placed a second Mickey Mouse doll on the stage. Wynn was able to remove one of the dolls through a trap-door on the puppet stage. Sometimes she did. When she raised the curtain, there might be two Mickeys on stage or just one. If the infants could add the number of dolls, they should expect that two Mickeys would be on stage when the curtain was raised. But if only one doll was there, they should be surprised and stare longer at the one doll than at the two. And that is just what happened, over and over again. Siena could not calculate past the number 3, or perhaps 4, but she apparently could do what she could do because her mind produced precise mental images.[154]

Later in that year, Siena looked where I was looking, imitated more complicated movements, and played with a book alongside me. She categorized her toys in simple ways, used very simple tools, estimated quantities of Cheerios, and commenced a simple understanding of how events relate to each other.[155] She moved her toy stove out of the way in order to reach her plastic hammer. If I covered the toy stove, she realized that it continued to exist even when she was unable to perceive it (Piaget's "Stage 4 object permanence task") and removed the cover.

She began to "learn to learn." After being rewarded for choosing one kind of stimulus over another, she learned a general rule that she could apply to an infinite number of situations. Psychologists think that she was not merely associating a stimulus with a response but was learning something more abstract. Between the ages of eight and ten months, she began to understand a few of my words.[156]

Near the end of this year, Siena began to show signs that she was acting intentionally and thought that I was, too.[157] I knew that the foundation for her theory of mind was being laid when her capacity for "joint attention" appeared.[158] Not only did she interact directly with objects, me, and other animals, but she began to "share the world." She looked at me when I read *Goodnight Moon.* She was beginning to understand, at least implicitly, that when I look or point to something, the attention I pay connects me to it.[159] She followed my gaze toward places she could not see. She began to "socially reference" by looking to me for emotional cues about how to respond to objects or events. She began to point.[160]

Her pointing may seem no big deal, but it was. Scientists say there are two kinds of pointing. They call the simpler "proto*imperative* pointing." When her favorite videotape, "The Best of Elmo," reached its conclusion, she pointed to the television screen, then alternated her gaze between me and it.[161] If I didn't rewind the tape, she pounded on the blank screen. A more complex kind of pointing is "proto*declarative* pointing." Siena might have pointed not because she wanted me to do anything for her, like rewind "The Best of Elmo," but simply to call my attention to a pull toy that was making a funny noise. Some scientists think that only humans can proto*declaratively* point. I think they're wrong. But we'll save that for Chapter 9.

The Second and Third Rings

Early in her second year, Siena watched me place a marble under one cup, saw me move it to another cup, and understood that it was now under that second cup (Piaget's "Stage 5 object permanence task"). Thus, she knew that an object could move independently of her actions, even though she may not yet have been conscious of her mental representations.[162] She related objects to each other by stacking blocks or putting cups inside each other to see what would happen.[163]

Later this year, her mind will explode. She will maneuver as if she is toting "cognitive maps" in her head that tell her where things are hidden.[164] Other abilities will arrive in a bundle. Simultaneously, she will recognize herself in a mirror and be able to use it to examine marks on her body. She will begin to exhibit altruism and self-conscious emotions, such as embarrassment.[165] She will start to cooperate with her twin.[166] She will begin to act and plan and comment on her plans.[167] She will learn symbols by imitating me and begin the kind of symbolic play that will lead to the kinds of tea parties that her older sister, Roma, ran a decade before.[168] She will imitate not just what I did but what I am trying to do, and she will understand something of my state of mind when I say that I'm looking for something, then find it.[169] She will respond to my request to show me her toy dog and will begin to understand that events in the world have physical causes.[170]

Sometime this year, Siena will pass Piaget's "Stage 6 object permanence task." This requires her to be able to show that she mentally represent objects. If I place her toy dog under one of the three pots that she is forever dragging from the kitchen cabinets, she will systematically search for the dog until she finds it.[171] Her inability to pass the Stage 6 test until this time may not have occured because she couldn't mentally represent objects—remember, before she was six months old, she could mentally add and subtract small numbers—but until now she couldn't show us that she did; the parts of her brain that allow her to *reach* were too immature.[172] By midyear, she may be able to begin to distinguish intentional from accidental behavior.[173] By the end of the year, she will have learned dozens of words and will be using them to comment about things after she has grabbed my attention. While I'm feeding her, she might hold up her cup, see that she has my attention, and then say "cup" or "juice." Her understanding that she has my attention and that we are both focused on the same thing may be the most obvious demonstration that she is beginning to understand the intentions of others.[174]

The third ring is of particular interest, as most primatologists agree that "barring inappropriate rearing conditions, great apes display symbolic cognitive skills similar to . . . human children" of this age.[175] Siena will develop what is called Level 1 visual perception. This means that she will understand that she can see her toy dog when I can't see it and that I might be able to see it when she can't. But she won't yet realize that it will appear differently to each of us depending upon the angle from which each of us views it or from what distance.[176] She will begin to evolve a desire psychology, the cognitively simpler half of the "belief/desire" psychology or "folk psychology" (I "want," I "wish," I "hope," I "should") that we all share.[177] She will know that if I *want* an apple, I may try to find one and that I may emotionally react when I either succeed ("Found it!") or fail ("Where *is* that damn apple?").[178] She will begin referring to herself ("Siena wants") and talking simply about her emotions and desires, as well as my feelings, and ascribing feelings to her dolls or her cat, Alice.[179] She will start to combine words to talk about events.[180]

The Fourth Ring

The fourth ring is of particular interest to us because of the oft-repeated observation of Professor Premack that with occasional exceptions, usually in favor of the chimpanzee, "a good rule of thumb has proved to be: if the $3\frac{1}{2}$ year old child cannot do it, neither can the chimpanzee."[181] Siena will start to understand that that her toy dog will appear differently to each of us depending upon angle and distance of viewing.[182] She will begin to layer a belief psychology (I "believe," I "expect," I "know") atop her desire psychology. If I believe something is true and she knows I believe it is true, she will be able to predict how I will act. She will understand "true belief" but will not comprehend "false belief" for at least another year, maybe two.[183] People have a false belief when they think that something is true because they are missing the information that would lead them to believe that their belief is actually false.[184] Siena will even be unable to acknowledge that she once had a false belief.[185]

For the approximately 0.1 percent of children who are autistic, Siena's developing folk psychology is unlikely ever to come together, except in bits and pieces.[186] Many autistic children seem to be, as Professor Simon Baron-Cohen says, "mindblind." Although they may recognize themselves in mirrors, many can't engage in vital "joint attention" behavior. They don't monitor gazes. They don't use protodeclarative pointing. Their seeing doesn't necessarily lead to their knowing the meaning of what they see. They may gesture to manipulate others' behavior ("Go away") but not manipulate their own mental states ("Everything will be all right"). If shown a video in which one person looks into a box and another just touches the box, most will say that the toucher, not the looker, knows what's inside it. They are impaired in their ability to pretend-play. After listening to a story about one character with a real bicycle and another who just thinks about a bicycle, most autistic children will claim that the thinker is the one who can touch the bike. Normal five-year-olds think that legs make them walk, eyes make them see, and brains help them think.[187] But when asked what the brain is for, autistic children usually answer in mechanical terms. Brains help them walk or see. Autistic children don't seem to understand that other minds can have beliefs different from their own. Their social awareness and behaviors are correspondingly severely impaired.[188] Autistic children will usually fail any test that requires them to know that other minds exist that hold beliefs. Beginning at this age, the likelihood that Siena will pass any such test rapidly becomes much greater. If she doesn't just observe but is actively involved in the deceit, her likelihood of passing the tests rises, though she won't become adept at deception for some time.[189] Here are a few examples.

Siena and Wally, an autistic child, watch a puppet show. A puppet boy places some chocolate into a cupboard, then leaves the room. While he is gone, his puppet mother moves the chocolate to another cupboard. When Wally is asked where the puppet boy will look for the chocolate when he returns, he will probably say that the boy will look in the cupboard in which his mother placed it. Siena may say it was the second cupboard. About half of four- to six-year-olds and most six- to nine-year-olds do.[190] In the "Smarties test" (English Smarties are like American M&Ms), Siena and Wally see that a candy box contains pencils and not candy. When asked what someone else will think the candy box holds, Wally will say "pencils." Siena may say "candy."[191] Perhaps best known is the "Sally-Anne test" (also called the Location Change task). Siena and Wally watch Sally put a marble in a basket, then leave the room. While she is gone, Anne moves the marble to a box. Wally will say that when Sally returns, she will look for the marble in the box and not the basket.[192] Siena may say the basket. Afterward, Wally may even claim that when he first saw the candy box, he thought it contained pencils.[193]

Siena and Wally will begin to diverge in their abilities to distinguish between appearance and reality. She, but not he, will be able to distinguish between someone thinking about a book and someone actually having it.[194] She will be able to form a "dual orientation," such as mentally representing a scale model both as a real object and as a symbol for something else. If I build a scale model of our family room and let her watch me hide a miniature stuffed bear in the scale model, she will be able to find the real bear in the real family room, whereas six months before, she would have been unable to find it.[195] She will learn to cooperate in complicated ways, mentally switch places, and understand roles played by others. She will burnish her ability to behave in ways that I am likely to misinterpret. When she uses this "tactical deception," she may gain an advantage over her unsuspecting dad.[196]

The Fifth Through Ninth Rings

In her fifth year, Siena will say that she intends to go to the bathroom and describe what she intends to do.[197] Given the opportunity to handle a sponge painted to look like a rock, she will recognize that it is a sponge made to look like a rock. Autistic children and normal three-year-olds will claim that the painted sponge is a rock.[198] She now possesses a rudimentary theory of mind and will often believe that the thoughts and beliefs of others are different from her own.[199] She may begin to make predictions based upon false beliefs.[200] She will teach her twin, Christopher, how to build with Legos.[201] Either this year or the next, she will begin to count accurately and reliably and understand what counting is for.[202]

Between the fifth and seventh rings, Siena will begin to understand that lengths, weights, quantities, and numbers remain the same even when they are packaged differently. I might show her two rows equal in length that contain the same number of pennies. If I push the pennies in one of the rows together and ask her if the two rows still have the same number of pennies, she will say "yes." She would have said "no" in her fourth year. If I show her two glasses filled with equal volumes of water, pour the water from one glass of water into a taller and thinner glass, and ask Siena whether the glasses still hold the same amount of water, she will say "yes." She will start "conserving" both number and quantity by understanding that one property of an object can stay the same even as other properties change.[203] At about six to eight or nine years old, she will exhibit second-order beliefs ("*I* think that *you* believe that today is Monday"). She won't be able distinguish my lies that are intended to deceive from my lies that are jokes or exaggeration or sarcasm until she reaches nine or ten.[204] She has been heading toward a fully developed "theory of mind" for most of a decade.[205] Now she's almost arrived.

Kanzi and COG

Harvard biology professor Edward O. Wilson relates an encounter with a young bonobo named Kanzi on a visit to the Language Research Center at Georgia State University run by the husband-and-wife team of Sue Savage-Rumbaugh, professor of biology and psychology, and Duane Rumbaugh, professor of psychology. Wilson, tall, imposing, and unfamiliar, scared the little bonobo. After accepting a cup of grape juice, Kanzi could not restrain his curiosity and

> drifted back over to me. This time, having been coached by Sue Savage-Rumbaugh, I imitated the flute-like conciliatory call of the species, *wu-wu-wu,wu,wu* . . . with my lips pursed and what this time I believed to be a sincere, alert expression on my face. Now Kanzi reached out and touched my hand, nervously but gently, and stepped back a short distance to study me again. (Wilson accepts a cup of grape juice.) I flourished a cup as if offering a toast and took a sip, whereupon Kanzi climbed into my lap, took the cup, and drank most of the juice. Then we cuddled. Afterward everyone in the group had a good time playing ball and a game of chase with Kanzi. The episode was unnerving. It wasn't the same as making friends with the neighbor's dog. I had to ask myself: was this really an animal? As Kanzi was led away (no farewells), I realized that I had responded to him almost exactly as I would to a two-year old child—same initial anxieties, same urge to communicate and please, same gestures and food-sharing ritual. Even the conciliatory call was not very far off from the sounds adults make to comfort an infant. I was pleased that I had been accepted, that I had proved adequately human (was that the word?) and sensitive enough to get along with Kanzi.[206]

I know how Wilson felt. Twice I have visited Kanzi and his half sister, Panbanisha, at the Language Research Center. I listened to Sue Savage-Rumbaugh speak to Panbanisha in English. I watched them converse by means of a folding board that contained a dizzying array of over two hundred and fifty brightly colored "lexigrams." I met the lexigram-using chimpanzees, Sherman and Austin. At the Chimpanzee and Human Communications Institute run by Professors Roger and Deborah Fouts at Central Washington University in Ellensburg, Washington, I have observed Washoe and her family of signing chimpanzees, Tatu, Moja, Loulis, and Dar. I watched Washoe ask the woman standing beside me to remove her shoe, to the delight of the Foutses. At the Primate Cognition Project, run by Professor Sally Boysen at Ohio State University, I was introduced to Abigail, Darrell, Kermit, Sheba, Bobby, and Sarah, who was the subject of Premack and Woodruff's original theory of mind work in the 1970s. As I watched, Sarah delicately stripped grapes from a stem while she patiently waited for Boysen to begin a number experiment. Afterward, lost in the mental image of the stately matriarch, I forgot Boysen's earlier warning about the juvenile males. As Boysen led me through the door of the chimpanzees' quarters, Bobby filled his cheeks with water and whistled a jet right between my eyes from thirty feet away.

Gerald Edelman once observed that "[e]xtraordinarily silly things have been proposed about the capacities of machines to think."[207] The vortex of much of this silliness is MIT. There, the "founding father of artificial intelligence," Marvin Minsky, asserted in 1993 that certain computer programs with memory features were "extremely conscious."[208] Daniel Dennett, who slashes away at the idea that any nonhuman *could* be conscious, has been advising a group building a robot named COG at MIT's artificial intelligence laboratory. Dennett claims he gets "an almost overwhelming sense of being in the presence of another conscious observer" when COG's camera eyes track him as he moves about a room. He predicts that COG "will be conscious if we get done all the things we've got written down."[209]

Interested, I made an appointment to visit the artificial intelligence lab. There I "met" COG. COG is a legless humanoid robot from the waist up. It was resting on a table that sets its camera eyes at human eye level. The friendly Ph.D. student who led my group of three about the lab told us that the long-term goal was to make COG "as smart as a six-month-old." Then he showed us YUPPEE, an artificial intelligence "dog," and KISMET, a "robot baby," and videos of all three in action. But he couldn't activate COG in the "flesh" for me. After listening to him use a raft of words that denote mental states to describe how the three robots were "feeling," such as "likes," "dislikes," "cries," "gets upset," "wants," "lonely," "interested," "bored," and "intimidated," I asked him if he expected COG to achieve consciousness within, say, the next fifty years. He plainly hoped it would. But he shook his head and said that thirty years ago, people thought that artificial intelligence was "just around the corner and we're still not even close."

The journalist Stephen Budiansky, no friend to the idea of animal consciousness, has labeled COG a "parlor trick" that Dennett and "his colleagues at the Artificial Intelligence Lab at MIT cooked up."[210] Professor Ben Schneiderman, who directs the Human-Computer Interaction Laboratory at the University of Maryland, thinks that work on COG might lead to "better animatronic dolls for Disney World or better crash-test dummies," but nothing else.[211] The "overwhelming impression" I got while observing COG, its "baby," and its "dog," is the same one I get when Roma and I visit the Hall of Presidents at Disney World or when I watch *Star Wars*, or when George Washington's eyes stare at me from no matter what angle I view his engraved picture on the one-dollar bill. But I neither think nor believe that the Disneyfied Lincoln, the *Star Wars* C3PO, or the Washington on the dollar bill are conscious or even on the road to it.

Dennett thinks that similarity of behavior between humans and nonhumans, whether they are robots or bonobos, suckers "the gullible or generous-hearted."[212] If so, they are in good company. Holmes (Oliver Wendell, again), not known for being either, wrote that "even a dog distinguishes between being stumbled over and being kicked."[213] He was hardly the only judge who thinks so. In hundreds of cases, thousands of judges have agreed with Holmes. In Dennett's view, not only do my cat, Alice, and my dog, Marbury, lack consciousness, but so do my twins, Siena and Christopher. Apparently they will be blinded by the light of consciousness just about the time they learn to talk back. Yet Dennett thinks that computer programs and robots are conscious, and he has nothing but scorn for those who "baulk at the *very idea* of silicon consciousness," though he concedes that "[t]his idea has been dismissed out of hand by most thinkers."[214] Certainly neither scientists nor philosophers are flocking to sign his dance card. Instead, some call his ideas "bizarre" and "intellectual pathology."[215]

De Waal's Rule of Thumb ("Strong arguments would have to be furnished before we would accept that similar behaviors in related species are differently motivated") can only be invoked for animals who have traveled similar evolutionary tracks. This is precisely why Dennett's plea that we open our minds to "silicon consciousness" but close them to the consciousness of nonhuman animals is so weird. Silicon COG and his family are continuous with nothing but aluminum and phosphorus on the periodic table. Estimates made by Greenfield, Gazzaniga, Fodor, and Pinker that nobody has the first idea about how consciousness works makes me wonder if Dennett has written anything down other than "Note to file: Make COG Conscious."

One day science may prove what my wife claims is impossible—that I am the only conscious being in the universe—or that our twins are not conscious, that neither are Marbury, Alice, chimpanzees, or bonobos. I don't know how. But if that day arrives, I will think of Bertrand Russell, a famous atheist, who rehearsed what he would say if God stopped him at the pearly gates and demanded to know why Russell hadn't believed in him when he was alive: "You didn't give me enough evidence!"

Wordless Thought

The last redoubt in the fight against animal consciousness is this syllogism: Language is necessary for consciousness; only humans have language; therefore only humans are conscious.[216] Here, I'll discuss why the first premise is wrong and save my attack on the second for the next chapter.

Today hardly any scientist believes that language *creates* minds. Even Budiansky, who acts as a flanker for those who want you to believe that nonhuman animals are not conscious, has been reduced to claiming not that language causes consciousness but rather that "whether or not language causes consciousness, language is so intimately tied to consciousness that the two *seem* inseparable."[217]

Seem inseparable to whom? To begin with, language is more than just *tied* to consciousness. Without consciousness, there could be no language. Few doubt that a human mind that bubbles with language is different than a languageless human mind. In the next chapter, we will see that even an ape mind with language differs from the mind of an ape without it. But it is a logical error to conclude that just because consciousness is necessary for language, language is necessary for consciousness. Though the John Hancock Building could exist without the observation deck, there could be no observation deck without the building.

The evidence that language is unnecessary for consciousness is powerful and converges from many directions. There is little evidence that humans think only in the language they know. If we did, we would expect our thinking to differ as our languages differed. But it doesn't.[218] Language and consciousness haven't seemed inseparable to many brilliant and creative people, from Einstein to the poet Samuel Taylor Coleridge, who both swore that their greatest ideas came to them wordlessly as pure images.[219] Language and consciousness don't seem inseparable to some highly functioning autistic adults who tell us that pictures dominate their thinking.[220] Language and consciousness weren't inseparable to Siena, who could add and subtract the numbers 1, 2, 3, and maybe 4 by the age of six months without knowing a single word. After months of chaotic dinners, my wife and I began showing thirty-minute videos to keep the twins occupied while we wolfed down our food. If the videos ran out, Siena would toddle into the kitchen, take my wife by the hand, lead her to the television set, point with her whole arm to the screen and grunt, all the while alternating her gaze between my wife and the screen.

After she learns how to speak, Siena isn't going to remember doing that. But hundreds of humans have lived languageless lives for years, sometimes for decades, before they learned language. Then they could tell us what it was like to think without language without waiting to be asked.[221] Psychology professor Harlan Lane described how, in the nineteenth century, Ferdinand Berthier indignantly rejected the "premise that thought depends utterly on language" in a speech to the French Academy of Medicine and the Second Class of the French Institute. Berthier was a deaf-mute and a teacher of deaf-mutes. He relayed an account in which another deaf-mute, Allibert, had been "asked at a dinner if he had any ideas before receiving [language] instruction at the institute; to general amazement he said that he had, and proceeded to describe a few of the ideas of his childhood." Berthier, who should have known, thought it "an uncontestable and uncontested fact that thought exists before language."[222]

The neurologist Oliver Sacks tells the story of Joseph, an eleven-year-old deaf and languageless boy whom Sacks knew. Sacks thought that Joseph could not "hold abstract ideas in mind, reflect, plan, play." But he "was like an eleven-year-old in most other ways."[223] He liked to draw, could easily solve visual puzzles and problems, and was proficient with problems that required perceptual categorization or generalization. Sacks concluded that "it was not that he lacked a mind, but that he was *not using his mind fully*."[224]

Sacks narrated the story of how in 1893, William James published a letter from Theophilus d'Estrella. D'Estrella, born deaf, had acquired his first formal language at age nine. He told James that before he learned language, "I thought in pictures and signs." He "considered that he *did* think, that he thought widely, albeit in images and pictures, before he acquired any formal language; that language served to 'elaborate' his thoughts without being necessary for thought in the first place," and James agreed.[225]

Sacks also told of Jean Massieu, born deaf in the eighteenth century, who did not acquire a language until he was nearly fourteen. After he mastered written French, he wrote that, until the dawning of language, he had communicated only with manual signs and gestures. As a child, he "saw cattle, horses, donkeys, pigs, dogs, cats, vegetables, houses, fields, grapevines, and after seeing all these things remembered them well." He "had a sense of numbers, even though he lacked names for these: 'Before my education I did not know how to count; my fingers had taught me. I did not know numbers; I counted on my fingers, and when the count went beyond ten I made notches on a stick." He "envied other people going to school; how he took up books, but could make nothing of them; and how he tried to copy the letters of the alphabet with a quill, knowing that they must have some strange power, but unable to give any meaning to them."[226] To Sacks, it was

> clear that thought and language have quite separate (biological) origins, that the world is examined and mapped and responded to long before the advent of language, that there is a huge range of thinking—in animals, or infants—long before the emergence of the language. (No one has even examined this more beautifully than Piaget, but it is obvious to every parent—or pet lover). A human being is not mindless or mentally deficient without language, but is severely restricted in the range of his thoughts, confined, in effect, to an immediate, small world.[227]

Then there is the remarkable story of deaf Ildefonso. Susan Schaller, a teacher of American Sign Language (ASL), relates it in *A Man Without Words*. When she met the twenty-seven-year-old man in Los Angeles, he "not only lacked any language but lacked any *idea* of language."[228] But he had herded goats and sheep, planted and harvested sugarcane, slaughtered chickens, begged, and worked on airplane parts! ("Hello, FAA? Your airplane parts manufacturers are hiring workers who aren't conscious!")[229] Yet Ildefonso, and other languageless men and women whom Schaller met, "anxiously began expressing their personal histories," indeed "whole autobiographies" as they learned language.[230]

Schaller was stymied in her efforts to learn almost anything useful from scholars about how the languageless thought. "Everything I read or heard was purely abstract, hypothetical, and speculative," she wrote.[231] She was assured by a graduate student working with two linguists who were following the progress of a languageless deaf adult as she learned language that it was a "once-in-a-lifetime experience."[232] Sitting in a class for languageless adults in the same city, she later wondered, "How could a researcher consider a prelingual deaf adult a once-in-a-lifetime happening when four were sitting at the same table only a few miles away?"[233] Frustrated that "people in universities seemed ignorant of languagelessness," she plunged into the deaf community, searching for languageless adults.[234] Soon she found herself surrounded by a roomful of them and watched in amazement as they acted out their lives for her.[235]

Maybe Ildefonso, d'Estrella, Massieu, and others who learned language very late and related their prelingual life histories just imagined they had had minds in their earlier lives. But I doubt it. Maybe Sacks and Schaller and James are just overly "gullible or generous-hearted" souls. But I doubt it. Normal humans report that they mentally represent in a combination of language, pictures, and pure thought. More likely Stephen Jay Gould was correct when he observed that "[p]rimates are visual animals, and we think best in pictorial or geometric terms. Words are an evolutionary afterthought."[236]

Gerald Edelman does not think that language is necessary to ignite higher-order consciousness. But as an incredibly efficient and superb sort of symbolic memory, language makes it burn.[237] In an 1988 book, *Thought Without Language*, more than a dozen scholars presented evidence that infants and nonhuman animals can think without language and that human adults with language often think without it. Its editor, Lawrence Weiskrantz, the experimental psychologist who named "blindsight," concluded, "That there can be mental content, a capacity to 'think of' in the absence of language, and in a form that has external reference, would not seem in doubt from a large variety of evidence presented here."[238] Ten years later, he remained firmly convinced. "If it were language that especially endowed one with consciousness, however, it would imply that only *Homo sapiens* were conscious, which few—and certainly not I—would wish to accept."[239] He concedes that without language, the consciousness of blindsighted monkeys cannot be proven and that we must fall back on something like de Waal's Rule of Thumb. But such an "analogy is strong if changes in awareness, as in blindsight, follow similar courses in animals to those in man, as we have seen actually happens, and if the brain systems revealed by brain imaging show such parallels (which they do)."[240]

I'll leave the last word to Kanzi's friend, Sue Savage-Rumbaugh:

> It requires but a moment's reflection to recognize that humans engaged in complex nonverbal activities—such as in dance, music, sculpting, and athletic skills—depend on wordless thought. To suggest otherwise is "a notion that only a college professor or other professional wordsmith could have ever taken seriously."[241]

9

Seasons of the Mind

THE EFFECTS OF SOCIALIZATION

After primatologists noticed that primates thrive in social environments filled with complexities and rife with intrigues, speculation arose that primate intelligence developed precisely from the demands of living such rough-and-tumble social lives.[1] In the late 1970s, Frans de Waal climbed into a ringside seat high above an island at the Arnhem Zoo in the Netherlands. For thousands of hours, he watched the lives of twenty-five chimpanzees in the world's largest captive colony unfold in a way that even Jane Goodall had not been able to observe at Gombe, where chimpanzees live and travel in dense, often impenetrable, forest. But at Arnhem, they were confined to a nearly treeless island where de Waal was able to assume an unobstructed God's-eye view of what transpired below.

The jockeying for advantage and power that he witnessed reminded him of nothing so much as the vivid word pictures that the Florentine political philosopher Niccolò Machiavelli painted of the machinations and infighting of seventeenth-century Italians.[2] When Newt Gingrich ascended to the speakership of the U.S. House of Representatives in 1995, he placed de Waal's book *Chimpanzee Politics* on the reading list for new members. By 1998, de

Waal was convinced that "the general principles I uncovered apply not only to apes on an island but to jockeying for power everywhere."[3] Some had even begun to call this social intelligence "Machiavellian."[4]

But primate social life was not all fighting for advantage and power. De Waal witnessed so much sensitivity to others, conflict resolution, and reciprocal exchange that he became "uncomfortable" with the term "Machiavellian intelligence" to describe it.[5] In *Peacemaking Among Primates*, he explained how primates also reconcile and forgive.

> Forgiveness is not, as some people seem to believe, a mysterious and sublime idea that we owe to a few millennia of Judeo-Christianity. It did not originate in the mind of people and therefore cannot therefore be appropriated by an ideology or religion. The fact that monkeys, apes, and humans all engage in reconciliatory behavior means that it is probably over thirty million years old.[6]

We will see in Chapter 10 that every normal chimpanzee and bonobo appears to have the biological capacity to achieve and express complex mental abilities. But the evidence strongly suggests that if they are deprived of adequate socialization, they may never be able to attain these abilities, certainly not in ways in which we humans understand. Although researchers have seen chimpanzees and bonobos do amazing things in the wild, none have observed two chimpanzees discussing whether termites are tastier than ants or bonobos counting *Treculia* fruits to eat. It is very hard for us accurately to understand what cognitive abilities chimpanzees and bonobos have in the wild or semiwild. We cannot subject them to strictly controlled studies; we can only scrutinize their behavior as closely as possible, then draw our best conclusions from what we observe. Some of these observations would lead a reasonable person to conclude that wild chimpanzees and bonobos possess the capacities for complex cognition. But we will see that when we drag them captive into a laboratory or raise them in a cage from birth, we may mute or even destroy their capacities. When we "enculturate" them in a socially and linguistically rich human environment, we appear either to awaken dormant humanlike cognition or may simply make them more culturally like us, which allows us to understand them better.

I live on an acre of New England land, half of which is wooded. In winter I stroll easily over bare ground that surrounds the dormant trees with their naked branches. With the arrival of spring, the vegetation slowly greens, thickens, and sprouts. By early summer, I can scarcely fight my way through the tangled undergrowth and thickly leafed boughs. "Wolf-children," such as the Wild Boy of Aveyron and Kaspar Hauser, both said to have been raised by wild animals, appear to have achieved much lower cognitive levels than normally socialized children do.[7] Their cognition and that of "wolf-chimpanzees" and "wolf-bonobos" who are raised in the intellectually bare, socially arrested, and linguistically deprived "winter" environments typical of most laboratories may exemplify a starker and sparser "winter mind." Bonobos and chimpanzees raised in the comparative richness of normal chimpanzee and bonobo society or human children who grow up in the more socially deprived human environments may exemplify a "spring mind," whereas humans and other apes exuberantly enculturated in a human society develop a luxuriant "summer mind."

A leader in the understanding of the effects of socialization upon the minds of apes is the primatologist Michael Tomasello, codirector of the Max Planck Institute for Evolutionary Anthropology in Leipzig. Tomasello argues that "the understanding of relational categories in general . . . is the major skill that differentiates the cognition of primates from that of other mammals."[8] He finds "just one major difference [between humans and other primates], and that is the fact that human beings 'identify' with conspecifics [other humans] more deeply than do other primates."[9] Humans develop their capacity to understand that other humans are intentional mental agents through a social or cultural process.[10]

Whether there is just one major cognitive difference between humans and other primates, as Tomasello believes, or whether multiple differences exist, general agreement exists that humans are "fish in the water of culture."[11] Tomasello emphasizes that social processes "do not *create* basic cognitive skills. What they do is turn basic cognitive skills into extremely complex and sophisticated cognitive skills."[12] Before we peek into the minds of chimpanzees and bonobos in Chapter 10, we need to examine the possible impacts of socialization on their development.

Winthrop and Luella Kellogg were among the first to argue that humanlike rearing was critical to humanlike cognitive development. In the early 1930s, they reared a chimpanzee, Gua, alongside their son, Donald, and concluded that if a chimpanzee is

> kept in cage for a part of each day or night, if it is led about by means of a collar and chain, or if it is fed from a plate upon the floor . . . these things must surely develop responses which are different from those of a human. A child itself, if similarly treated, would most certainly acquire some genuinely unchildlike reactions.[13]

Doubters might consider "breakfast with Lucy." For months, Roger Fouts, Lucy's home American Sign Language tutor, arrived at the chimpanzee's home at 8:30 every morning. Lucy, Fouts said,

> would greet me at the door, give me a hug, and show me into the house. While I sat in the kitchen, six-year-old Lucy would go to the stove, grab the teakettle, and fill it with water from the kitchen sink. She did all this chimpanzee-style, by jumping from counter to counter. After getting two cups and two tea bags out of the cupboard, she would brew the tea and serve it like the perfect hostess. Then her ASL lesson would begin.[14]

Although these social processes are not well understood, along with the pioneering Russian psychologist Lev Vygotsky, Tomasello and his colleagues argue that the power of socialization is so strong that if human infants mature outside it, their "winter minds" will not mentally develop in the usual human way. But a wolf-child can spring up even among humans who live like wild animals. Primo Levi, the Italian survivor of that most abnormal of human societies, Auschwitz, recalled Herbinek, a pathetic boy who may have been born in the camp and who lived out three languageless years among thousands of uncaring inmates before he died.[15] Tomasello argues that wolf-children would be able to engage in very little thinking about cause and effect, mathematics, or other minds, and they wouldn't be able to use symbols or language.[16] The development of language, and probably other complex cognitive abilities in human and ape, seem exquisitely sensitive to socialization.[17]

Twenty years ago, researchers reported that chimpanzees raised in socially and environmentally deprived conditions suffered severe and irreversible cognitive deficits.[18] Almost ten years before that, studies showed that chimpanzees deprived of social contact with both humans and other chimpanzees failed the mirror self-recognition test that, we will see in Chapter 10, is a standard test for visual self-consciousness. But if they were later allowed to socialize with other chimpanzees, some of this mental damage could be repaired.[19] With English understatement, psychologist Andrew Whiten and his colleagues have suggested that even the basic cognitive abilities of chimpanzees to imitate and transmit culture may atrophy "in the relatively sterile lifetime experience and opportunities" suffered by captive apes.[20]

Many scientists believe that a critical window in brain development opens both for humans and other apes during their first year of life. When Sue Savage-Rumbaugh exposed three and one-half year old Tamuli, a mother-reared bonobo, to the same rich social and linguistic environment to which Panbanisha, a language-using bonobo, had been exposed from the age of six months, Tamuli's capacity for language comprehension all but failed to materialize.[21] It seems that if humans and other apes are not exposed early enough or well enough to "conversation" about language, and perhaps other complex mental states, these cognitive abilities may well be delayed or permanently stunted.[22] This may explain why congenitally blind or deaf children raised in a "late-winter" or "early-spring" environment in which they have little or no access to "conversation" about the mental states of others often fail an equivalent of the Sally-Anne test (which tests the highly complex ability to have knowledge of another's false beliefs) for many years, well after normally hearing or sighted children pass. Deaf children have also been shown to suffer cognitive delays, and not just in acquiring language.[23]

A SCALE OF SOCIAL INTERACTION

Utter social deprivation anchors the most impoverished end of a "scale of social interaction." At the richest end of the scale lies "enculturation" with a human culture. While some scientists suspect that great apes enculturate their own offspring in the wild, almost nothing is actually known about whether, or how, they do.[24] I will therefore use the term as primatologist Lyn Miles does, to mean "the deliberate process of raising an individual in a human setting, with the intention of transmitting cultural models and symbolic forms of communication."[25] The "summer" minds of children and

apes are "enculturated" by incessant exposure to those who intentionally treat them *as if* they possessed complex mental abilities.[26]
With two exceptions, only humans are known to have enculturated
a nonhuman ape. The exceptions, of course, would themselves
have to have been enculturated. One exception was Washoe, who
enculturated her infant foster son, Loulis (for whom Fouts had to
pay sales tax). The other is the bonobo Panbanisha, whose enculturation of her infant, Nyota, is in progress at Sue Savage-Rumbaugh's
Language Research Center. Savage-Rumbaugh reports that Nyota
is being received in a "bi-cultural *pan-homo* environment" consist-
ing of a small number of human caretakers and his linguistically
competent mother and uncle Kanzi, and today is far ahead of
where Kanzi and Panbanisha were at the same age.[27]

Josep Call and Michael Tomasello argue that both for children
and nonhuman apes,

> being treated intentionally by others, i.e., being enculturated into a cognitive
> community, is an integral part of the ontogeny [the course of a child's devel-
> opment] of certain sociocognitive abilities, especially the ability to under-
> stand behavior intentionally. It may be especially true to have such
> experiences early in development.[28]

After canvassing the scientific literature, they concluded that "do-
mains in which humans seem to have the greatest effect on apes
are intentional communication and social learning."[29]

In Tomasello's view, enculturation allowed the bonobo Kanzi to
demonstrate more complex cognitive abilities than his wild or captive brothers and sisters or even home-raised apes who were not
actively taught about intentional states appear to show. Human beings "on a daily basis encourage Kanzi to share attention to objects
with them, to perform certain behaviors they have just performed,
to take their emotional attitudes toward objects, and so on and so
forth. *Apes in the wild have no one who engages them in this way—no
one who intends things about their intentional states.*"[30] Tomasello and
Call argue that by *treating* apes as intentional beings—for example,
by pointing, explaining, and actively teaching them—enculturators
appear to be able to "socialize their attention," which primes them
to acquire deeper and more sophisticated mental skills in precisely
the manner that my wife and I "socialize the attention" of our
young twins every day. This stimulating involvement demon-

strates not just "the impressive learning skills of the great apes" but "the power of cultural processes."[31]

Enculturated chimpanzees and bonobos thereafter exhibit greater abilities to "socially reference" (to monitor the reaction of another to an object or event, then use that information to respond), to engage in more frequent and more extended episodes of "joint attention" with humans concerning an object, to understand the meaning of human pointing and point themselves both to request (protoimperative pointing) and to comment (protodeclarative pointing), to imitate, cooperate, understand the mental states of others, and learn the symbols necessary for both language and simple mathematics.[32] The following studies seem to bear this out.

When Tomasello and his colleagues compared the abilities of two-year-old children and unenculturated chimpanzees to imitate, a skill that seems to require some understanding of other minds, the children outshone the apes.[33] But when he and other primatologists, including Sue Savage-Rumbaugh, compared the imitation abilities of mother-raised chimpanzees, two-year-old human children, and enculturated chimpanzees, they found that the children and enculturated chimpanzees scored at about the same levels, each far beyond those of the mother-raised chimpanzees.[34] In a broader experiment, Tomasello and Call compared the more complex abilities of two-year-old and three-year-old human infants, unenculturated adult and juvenile chimpanzees, unenculturated orangutans, as well as the enculturated orangutan Chantek (who had been the subject of a decade-long sign-language experiment at the University of Tennessee), to distinguish an experimenter's intentional actions from her accidental ones. Chantek's scores were strikingly higher than the average human two-year-old and the unenculturated apes and higher even than the average human three-year-old.[35] Finally, Sue Savage-Rumbaugh reported that Panbanisha, an enculturated language-using bonobo, passed a version of the Sally-Anne test, whereas Tomasello and Call said that two unenculturated chimpanzees, who did not know language, and Chantek failed its nonverbal equivalent.[36]

Like Tomasello's, Sue Savage-Rumbaugh's research over thirty years has left her with little doubt about socialization's power to stimulate the development of complex cognition. She is convinced that "the ability to think and reason as a cultural and linguistic being in a given society is constructed anew by each individual during the process of coming to behave as a competent member of that society."[37] The very *process* of enculturators treating children and apes *as if* they have a self-concept, *as if* they can engage in joint attention, *as if* they can intentionally communicate, *as if* they have desires or beliefs, *as if* they have a theory of mind, and *as if* they can use language actually *causes* these abilities to blossom into a "summer" mind. When enculturators of children or apes consistently assume that what the child or ape is saying or doing *makes sense* and treat them *as if they were making sense*, eventually they do make sense.[38] "Had we not *treated* apes we have worked with as though they understood our language, it is unlikely that they would have come to do so. It was the treating them 'as though' the possibility existed that made the emergence of language possible."[39] Savage-Rumbaugh recently wrote she doesn't necessarily think that enculturation makes apes "smarter." More likely, chimpanzees and bonobos already have capacities for at least some of the elements of theory of mind.[40] What makes them "smarter," she thinks, is development within a stable, complex, social environment in which the nonsocial environment places demands on the group. Enculturation, whether by human, chimpanzee, or bonobo, makes any child—ape and human—more culturally resemble his or her enculturators. The enculturators can then better understand them, because the human and ape children respond in ways in which they both understand.[41]

Tomasello, Bruner, Vygotsky, and others have emphasized the way socialization sparks both children and other apes to develop such complex cognitive abilities as a full-blown theory of mind, or at least some of its elements, and how language helps explain studies in which it was found that just talking to young children about false beliefs helped them better understand what false beliefs were. It also helps explain why children who grow up in large families appear to perform better on false belief tests and why the children of parents who discuss differing viewpoints with them are more socially competent than children of parents who don't.[42]

An infinite number of possible social environments can be positioned along the "scale of social interaction," from utter social deprivation to enculturation. But the scale should be envisioned not as a straight line but as a forked stick. Along much of the stick's length, socialization affects humans, chimpanzees, and bonobos in much the same way. But the more that any are treated as *sense-less*, or as mere tools, and the less their keepers care about their mental abilities, the closer their socialization will approach utter social deprivation. The mental developments and abilities of Jerom and the other AIDS-experimental chimpanzees at Yerkes were of little or no consequence to their captors. Consequently, their minds were almost certainly terribly stunted, just as the minds of their captors' children would have been stunted had they been treated so brutally. Apes imprisoned in "winter" laboratories or zoos by captors who care little about their mental abilities and much about exhibiting or testing them or vivisecting them are no more likely to develop complex mental abilities than would the captors' children living in the same society. Deprived even of mother-rearing and normal chimpanzee or bonobo society, ape youngsters may not even unfurl the basic mental abilities normally seen in their wild or semiwild brothers and sisters.

At the fork in the scale of social interaction, one direction leads to enculturation and the development of more humanlike "summer" minds. Closest to enculturation lies the "early-summer" social and cultural environments in which chimpanzees and bonobos are intensively exposed to the teaching of language and mathematical symbols by humans who want the apes to learn them. The other direction leads to whatever mental abilities are developed by a "springlike" full and vigorous immersion in wild chimpanzee or bonobo society, such as those at Gombe and Wamba (a village in the Democratic Republic of the Congo near where wild bonobos live). Hard by full-scale immersion in society on the scale of social interaction are the mental abilities that we see in apes living in a semiwild ape society, such as the Arnhem Zoo.

Only a handful of apes have ever been enculturated by humans. The chimpanzees and bonobos who produce almost all the data for almost all the studies of ape cognition that have ever been done are usually raised and maintained in "winter" concrete-and-steel laboratory settings that are stressful, sterile, frightening, isolating, and mentally and emotionally unhealthy. These prisons exist for the benefit of researchers who are not primarily interested in the welfare or cognitive development of the apes, apart from wrenching whatever data they can from them. The writer Eugene Linden documented the sad fates of some of the early signing chimpanzees, who, having been "transformed from commodities into personalities," reverted to being treated as commodities again when the research concluded or the researcher decided to move on. Subsequently, the chimpanzees vanished into the gulag of biomedical research laboratories.[43]

In contrast, nearly all human child–research subjects are heavily enculturated. Any exception would be rare and prominently noted, and no child-development researcher would ever think of comparing data obtained from the stunted minds of unenculturated, socially deprived, intentionality deprived, and linguistically deprived children or from a "wolf-child" to "children" everywhere. Yet it is second nature for many ape researchers to generalize negative results obtained from a small number of these socially deprived and cognitively stunted "winter" chimpanzees and bonobos to "chimpanzees" and "bonobos" everywhere. This flawed "apples-and-oranges" methodology is so much the scientific norm that I will assume that any data obtained came from these populations, unless I note differently. Here are two prominent examples.

NIM

In "Project Nim," Professor Herbert Terrace, a lifelong bachelor, attempted to teach a single chimpanzee, Nim Chimpsky, a signed language in the late 1970s.[44] Terrace understood that "human language develops out of complex social interactions between a child and its parents."[45] And Terrace's proclaimed goal was for Nim to learn sign language the way a child does.[46] One must read Terrace's scientific article, "Can an Ape Create a Sentence?" which appeared in *Science*, together with his popular book, *Nim: A Chimpanzee Who Learned Sign Language*," to understand how Nim's socialization went wrong.

In his *Science* article, Terrace briskly related that Nim had been "raised in a home environment by human surrogate parents and teachers who communicated with him and amongst themselves in ASL." That was it concerning Nim's socialization. Only in *Nim* does Terrace's admirably frank discussion of it appear. Unlike every successful ape-language researcher before or since, Terrace immersed Nim in a teaching world that was as far as Terrace could arrange from the rich social and linguistic world in which human children normally acquire language. Despite his goal of having Nim learn sign language the way a child learns it, Terrace instructed his teachers *not* to treat Nim as a child, held up as models those instructors who refused to treat Nim as a child, and refused even to allow them to comfort Nim when he cried out during the night.[47] Over forty-six months, Nim's crowd of caretakers "cycled" through Nim's life in what Terrace described as "a revolving-door manner."[48] In what Terrace thought "must have struck Nim as an endless stream," upwards of 240 prospective teachers taught Nim for at least one session. More than sixty, ranging in age from nine to fifty-three, actually taught Nim sign language.[49] Although it became apparent that Nim responded very positively to fluent ASL signers, almost all of his teachers first had to be taught ASL before they could, in turn, teach Nim.[50] Most never became fluent and Terrace reported that "[a] good nonfluent teacher could manage one, two, or three sentences on a given topic before going on to sign about the next topic."[51]

Nim's teachers "proved themselves," Terrace wrote, "not by how well they played with Nim or by how well Nim liked them, but by how well they taught him to sign."[52] His first main signing instructor, a teacher of retarded children, used the same techniques of behavior modification that she used to teach sign language to retarded children and the same techniques to punish Nim when he bit someone or screamed too loudly as she had used to punish hyperactive children.[53] Other teachers treated Nim as if he were a child with a severe learning disability.[54] Terrace, who sometimes slapped Nim when he bit or made a bad mess in the kitchen, advocated using a cattle prod on him.[55]

In what Terrace described as "a rather grueling schedule," Nim's teachers drilled him one-on-one, six hours a day, in the teaching environment in which I feel confident that Terrace would not have wanted his children to be educated—had he had any—alone, in a sterile, empty, windowless 8 by 8-foot white concrete cinder-block cubicle.[56] Eventually the sessions grew to seven, and finally, thirteen "backbreaking" hours a day.[57] After Nim was gone, Terrace returned to the classrooms and "wondered how I (!) and the other teachers could have spent so much time in these oppressive rooms."[58] Although Terrace claimed in *Nim* that "[a]ll communication was based on American Sign Language," in his *Science* article, he admitted it was not. Nim was taught "Pidgin Sign English" (PSL), which is not even a language.[59] Not having been exposed to a real language, Nim never learned one. Terrace claimed that PSL was what Washoe had learned, but here he was mistaken.[60]

Unsurprisingly, Terrace produced a sometimes angry and increasingly aggressive student.[61] Nim's signs were usually unspontaneous, laboriously acquired, and almost exclusively requests ("Nim banana hurry"). Terrace wrote revealingly that on occasion, Nim did spontaneously name things or people, "but only when he was with people he liked and trusted. Those teachers constituted too small a fraction of the large group of caretakers who looked after Nim."[62] He concluded that "[w]ithout the intensive and intimate bonds that only a stable group of caretakers could create and maintain, I doubt that there would be enough social motivation for a chimpanzee to share its world with its caretakers by signing about it."[63] Lucy the tea hostess initiated more than three-quarters of her conversations with Roger Fouts.[64] But taught to imitate by his gaggle of trainers, Nim imitated.[65] His "sentences" resembled Thomas Hobbes's description of the life of man in his natural state: "solitary, poore, nasty, brutish, and short." ASL-using humans and other signing chimpanzees use the space in which they sign, facial expressions, and gaze direction to modulate the meaning of their signs. Nim did not. In the opinions of Thomas Van Cantford and James Rimpau, who worked with Washoe and other signing apes, "[e]ither Nim's rearing conditions were so aberrant that these behaviors did not occur or these behaviors did occur but Nim's teachers, novices with ASL, did not report them."[66]

It was never clear that Nim understood what he produced. Communication was usually a one-way street. Humans "listened" to Nim, but not the other way round.[67] Having assumed from the outset that Nim would need "much more explicit instruction" in language than would a child, Terrace would later express surprise that "special training was needed."[68] And Terrace freely admitted that Nim's loss of his original "family" of teachers "at a critical stage of his growth had a permanent adverse effect on his social, linguistic, and emotional development."[69]

The linguist Philip Lieberman, after reading the reports of Project Nim, snorted that Terrace had created a "wolf-ape."[70] Terrace concluded that Nim was unable to learn important elements of language. But what he actually proved was that with enough chaos and with poor enough teaching, a chimpanzee can nearly be prevented from learning ASL. I say "nearly," because Nim's language improved once he was freed of "Project Nim."[71]

THE NEW IBERIA SEVEN

For several years, Daniel J. Povinelli has been reporting that the minds of "chimpanzees" are virtual nonstarters. This claim appears all the more remarkable in light of the observation of one psychologist that the seven captive New Iberia Research Center chimpanzees that Povinelli has used reveal an "impressive quality" of mind.[72] Povinelli bases his conclusions primarily on a long series of experiments in which the New Iberia Seven were just as likely to gesture for food to a researcher who could not see because she was blindfolded, had her eyes closed, had a bucket on her head, or had averted her eyes as they were to gesture to a human who could see. From this he concluded that chimpanzees do not understand that eyes connect internal states of attention to the world.

In Chapter 10, I will argue that Povinelli's methodology was deeply flawed. For now, I will assume that it wasn't and that he accurately described the minds of the New Iberia Seven. But his descriptions of their minds still fit the minds of such enculturated apes as Kanzi and Panbanisha like Cinderella's slipper. Tomasello's remark that Povinelli's findings will "be surprising to most researchers who have worked extensively with apes" and Povinelli's own observation that "[i]f you have ever experienced a chimpanzee following your gaze, you know that it is nearly impossible to resist thinking that the chimpanzee is trying to figure out what you are looking at" tip us that something about his methods or his subjects went terribly wrong.[73]

As Povinelli has not written his *Nim*, we know little about how the New Iberia Seven were socialized. All he tells us is this: They were captive-born, four were taken from their mothers at birth, and two were removed a year later.[74] Their nursery consisted "of several large playpens with toys, blankets, and stuffed animals, sleeping cribs, and several human caretakers."[75] At the age of one year, the youngsters were transferred to a "transition nursery" with "an outdoor play enclosure with access to an older group of chimpanzees through a wire fence."[76] When they were four or five years old, the chimpanzees were "transferred to a specialized living and testing complex" that consisted of "three indoor-outdoor units." That's it.[77]

Povinelli gives us next to no information about the chimpanzees' socialization for a very good reason. He doesn't *care* about it. He doesn't think it's *important*. In opposition to Tomasello, Savage-Rumbaugh, and most other primatologists, Povinelli doesn't think that apes reared in a socially or linguistically rich environment display more complex mental abilities than do apes raised in a socially deprived environment.[78] Part of what makes Povinelli's airy dismissal of socialization so odd is, of course, that most psychologists think that enculturation is precisely the method by which the *children* whom Povinelli tests and whom he compares to the chimpanzees are able to display their advanced mental abilities. It is no wonder that the New Iberia Seven often produced negative outcomes. But what do the negative outcomes of Povinelli's work with the New Iberia Seven actually teach us about the minds of chimpanzees?

"The inadvisability of making positive claims on the basis of negative results is well-known."[79] I might test my daughter, Roma, on whether she knows the product of 12 times 11, and she might answer "144." That may mean that she doesn't know what 12 times 11 is. But it may mean many other things; she might know the answer but simply have made a mistake; she might not have been paying attention; she might not have heard me correctly; or she might be tormenting me because she is sick and tired of my asking what 12 times 11 is. During the months I waited to learn whether I had passed the Massachusetts bar exam, I taught chemistry and biology at the Cambridge Rindge and Latin High School. One of my quieter chemistry students, a Haitian immigrant, received a zero on her first chemistry test. Had she earned a "50," I might have assumed that she wasn't studying hard enough. But a zero? I was intrigued. When I asked her to stay after class to chat, I quickly realized that she might or might not have known any chemistry, but she certainly knew no English, which was the language in which I administered the test and conducted my classes.

Although we don't know much about how the New Iberia Seven were raised, we do know how they were *not* raised. They were not enculturated. They were not treated *as if* they had a self-concept, *as if* they could engage in joint attention, *as if* they had desires or beliefs, *as if* they had a theory of mind. Povinelli acknowledges that "there are certainly other apes which have received far more extensive exposure to the social and material culture of humans." But he could truck the seven to New York City, have them panhandle the length of Fifth Avenue each morning, shepherd them to the New York Public Library every afternoon, sit them down to a hundred performances of La Traviata, and soak up the sun in the bleachers of every Yankees home game for the next five years and it wouldn't matter. Mere "exposure to the social and material culture of humans" is *not the same thing* as "enculturation."[80]

If enculturation is as important to the development of complex humanlike mental abilities, or even the ability to express them in ways that humans understand, as much of the data I have discussed in this chapter strongly suggest, then we may have a diagnosis for the comparative failures of the New Iberia Seven: unfair comparisons. It is unfair to compare the mental abilities of enculturated children, lovingly raised and nurtured in a language-drenched and socially rich natural environment, to chimpanzees raised in an apish Sing-Sing. In the words of Tomasello and his colleagues, such apes and children have "vastly different" social experiences.[81] Enculturated children should be compared to enculturated apes, not "wolf-apes." We will see in Chapter 10 that Savage-Rumbaugh carefully compared the language comprehension skills of a human child, Alia, not with Sherman or Austin, two unenculturated, though symbol-using, chimpanzees, or with one of the Arnhem chimpanzees or with a wild bonobo from Womba, but rather with Kanzi, an enculturated bonobo. Kanzi would have come off like Shakespeare had he been paired, not with Alia, but with the Wild Boy of Aveyron or a little girl led blinking from the dark cellar in which she had been chained alone for her entire life. Kanzi's half sister, Panbanisha, the enculturated bonobo who has passed a Sally-Anne test for theory of mind, would appear to mind read like "The Amazing Kreskin" compared to a severely autistic child. To be both fair and accurate, one should compare enculturated apes to enculturated children, partly enculturated captive apes to partly enculturated captive children, and "wolf-apes" to "wolf-children." The failure of researchers like Povinelli to do this severely compromises the validity of any negative findings.

On the other hand, any positive findings that emerge from chimpanzees and bonobos kept in such destructive conditions can be taken as nearly assured. That I can play man-to-man with Michael Jordan wearing a full hip cast after he breaks his leg and hold him scoreless for a quarter says nothing about whether he will score

100 points against me when the cast comes off. But if Michael beats me by scoring 100 points while he is wearing the hip cast, it says a lot. Is there any doubt what would happen if I went man-to-man with a healthy Michael Jordan or batted head-to-head with Mark McGuire in a home-run derby, or duked it out for a round with Mike Tyson? Povinelli can pooh-pooh enculturated apes and refer to them as "abnormally reared," as if apes are "normally reared" in concrete and steel prisons.[82] One round with Tyson might help him understand.

Of course, Povinelli will never find a significant number of un-enculturated children to compare with unenculturated apes, nor will he find partly enculturated captive children to compare with partly enculturated captive apes. He couldn't use them if he did, and I'm sure he wouldn't want to. That leaves him one theoretical option if he wants to do valid research. But even this option isn't actually open to him. Enculturation demands an enormous emotional commitment on the part of the enculturator to make it work. He must treat the apes *as if* they have a self-concept, *as if* they can engage in joint attention, *as if* they can intentionally communicate, *as if* they have desires or beliefs, *as if* they have a theory of mind, *as if* they can use language, and, finally, *as if* they were his own children. I wouldn't hire a nanny for Siena and Christopher who didn't fervently believe that they can act intentionally, share attention, point protodeclaratively, and learn language and who didn't treat them as if they could. I would be out of my mind to hire a nanny whose career was based on a claim that children can't possibly do any of these things. After she finished with them, they probably couldn't! Povinelli recently told *Science* that "[h]umans have a whole system that we call theory of mind that chimps don't have."[83] It is about as likely that Povinelli will find theory of mind in chimpanzees as the pope will conclude that God is dead!

Enculturating any primate is a time-consuming, lengthy, incredibly expensive, and weighty burden assumed only by the most dedicated and caring enculturators. The enculturators change the lives of the apes. But as the subtitle of Roger Fouts's *Next of Kin: What Chimpanzees Have Taught Me About Who We Are* suggests, the apes change the lives of the enculturators. As we will see in Chapter 10, it is no coincidence that the richest cognitive data emerge from these "summer" collaborations.

10

Chimpanzee and Bonobo Minds

Frans de Waal once complained of our "tendency to compare animal behavior with the most dizzying accomplishments of our race and to be smugly satisfied when a thousand monkeys with a thousand typewriters do not come close to Shakespeare."[1] If fundamental rights turned on the ability to write prose like Shakespeare or poetry like Dante, to do science like Einstein or mathematics like Newton, to sculpt like Michelangelo or paint like Leonardo, to demonstrate the insight of Freud or the political skills of Lincoln, few humans would have them. But because fundamental liberties and equality are universal among humans, they must turn on something else. We will see that this "something else" is not their membership in the species *Homo sapiens* but their autonomy. The fundamental rights of a chimpanzee or bonobo should turn on the same thing.

Minds are critical for legal rights. It would be hard to persuade a reasonable man that a chimpanzee with the mind of Aristotle should be denied every legal right. It would also be hard to convince a reasonable woman that a bonobo with the mind of a toaster should be entitled to any rights at all. This chapter is about the kinds of minds that chimpanzees and bonobos have. Of course, we can't look directly into them. But we can show their complexities, then apply de Waal's Rule of Thumb from Chapter 8 ("Strong arguments would have to be furnished before we would accept that similar behaviors in related species are differently motivated") to shift the burden to the skeptic to show that chimpanzees and bonobos don't measure up.

After twenty years of working with Ai, a chimpanzee, on the study of form and color perception, pattern perception, face recognition, auditory perception, and more, Tetsuro Matsuzawa and his colleagues at the Primate Research Institute at Kyoto University in Japan have concluded that chimpanzees generally perceive the world the way we do.[2] In a 517-page book, primatologists Michael Tomasello and Josep Call recently sought to review "all that is scientifically known about primate cognition."[3] They concluded that "[t]here is no question that primates behave intentionally" in that they choose their goals and the means to achieve them and monitor their progress toward success, and I have no reason to argue with them.[4] Whole books, chapters, and articles have been written on chimpanzee culture.[5] I won't add to them, except briefly to note the following: First, Christophe Boesch and Michael Tomasello, who have clashed throughout the last decade on issues of chimpanzee culture, now agree that "there is no question that much of chimpanzee behavior is culturally transmitted" and that "many deep similarities" exist between chimpanzee and human cultures.[6] Second, in a June 1999 article entitled "Cultures in Chimpanzees" published in *Nature*, nine prominent primatologists and anthropologists, including Jane Goodall, William McGrew, Toshisada Nishida, Richard Wrangham, and Christophe Boesch pooled 151 years of wild-chimpanzee observation at the seven longest-established African field sites that highlighted more than three dozen cultural variations in such chimpanzee behaviors as nutcracking, termite fishing, rain dancing, and object throwing.[7] Frans de Waal concluded that "the evidence is overwhelming that chimpanzees have a remarkable ability to invent new customs and technologies, and that they pass these on socially."[8]

I will focus on seven areas of cognition. One is primitive, but very important: *the capacity to feel pain.* The other five, often heavily overlapping, categories will ascend to some of the most complex cognitive abilities that chimpanzees, bonobos, and human beings possess in whole or in part: *mental representation, self-conception, logical and mathematical abilities, tool use, the knowledge that minds exist, and nonsymbolic and symbolic communication, including language.* Embedded within each may be numerous, less-complicated cognitive skills. Here's an example: Lincoln delivered a three-minute speech at Gettysburg. Imagine the cognitive skills embedded in that!

There is no agreement as to when human children develop such complex skills as self-conception, theory of mind, and language or even on how to define them precisely, let alone how they develop. But no one requires these answers to entitle children to fundamental liberties or equality. Chimpanzees and bonobos don't need a full-blown self-conception or theory of mind or language to be entitled to them, either. Even those who claim that they lack complex mental capacities in full are hard-pressed to argue that they don't possess a sizable chunk of each.

If You Prick Us, Do We Not Bleed?

Like Daniel Dennett, P. T. Barnum sized us up as suckers. But even if one *is* born every minute, according to my calculations, most of us are still not that gullible. And they can't *all* be sitting in our legislatures. Every jurisdiction of which I'm aware has enacted an anticruelty statute, sometimes dozens of them. Hundreds of these statutes around the world prohibit nonhuman animals from being tortured, tormented, abused, cruelly treated, overworked, starved, deprived of water and shelter, and having unnecessary pain inflicted upon them.[9] Great Britain has forbidden using great apes in biomedical research; New Zealand, all nonhuman hominids. On the other hand, no statutes prohibit cruelty to robots or washing machines, and for good reason. They're not conscious. They don't suffer. They can't feel pain. They can't feel a thing.

For the ten years I have been teaching the course "Animal Rights Law," I have assigned one of Shylock's speeches from *The Merchant of Venice.*

Hath not a Jew eyes? Hath not a Jew hands, organs, dimensions, senses, affections, passions; fed with the same food, hurt with the same weapons, subject to the same diseases, heal'd by the same means, warm'd and cool'd by the same winter and summer, as a

Christian is? If you prick us, do we not bleed? If you tickle us, do we not laugh? If you poison us, do we not die?

No, I'm not comparing Jews to chimpanzees and bonobos. I'm comparing *all* of us to them. Vast numbers of nonhuman animals *can* feel pain; they *can* suffer. It is almost certain that chimpanzees and bonobos do. Although pain and suffering are elusive concepts in any species, including our own, a model used in standard medical textbooks holds that pain and suffering have three components.[10]

First, there is nociception. Primate bodies detect and signal noxious events through specialized nerves that begin to develop before the midpoint of gestation.[11] Nociception appears to exist in all vertebrates and may exist in some invertebrates as well.[12] Second, there is pain. This is our conscious perception of the nociceptive stimulus. Nociceptors transmit signals to our central nervous system where "pain" is recognized and acted upon by other neurons located in our brains.[13] All vertebrates have these neural connections.[14] A primary somatosensory cortex, a gyrencephalic or furrowed cortex, and a prefrontal cortex, which all normal primates have, are strongly associated with this ability to feel pain.[15] Third, there is suffering, which is not the same thing as pain. Pain is just one of many things that can cause suffering, although greater pain is believed to cause greater suffering. Suffering is the emotional response to perceived pain. It's exceedingly difficult to analyze or measure suffering, even in humans. But emotions affect it, especially fear—a primitive state of mind, indeed—and anxiety, a fear of what the future may bring.[16] Dr. Eric Cassell, an expert on suffering, has reported that humans "frequently report suffering from pain when they feel out of control, when the pain is overwhelming, when the course of the pain is unknown, when the meaning of the pain is dire, or when the pain is apparently without end."[17] This suggests that the better able any mind is to experience them, the greater will be the suffering.

Arguments that chimpanzees and bonobos cannot feel pain or suffer recall that until recently, many doctors believed that human newborns couldn't either and mercilessly operated on them without anesthesia.[18] One 1977 study revealed that more than half the children between the ages of four and eight months who had undergone major surgery were given no medication for postoperative pain.[19] Drawing upon newborn anatomy, neurology, and behavior, as well as research on nonhuman animals, the authors of a 1987 paper in the *New England Journal of Medicine* argued "that marked nociceptive activity clearly constitutes a physiologic and perhaps even a psychological form of stress in premature or full-term" infants that could be alleviated by anesthetics. They concluded that "[c]urrent knowledge suggests that humane considerations should apply as forcefully to the care of neonates and young, nonverbal infants as they do to children and adults in similar painful and stressful situations."[20]

Mental Representations

Few who work with chimpanzees and bonobos doubt their ability to form mental representations of at least the first-order. That means they can mentally represent facts or things, if not other minds. The evidence for these first-order mental representations is strong. We'll begin with cognitive mapping. But keep in mind that many of the seven cognitive abilities that we're discussing heavily overlap, as many of them depend upon mental representations.

Remember that cognitive maps allow children to know where places are or where objects have been hidden. This ability appears late in their second year. Chimpanzees and bonobos seem to carry maps around in their minds as well. Jane Goodall reported that chimpanzees at Gombe could "easily locate food patches within the 8 to 24 square kilometers of the Gombe home range. Their spatial memory is rich in detail; they know not only the position of major foods—great stands of richly fruiting trees, for example— but also the whereabouts of solitary trees and individual termite mounds."[21]

In one series of experiments, a captive chimpanzee observed a scientist walk about hiding foods within a one-acre enclosure. The other five chimpanzees in the colony were not permitted to watch. The observing chimpanzee (each had a turn) could find most of the foods and typically recovered her favorite foods first, no matter in what order the food had been hidden, and she did so in an efficient pattern, while the nonobserving chimpanzees usually located just one food after a random search.[22]

Fifty-five acres of forest surround the Language Research Center. Savage-Rumbaugh and her colleagues set up seventeen locations throughout this forest, gave them such names as Tree-house, A-Frame, Midway, Flatrock, Lookout Point, and Criss-Cross Corners, and stocked them with different kinds of foods in coolers. They started bringing Kanzi into this woods when he was two and a half. Within five months, he knew every location. He chose where to travel by picking out the location or the type of foods stocked there, in a photograph. If he pointed to blackberries, cheese, or orange juice, off the party cruised to Criss-Cross Corners; bananas or juice sent them trudging to Tree-house. Once he announced which location he intended to visit, Kanzi would pass through any of the other sixteen locations until he reached the chosen one. He would hop onto the shoulders of visitors and guide them to any location by gesturing in the proper direction when they came to a trail fork or by jumping down and leading them. While humans followed the marked trails or carried maps, Kanzi could bushwhack straight to any location.[23]

Piaget's Stage 6 object permanence task, which infants begin to pass late in their second year, is thought to require mental representation. Recall that an experimenter places an object under one of several possible covers and the infant systematically searches until she finds it. Although several experimenters have claimed that chimpanzees have passed formal Stage 6 tasks, others have criticized these experiments as not being controlled strictly enough.[24] But at least one capuchin monkey has passed this formal Stage 6 test. So has a gorilla, a dog, cats, and an African Gray parrot.[25] Even more compelling is the fact that many of the abilities that chimpanzees and bonobos demonstrate are more complex than those required for Stage 6 abilities. It is therefore highly likely that chimpanzees and bonobos can not only pass this test but that they achieve these abilities at about the same time as human infants.[26] Data exist that show that chimpanzees achieve Stage 6 abilities for being able to use a tool to solve a problem without trial and error and to use spatial knowledge.[27]

The psychologist David Premack asked the language-trained chimpanzee Sarah to reconstruct a chimpanzee face from a cut-up puzzle. When Sarah had been wearing a hat or glasses or necklaces, she would add clay to the top of the chimpanzee's face or around the eyes or to the neck or bottom of the face.[28] Chimpanzees easily navigate video mazes using joysticks. They can locate objects outside their line of vision by following a closed-circuit television that shows their hand, which is hidden behind a screen, and the object for which they are searching.[29] In one experiment that suggests that baboons can rotate objects in their minds, they were shown the letter "F" on a television monitor. Then they were shown both the letter "F" and its mirror image, each of which was rotated the same 60°, 120°, 180°, 240°, or 300°, and were asked to choose the image that matched the letter "F." Most of them succeeded.[30] Chimpanzees given a set of five nesting cups which, rather like Russian dolls, progressively fit into one another, quickly learned to put the set together.[31] Scientists believe that this "may involve mental representation in the form of planning by imagining potential outcomes of specific actions with the cups."[32]

One experiment seemed to require a research subject to hold not one but two mental representations simultaneously to compare them, so called "dual representations." In that test, chimpanzees were shown a Lilliputian can of soda hidden in a scale model of a real room with which they were familiar. Unlike most three-year-old children, the chimpanzees were able to go straight to the real can of soda hidden in the real room.[33]

Our confidence that we are observing a pretend mental representation should increase the further from reality it seems to be.[34] Wild chimpanzees have been spotted "fishing" for ants with a stick where there were no ants. The anthropologist Richard Wrangham reported that he watched an eight-year-old wild chimpanzee, Kakama, play with a small log as a child would play with a doll for four hours. Four months later, two field assistants watched him at it again for three hours.[35] But most elaborate symbolic play has been observed in enculturated chimpanzees and bonobos. They keep, bathe, feed, and tickle dolls, use imaginary food in a wide variety of ways, and engage in such imaginary play as pulling a pretend pull toy, and picking up a toothbrush while signing "hairbrush," then using it to brush hair.[36]

Kanzi plays with imaginary objects in a remarkable number of ways. He pretends to eat them, feeds them to others, hides them, takes them from others, returns them, plays keep-away with them, and feeds them to his toys. He may place an imaginary morsel of food on the floor and pretend to ignore it until someone else pretends to reach for it. Then he will snatch it away. He may pretend to hide a pretend object, then at a later time take it out and pretend to eat it or spit it out and comment "Bad" on his computer. He may pretend to grab and munch the peach that decorates every Georgia license plate. He sometimes puts on a monster mask and pretends to chase and bite people or other bonobos. He pretends that a toy is biting or chasing him. He may ask others to pretend to be monsters and chase him.[37]

LOGICAL AND MATHEMATICAL ABILITIES

Many American high-school seniors dread the analogies section of the SAT. When one analogizes, one understands and compares relations. Once chimpanzees are taught a relationship such as physical identity (matching two designs or objects to a third), oddity (not matching two designs or objects to a third), or same-different (two pennies are like two nickels but not like a nickel and a button) with specific objects or designs, they can compare, classify, and sort objects.[38] Sameness and difference support the ability of chimpanzees and bonobos to classify or sort objects. Kanzi can classify by size, color, shape, and material.[39] Sarah could sort small bells and cars as well as children older than three and one-half to five and one-half years old.[40] The home-raised chimpanzee, Viki, once sorted a mound of screws, bolts, washers, paper clips, nuts, and nails to near perfection.[41] But even chimpanzees who have not been enculturated, who do not know language, and who have not been home raised have sorted based on physical identity or previous familiarity.[42] But no senior high-school student squirms at the prospect of comparing nickels and buttons. They dread the more difficult task of comparing "second-order relations." In second-order relations, one doesn't compare objects but relationships *between* objects. To do that, one must have some understanding of analogies.

Chimpanzees can understand second-order relations involving "same" and "different." That Sarah could match a pair of apples with a pair of oranges and not with a banana and apple shows that she understood the relationship of sameness.[43] But this was not all. She could call the sentences "apple is red" and "red color is apple" the same and the sentences "apple is red" and "apple is round" different.[44] At the time, Premack thought that it might be the language training itself that allowed Sarah to express this ability. But based on results obtained with five chimpanzees at Sally Boysen's Primate Cognition Project, Sarah, Sheba, Darrell, Kermit, and

Bobby, Boysen and her colleagues concluded that the learning of symbols by chimpanzees may result in a broad attentional shift in the way their minds generally solve problems. They become able to encode an abstract proposition, like a second-order relationship, onto a symbol. This symbol then acts much like a word, so when they think of the symbol, they can retrieve the information that it represents.[45] When chimpanzees choose keys over other objects to match with locks or match bottles with caps, they show an understanding of another second-order relationship called "complementarity." Unenculturated, non-symbol-trained chimpanzees who were experienced in assembling and disassembling objects have also demonstrated abilities to sort on that basis.[46]

Professor Jonas Langer discovered that human children engage in what he called "first-order" logical-mathematical operations from about the ages of six to eighteen months. For example, infants might substitute individual objects within one set of objects. They might put two sticks into a cup, remove one of the sticks, then replace it with a spoon. Over the next year and a half, they progress to "second-order" manipulations. They might substitute individual objects between two sets of objects by placing a stick in a cup that is near to a second cup that already has a stick in it, then swap the sticks. At the age of about thirty-six months, they advance to "third-order" manipulations that involve three sets of objects. This opens the gateway to an ability to engage in complex hierarchical classifications. Only a small number of chimpanzees and an even smaller number of bonobos of various ages, degrees of exposure to human culture, and instruction in the use of such symbols have been tested on their abilities to engage in logical-mathematical operations. Young unenculturated chimpanzees untrained in the use of symbols have attained a rudimentary "second-order" ability by about fifty-four months of age, while some older, more enculturated, and language-using chimpanzees and bonobos have displayed the beginnings of "third-order" operations.[47]

COUNTING

Counting, as Westerners understand it, is not universal in human societies. The Bemba of Zambia and the Weddah of Sri Lanka don't count, whereas the Ponam of Papua New Guinea designate just first, middle, and last.[48] The primitive abilities of human infants to add and subtract whole numbers of 4 and less appears innate and involves specific mental representations. But it does not necessarily require them to understand symbols.[49] More complex mathematics, however, requires the learning of symbols and appears somehow related to the capacity to learn another kind of symbol: language.[50] It is also a more complicated mental process than most of us think. That's why mathematics can be so hard to learn.

My son, Christopher, will have to acquire five separate abilities before he will be able count accurately, sometime between the ages of four and six. The first is the "one-to-one principle." He will learn to place one mental marker, but only one, on each counted object. That way he will know which items he has counted and which he hasn't. Second is the "stable-order principle." He will have to use his markers in the same order (1 comes before 2, 2 comes before 3, and so on). Third is the "cardinal principle." He must realize that his last marker refers to the total number of items counted. Fourth is the "abstraction principle." He will have to understand that the first three principles apply to anything—bananas, pieces of paper, chimes of a bell, anything. Fifth is the "order-irrelevance principle." He will have to know that the total number of bell chimes and bananas don't depend upon the order in which he counted them.[51] Studies have shown that chimpanzees can both learn abstract symbols and then use them to count.[52]

Chimpanzees who learn abstract symbols can engage in a kind of mathematics that is advance upon the primitive ability of human infants to add and subtract small integers. At Tetsuro Matsuzawa's Primate Research Institute, Ai first learned the symbols for the Arabic numerals 1 to 9, then how to order up to four of them from lower to higher (2 comes before 7).[53] Ai, and other chimpanzees, have learned to count to a number less than 10.[54] As children do, some chimpanzees have learned to help themselves to learn. When Sheba was learning to count candies at Sally Boysen's Primate Cognition Project, she exhibited so-called "indicating acts" while she counted. She might point to a candy without touching it or touch a candy without moving it or move a candy a short distance. The more candies Sheba had to count, the more indicating acts she used. When Christopher learns to count, he will do the same things. Scientists think that indicating acts help new counters organize their counting by marking which items they have counted and separating counted from uncounted items.[55]

That chimpanzees can both understand Arabic numerals (1, 2, 3, 4) as symbols and add them had been demonstrated in a series of investigations ongoing for more than a decade at Boysen's Primate Cognition Project.[56] Years into the project, after the chimpanzees had grasped number and simple addition, Boysen introduced them to a new problem. Sheba and Sarah were initially teamed. Sheba was shown two food dishes. One held six candies and one held a single candy (having innocently dipped my hand into the jar of candies Boysen reserved for the chimpanzees and suffering a withering chimpanzee glare, I can attest to their fondness for candies). The rules were simple. Whichever tray of candies Sheba chose, *Sarah* received. Sheba got the dish that she *didn't* choose. No matter how many times the test was run, Sheba always chose the tray with six candies. To her increasing agitation, she received just one candy and watched Sarah's delight as she was showered with candies. Boysen was baffled. Sheba was adept at simple counting and addition. How could she keep making such a mistake? But when they switched roles, the same thing happened to Sarah. Then it happened to Kermit, Bobby, and every chimpanzee in the colony. As sometimes happens with children, there was something about seeing real candies that interfered with their ability to choose the dish with the larger number of candies, even when Boysen knew that they knew that their choices were working to their disadvantage.

Boysen got an idea. She knew that each chimpanzee understood Arabic numerals between 1 and 6. Instead of presenting them with dishes that held real candies, she would show them dishes with only numbers in them and see which dish they chose. Otherwise the rules stayed the same. Sheba was first. If she chose the dish that contained the number 6 over the dish with the number 1, Sarah would get six candies and Sheba would receive one. "With absolutely no hesitation," Boysen reported, "and given the first opportunity to make such a choice, Sheba chose the smaller numeral 1."[57] All the chimpanzees now performed significantly better when Arabic numerals replaced actual candies. Their apparent ability mentally to represent a number as a *symbol* nearly eliminated the interference effect that the sight of the candies produced.[58] When one of the candy trays was replaced by a symbol, the interference effect was eliminated in just about the same way.[59]

But chimpanzees can do more. Sherman and Austin were shown two trays that contained two piles of chocolates each and were allowed to choose a tray. The first tray had one pile of four chocolates and one pile of three. The second had one pile of five chocolates and a single chocolate. They usually chose the tray that contained the larger *total* number of chocolates. This strongly suggests that they were mentally *adding* the piles in each tray and then *comparing* the numbers.[60] Chimpanzees have been able to add and compare fractions as well. After being taught the concept of fractions, Austin and Sherman were shown such items as a quarter of an apple and half a glass of milk, then asked to choose between one full disc and three-fourths of a disc. They chose correctly well above the level of chance.[61] The mathematician and cognitive neuropsychologist Stanislas Dehaene concluded that they "were obviously performing an internal computation not unlike the addition of two fractions: $1/4 + 1/2 = 3/4$."[62] The adding of fractions demands sophisticated reasoning abilities.[63]

Harder than counting is the ability to "conserve" number, weight, or volume. Remember that children usually don't attain this ability until they reach the ages of five to seven. Some of the most extensive ape "conservation" tests were performed with the young Sarah when she lived at David Premack's laboratory. In the late 1970s, she learned to use two plastic symbols that meant "same" and "different." Without being taught, she had no trouble conserving the volume of liquids no matter what shape glasses they were poured into. Nor did she have any problem judging that a lump of clay remained the same no matter how thinly it was stretched or how hard it was compressed.[64] In the early 1980s, her ability to conserve liquid quantities was replicated by other chimpanzees.[65]

TOOLS

In October 1960, Jane Goodall startled the scientific world when she reported seeing a Gombe chimpanzee, David Greybeard, fish for termites using a branch from which he had stripped the leaves. David Greybeard had made a tool. Tool using and tool-making had long been thought hallmarks of the human being alone. The anthropologist Louis Leakey famously wrote Goodall, "Now we must redefine tool, redefine man, or accept chimpanzees as human."[66]

The once-electrifying news from Gombe has become ho-hum. Now we know that nearly every wild and captive chimpanzee regularly uses tools and for a wide variety of purposes.[67] Tetsuro Matsuzawa has proposed a four-level system to describe the complexities of those wild chimpanzee tool-using behaviors seen so far. Picking a nut from the ground is Level 0; no tool is used. There is just one object involved, the nut. Using a branch to investigate by probing or to fish for ants or termites, as David Greybeard did, or using a rock to throw or club as a weapon is Level 1. So are using tools to clean themselves by wiping and dabbing at bleeding, as "stepping sticks" and "seat sticks" to protect their feet against thorns, and as "leaf cushions" to keep their bottoms dry when they sit on wet ground.[68] These are the simplest uses of tools and show that the chimpanzee recognizes that a causal relationship exists between the use of two objects.

Chimpanzees in the Tai National Park of the Ivory Coast and in Boussou, Guinea, often use a stone or a piece of wood as a hammer to crack nuts for food against another stone used as an anvil. This Level 2 use of tools demonstrates that they understand how three objects can relate. They may carry these nutcracking tools into the branches of trees to use high above the ground.[69] They may search for, select, and transport a variety of stone hammers of different weights. They appear to choose them according to the hardness of the nuts they wish to crack, even when they choose the stone before the nuts.[70] When they crack a particularly hard nut, they use stone, not wooden, hammers. They may adapt their tool use to new circumstances and use trees instead of stones to serve as anvils when no stones are available. Sometimes they create ladders by leaning logs against trees and open unfamiliar boxes with sticks.[71] Finally, when chimpanzees use a three-tool set by adding a wedge stone to stabilize an anvil stone and keeping it flat enough to allow them to crack nuts on it with a hammer, they have reached Level 3, the most complex wild chimpanzee tool use so far observed and the stage that a human child reaches between the ages of three and five.[72] Kate, a Gambian chimpanzee, has been photographed using a tool *set*. This means that she used several tools one after another to achieve a goal, employing a stout chisel, a fine chisel, a sharp-pointed bodkin, then a flexible dipstick to steal honey from a bees' nest.[73]

Tool use alone does not necessarily mean that the tool user "sees" mental representations. We can more readily infer this when chimpanzees show they understand that physical causes make things happen in the world by choosing, modifying, or making a tool to solve a new problem ("After thinking it over, this must be the way to go") without engaging in extensive trial-and-error strategies ("Let me try this. Darn. What if I try that? Shoot. Maybe this will work).[74] This reaches is at least Piaget's Stage 6 of sensory-motor development.[75] Gombe chimpanzees have been known to select a termite-fishing tool when a termite mound is not even in sight, and they may carry tools hundreds of meters to probe termite mounds.[76]

In the laboratory, chimpanzees use rakes to reach food. They have sought the missing parts of toys and turned their heads to see where a paper airplane came from. When handed a toy being operated with a cord that ran beneath a table, they have looked under the table.[77] A chimpanzee named Julia was presented with five transparent boxes, each of which held a key that could be used to open another box. Without trial and error, she determined which keys opened which boxes, then opened them.[78] Sheba and Darrell solved a problem involving a devilish little device called a "trap tube" that is so difficult that children usually can't solve it until at least their fourth, and sometimes their fifth, even their sixth, years. Sally Boysen and her colleagues concluded that the chimpanzees "represented beforehand the consequences of their actions and mentally organized a strategy for solution."[79]

The archaeologists Nicholas Toth and Kathy Schick are primarily interested in how such human ancestors as Australopithecines, *Homo habilis*, and *Homo erectus* made and used tools. They teamed up with Savage-Rumbaugh and Kanzi to see if the bonobo could learn to make stone tools. First, they showed Kanzi that sharp stone tools could be used to cut cords and allowed him to cut a cord to open a box that contained a treat. Kanzi was enthused. The first day he used stone flakes that Toth and Schick had cut to open the box. The second day he learned to choose the sharpest knife from an array of five. By the end of the first month, he was beginning to "knapp" stone tools by hitting rocks against each other.

Several months into the study, he startled Schick and Toth by inventing a method for producing sharp stone tools. Although he had shown little interest in throwing objects before, he began to throw stones hard against a concrete floor. They fractured into stones with cutting edges. Because Toth was more interested in investigating Kanzi's knapping than his throwing abilities, he asked Savage-Rumbaugh to discourage Kanzi's stone throwing. She covered the hard floor with a rug. Savage-Rumbaugh relates what happened when Kanzi first entered the newly carpeted room.

> [H]e threw the rock a few times and looked puzzled when it didn't shatter as usual. He paused for a few seconds, looked around to find a place where two pieces of carpet met, pulled back a piece to reveal the concrete, and hurled the rock. We have assembled a videotape of the toolmaking project, which I show to scientific and more general audiences. Whenever the tape reaches this incident there is always a tremendous roar of approval as Kanzi—the hero—outwits the humans again.[80]

In the relatively short period of time that Savage-Rumbaugh could allow her star language student (as we will see) to devote to stone toolmaking, he failed to match the technical skill of our human ancestors. But within nine months, Kanzi was producing stone objects similar enough at least to compare with the archaeological record. Generally they were not as good, though the psychologist Richard Byrne reports that Toth told him that some of Kanzi's stone tools were "Oldowan" (the kind our human ancestors made about 2.5 million years ago).[81] Kanzi, the team reported in 1993,

> made significant and rather startling progress in his stone-making ability, rapidly acquiring many of the basic skills required to produce sharp-edged cutting tools from stone cores. He has not only shown fairly rapid improvement in the hard-hammer technique demonstrated to him, but he has displayed ingenuity and insight learning in his innovation of the throwing and directed-throwing techniques.[82]

By 1997, Kanzi had amassed about 120 hours of stone tool making and tool using. It appeared that his preference for producing stone tools by throwing might be attributed to a bonobo's inability to grip a stone hammer and core with sufficient strength and coordination to produce stone flakes by using his hands. He seems to understand that the sharper the flakes he produces, the better they cut, and he even tests their sharpness with his tongue. He has learned to produce flakes by better aiming his throws and prefers the larger and heavier flakes as cutters. Because the stone cores themselves do not betray the fact that they have been thrown, Schick and Toth now wonder whether some of the stone flakes produced in Africa more than 2 million years ago by our ancestors might also have been produced by throwing.[83]

Finally, Lucy, the tea hostess, was born into a family of carnival chimpanzees, given away at the age of two days, then raised in the family of psychologist Maurice Temerlin. At the age of three, she placed a pocket mirror on the floor, squatted over it, and watched as she used a pair of pliers to pull apart her labia. She moved on to daily masturbation techniques that would have had Masters and Johnson beaming. She rubbed her clitoris with a pencil. This she abandoned as too "low-tech." One afternoon, she went from the Temerlin living room to the kitchen, fetched a glass from a cabinet, moved to a second cabinet where she found a bottle of gin and poured herself two or three fingers. Sashaying back to the living room, she picked up a *National Geographic*, plopped on the couch, and began to browse. Three to five minutes later, she placed drink and magazine on the floor, then headed to a utility closet in a hall twenty yards away. She pulled out a vacuum cleaner, hauled it to the living room, plugged it in, removed the brush attachment, flicked on the machine, and applied the sucking pipe to her genitals. After a while, according to Temerlin, she laughed, "looked happy," stopped, turned off the vacuum, and returned to her drink and magazine on the couch. Those readers over twenty-one years of age are invited to view the photographic evidence in *Lucy: Growing Up Human*. The Temerlins were also kind enough to keep *Playgirl* magazines around the house for Lucy to open, squat over, and rub against photographs of male genitals.[84]

THE KNOWLEDGE THAT MINDS EXIST

Remember from Chapter 8 that animals are commonly thought to have a theory of mind when they can metarepresent. That is, they can predict and explain the behavior of other animals by mentally representing their mental states.

Scientists often report behaviors of chimpanzees and bonobos that appear to possess many, if not all, the elements of a theory of mind. At least, it's not easy to describe what is going on in another reasonable way. We will see that these apes have an implicit, and perhaps explicit, theory of mind, are self-aware, engage in joint attention, imitate, point, teach, intentionally deceive, empathize, symbolically communicate without using language, and use language. The list is not exhaustive, for I don't discuss such an ability as ostension (a way of expressing and assessing the intent to communicate).[85] These behaviors appear to demand they understand that other minds may see things differently than they do.[86]

Professors Vittorio Gallese and Alvin Goldman may have provided the first neurological support not just for theory of mind in macaque monkeys but for the simulation brand (which claims that a "mind reader" reads minds by "slipping on the mental shoes" of another, imagining how she would feel in the same situation, then attributing her feelings to him). So-called "mirror neurons" have been located both in the macaque and human brains. These neurons fire both when an individual performs an act such as grasping, holding, or manipulating an object *and when she sees another individual performing those same acts.*[87] Gallese and Goldman write that mirror neuron activity "seems to be nature's way of getting the observer into the same 'mental shoes' as the target."[88] They speculate that mirror neuron activity in the macaque may either be a precursor to theory of mind or a primitive version of theory of mind itself.[89]

As we have seen, a sophisticated theory of mind emerges in a child only after years in which increasingly complicated mental skills have pieced together and layered one atop another. It begins when a child can impute any mental state to anyone, no matter how incompletely, and is suggested by such early behaviors as imitation, joint attention, and protodeclarative pointing (pointing to show). Next, she begins to understand desires and intentions, then "true belief." Around her fifth year, she starts to understand "false belief." As she moves toward ten years of age, she begins to have beliefs about beliefs and, near the end of that first decade, she realizes that deceitful lies are different from jokes, exaggeration, and sarcasm.

Mindblind autistic children often fail to advance beyond the first stage. Let's refresh our understanding of autism with cognitive psychologist Steven Pinker's description of autistic children.

> When taken into a room they disregard people and go for the objects.
> When someone offers a hand, they play with it like a mechanical toy

... They pay little attention to their parents and don't respond when called. In public, they touch, smell, and walk over people as if they were furniture. They don't play with other children.[90]

One supporter of Machiavellian intelligence realized that comparing the razor-sharp social intelligence of many primates with the mindblind autistic human child might help us understand how theory of mind develops.

[I]t suggests that with a deficit in mindreading, one primate at least is severely disabled in social competence. Given that so many non-human primates are highly socially competent, engaging, for example, in the manipulation of coalitions and negotiation of sexual partnerships ... one must ask how this might be managed without mindreading, when the case of autism appears to demonstrate that mindblind social behavior is so limited? (or to put it the other way round, if primates lack a theory of mind, yet are so socially expert, why need mindblind people be so severely constrained in their social abilities?).[91]

Implicit Theory of Mind

In Chapter 8, I said that some scientists believe that young human infants who can only mentally represent at a first-order (represent his mother's behavior but not her mental state) might still engage in "intersubjectivity." When face-to-face, this allows him to "feel" some of mom's mental states by experiencing them as inseparably linked to her behaviors in the same way that he might perceive the color red as not being an internal state of a red fire hydrant but as a property of it. Professor Juan Carlos Gómez thinks that such an infant possesses an "implicit" theory of mind. He uses it to acquire some of the same information as an adult who can mentally represent at the second-order level, and it allows him to understand another as a subjective entity with intentions. Gómez argues that the mental state of visual attention invariably requires a human mother to turn her mind to whatever she is paying attention to. But when she does that, she usually turns some parts of her body toward it as well. In other words, she *looks* at it. An infant who lacks the ability to metarepresent won't know that another mind is spinning inside his mother. He may, however, perceive her mental state of visual attention not as something coming from a mind that he doesn't know is there but as part of her outer behavior of looking. The same might go for the ape. Although a chimpanzee infant might lack the ability to metarepresent his mother's mind, he might still be able to interact with her and understand that she can interact with him in the same way that the human infant can "feel" some of his mom's mental states by experiencing them as inseparably linked to her behaviors.[92]

When two visual attentions contact, we call it a "gaze." Gómez says that infants and animals unable to metarepresent can, in that case, understand that another is behaving like a subjective entity with intentions as well and, indeed, that is what human adults usually do. A mutual awareness exists. Locked in a gaze, we rarely mentally represent the other's mental representation of us.

Gómez emphasizes that an implicit theory of mind is both a necessary building block of theory of mind and an exceedingly valuable ability in its own right that is not supplanted when an explicit theory of mind arises but is routinely used alone, even by human adults. It allows us to intentionally communicate with one another. For example, Gómez explains that a man sitting in a restaurant who would like salt brought to his table may make a salt-shaking gesture to the waiter while looking him in the eye and saying, "May I have some salt?" This may appear to be a simple communication. But Gómez says that the diner is actually engaged in this highly complex fifth-order level of request: "I *want* you to *understand* that I *want* you to *understand* that I *want* you to bring me some salt." However, we often short-circuit this process by using our implicit theory of mind and relying on the nonverbal behavioral cues of gestures and eye contact.[93] This can be so powerful that our gestures can override our words. If I say to the waiter, "May I have some salt?", but at the same time vigorously shake my head "no" from side to side, it is at least as likely that the waiter will either not bring me any salt, will come over to clarify what I mean, or will decide to attend to his saner customers, as that he will bring me some salt.[94] That there is truth in Gómez's argument was brought home to me as I sat in a taverna in Athens, unable to speak Greek, or a ristorante in Rome, unable to speak Italian, gesturing to the waiters to bring me the check. And they did.

Self-Awareness

When a bonobo closes his eyes or covers them with one hand to handicap his abilities to do gymnastics or chase through a tree, it suggests that he is *self-aware*.[95] Psychology professor Gordon Gallup Jr., who invented the benchmark test for visual self-recognition in 1970 that I mentioned in Chapter 8, proposed in 1983 that one is self-aware to the extent one can become the object of one's own attention.[96] Gómez argues that when human infants or apes incapable of metarepresentation but having an implicit theory of mind become locked in a gaze with another or share another's attention, they must become aware of themselves at some level, even if it may not be at the level of understanding that they themselves have a mind.[97] As with the chicken and the egg, scientists don't agree whether awareness that I have a mind preceded or followed my awareness that others have minds.[98] How can we know when others conceive their own mental states? Once again the answer is: We can't, for sure; we can only make our best guess based on the evidence.

Human and nonhuman animals who have learned at least some elements of language can reveal aspects of self-awareness directly. All symbol-using chimpanzees have been taught their names and the symbol for "me" and use both.[99] Allen and Beatrix Gardner, pioneers in the teaching of ASL (also called Ameslan), asked this question of Washoe as she looked into a mirror.

"Who that?"
"Me Washoe," came the reply.[100]

Don't be distracted by the seemingly pidgin quality of ASL on the printed page. ASL is not Pidgin Sign English (PSL), but a true language.[101] More than twenty years ago, the Gardners explained that "[b]ecause the structure of Ameslan is so different from the structure of English, it is difficult, perhaps impossible, to speak good English and sign good Ameslan at the same time." Attempts to do so resemble attempts to speak English and write German simultaneously."[102] Because an ASL sign is understood not only by the precise hand configuration used but by where it is placed in relation to the body and the way the sign is moved through space, ASL translations into English often appear crude. But ASL in action is anything but.[103]

One might choose to disbelieve Washoe or think that she was not actually referring to herself. But the ASL-using orangutan Chantek said the same thing. When drinking a soda, she replied *"Me Chantek"* when asked *"Who drink?"*[104] Terrace reported that before he learned his name sign, Nim would sign *"Me"* at his mirror image or pictures of himself. Later, he would sign *"Nim"* when a teacher pointed to them and asked *"Who that?"*[105]

A large body of nonverbal evidence supports chimpanzee and bonobo self-awareness, as well. At Tetsuro Matsuzawa's Primate Research Institute in Kyoto, the chimpanzee Ai was taught to use lexigrams that represent objects and their qualities. She uses such pronouns as "you," "me," "him," and "her," recognizes herself in photographs, and seems at least partially to understand that she might "possess" an object, in her case a green feeding bowl.[106] Matsuzawa is probably the laboratory primatologist most experienced in studying wild chimpanzees. He noticed that wild chimpanzees in Boussou, Guinea, not only made stone tools to use to crack nuts but lugged favorite tools around with them. Matsuzawa thought this might indicate that they, too, possessed "a rudimentary form" of object possession.[107]

Gallup's original test for visual self-recognition has been successfully repeated so often that it has become the "gold standard" for visual self-recognition in both nonhuman animals and human children. He gave chimpanzees an opportunity to familiarize themselves with mirrors. Then he placed them under general anesthesia. While they were unconscious, he marked an eyebrow ridge and the opposite ear with odorless and tasteless red dots. Gallup reasoned that chimpanzees who touched the red spots while gazing into a mirror recognized the mirror image as themselves.[108] Many did. Bonobos do as well.[109]

No child less than fifteen months old has ever passed the mirror self-recognition test. Children under twenty months usually fail as well, though the percentage of successful children increases as their ages increase from fifteen to twenty-four months.[110] Chimpanzees as young as two and one-half years old have also passed the "dot test."[111] At Sally Boysen's Primate Cognition Project, four-year-old Kermit displayed mirror-guided, self-directed acts within five minutes of seeing his image in a mirror for the first time.[112]

Set up a video camera and monitor in any public place and it will quickly become apparent that watching oneself live on camera elicits similar responses to watching oneself in a mirror. On camera, Austin, the lexigram-using chimpanzee at Savage-Rumbaugh's Language Research Center, would watch himself pretend-swallow imaginary food that he pretend-scooped from imaginary bowls. Or he might shine a flashlight down his throat while he looked to see what was illuminated on camera.[113] Both Sherman and Austin differentiated between live videotapes and tape recordings of themselves. According to Savage-Rumbaugh, "When they saw a self-image, they would stick out their tongues, bob their heads, and so on to see if it were a live image. If it was live the behaviors I described above would begin to appear. If the image did not react, they typically went about their business."[114]

As with mirrors and live cameras, shadows can also test self-recognition. Nearly twenty years after he developed the mirror self-recognition test, Gallup invented a "shadow self-recognition test" for children. He found that "shadow permanence," or the ability to locate an object by seeing its shadow, emerges in the human infant at about the age of thirteen months. The ability to manipulate and play with one's shadow arrives about six months later. The ability to understand that your shadow is cast by you can emerge as early as twenty-five months but does not emerge in the majority of children until about forty months.[115] These results are consistent with the fact that full self-awareness does not begin to appear in children until about two years of age.[116] Sheba, a chimpanzee at Sally Boysen's Primate Cognition Project, passed the shadow self-recognition test at age eight.[117] On sunny Georgia days, teen-age Austin might manipulate his shadow in unusual ways. Watching movies of wild chimpanzees, he would sometimes make a shadow on the screen by placing his body between it and the projector, then use it to chase the wild chimpanzees.

Not all chimpanzees and bonobos appear to recognize themselves. In one mirror self-recognition test administered by Karyl Swartz and Siân Evans, only one of eleven chimpanzees passed the formal dot test, although three of them displayed other sorts of self-directed behaviors.[118] In another conducted by Daniel Povinelli and colleagues, 26 percent of thirty-five chimpanzees over the age of sixteen passed, 75 percent of twelve chimpanzees between the ages of eight and fifteen passed, 20 percent of ten six- and seven-year-olds, and just 2 percent of forty-eight one- to five-year-olds passed.[119]

These inconsistencies might be attributable to differences in rearing and socialization. Or they might stem from differences in the individual chimpanzees and bonobos under study. Or the apes might be shy or embarrassed by the dot, as human children sometimes are, and not touch it.[120] Povinelli reported that the mirror self-recognition abilities of chimpanzees varied with age and perhaps other factors.[121] Other studies show that differences in culture, social class, age, developmental levels, the values of the researcher, physical makeup, amount and type of education, and rearing history can cause researchers to draw incorrect conclusions about how even *humans* think.[122] Motivation, opportunity, and serendipity may also play their parts. Gorillas neither make nor use tools in the wild. But it would be a mistake to conclude that they can't use or make them, for they often do in captivity.[123] One hundred thousand years ago, human beings virtually identical to any reading this produced no known art, had no written, and perhaps spoken, language, and did not use agriculture. That we do all three today suggests that it would be a grave mistake to conclude that our ancestors of one hundred thousand years ago lacked the necessary cognitive capacities.[124]

Joint Attention

Remember that "joint attention" is the ability to "share the world." Autistic children lack it; normal children begin to develop it toward the end of their first year. Enculturated chimpanzees and bonobos have it. In one experiment, when asked in English "What's that?" enculturated chimpanzees could name an object referred to solely by gaze.[125] Even some unenculturated chimpanzees who have had a great deal of human contact appear to understand at least some aspects of joint attention. Some have assumed another's visual perspective by following a human gaze toward a spot above and behind them, even around a dividing wall.[126] Others have been able to follow the gaze of other chimpanzees.[127] Still others have figured out under which box raisins lay solely through the experimenter's cueing either by his facing the box and gazing at it or moving his eyes to glance at the box while keeping his head oriented straight ahead.[128]

In Chapter 9, I mentioned an experiment conducted by Tomasello and Call in which they compared the abilities of two-year-old and three-year-old human infants, unenculturated adult and juvenile chimpanzees, unenculturated orangutans, and Chantek, an enculturated orangutan to distinguish an experimenter's intentional actions from her accidental ones. I said that Chantek's scores had been strikingly better than the average human two-year-old and the unenculturated apes, and generally higher than the average human three-year-old as well. But even the unenculturated chimpanzees showed that they understood *something* about the difference between an experimenter's intentional and accidental acts, if not as much as a three-year-old does.[129] These chimpanzees may also have labored under the disadvantage of not having the experimenter's intentions be made as plain to them in terms they understood as well as the children did. The cues that an act was an accident included the experimenter making "a facial expression of disapproval" and saying "Oops!" which was used "because it is a natural behavior in English-speaking humans."[130] Especially for unenculturated chimpanzees, who were neither English-speaking nor human, neither the facial expression nor the "Oops!" might have communicated anything, and indeed there was no difference in the results obtained by chimpanzees who were told "Oops!" and those who weren't.[131] A hint that different experiences might have influenced the results was given when the children showed they understood the experimenter's act of placing much better than her act of dropping as intentional. Tomasello and Call attributed this to childhood experience with games in which dropping is often understood as accidental.[132] On the other hand, the chimpanzees understood dropping as more intentional than placing, which Tomasello and Call attributed to the fact that when a chimpanzee shares food, he drops it in front of the other chimpanzee.[133] Interestingly, the enculturated Chantek understood intentional acts as placing the same way the human infants did.[134]

As children do, beginning about the age of nine or ten months, chimpanzees may engage in the form of joint attention known as "social referencing," by which they seek emotional information from another by alternating looks between the individual and an object or event. They then use that information, whether it's fear or enthusiasm or whatever, to influence their own behavior toward the same object or event.[135]

Pointing can also help establish joint attention. No one has reported seeing a wild chimpanzee point, but they often point in captivity. We humans point to *refer* to something, to direct someone else's attention to an object or a place. Chimpanzees and bonobos with extensive human contact also often point to direct human attention.[136] How can we know that they point to refer to something? We can no more know for sure than we can know that infants do it. But for almost twenty-five years, child development psychologists have believed that children refer to something when they extend their finger or arm, try to get the listener's attention, then either monitor their listener's gaze or look back and forth between the listener and the object or place to which they are pointing.[137] Chimpanzees do the same thing. They wait for a human to face them, point, then maintain eye contact or alternate their gaze between the human and the object or place to which they are pointing.[138] So it's fair to say that if children point to refer, so do chimpanzees. Here are several examples.

Savage-Rumbaugh relates how she once purchased unusually shaped colored plastic drinking cups for Sherman and Austin. Austin could not have cared less about them. But Sherman was captivated. He carried his cup everywhere throughout the first morning. Savage-Rumbaugh put the glasses away during nap time. That afternoon, she returned with a strawberry drink in one new glass and one old. Sherman immediately rushed to the computer that he used to communicate with lexigrams and pounded out "Glass strawberry drink." Then he pointed to the new glass.[139]

When Kanzi was eighteen months old and riding through the forest on Savage-Rumbaugh's shoulders, he would point with an entire outstretched arm in the direction he wished to travel. To drive his point home, he might force Savage-Rumbaugh's head in that direction with his hands or even lean his whole body toward where he wanted to go.[140] Kanzi's stepsister Panbanisha also began pointing at the age of one.[141] Roger Fouts tells a similar story about the young Washoe. Riding piggyback around the Gardners' backyard, Washoe would touch Fouts's chest to make the ASL sign for *"You,"* then make the *"Go there"* sign by stretching out her arm and index finger. When they arrived "there," she would point again, and again, and again.[142] Chimpanzees point to direct the attention not just of humans but of other chimpanzees. Sherman and Austin learned to communicate with each other through symbols that lit up on a computer. But they would also point back and forth between a desired food and the symbol for that food.[143]

Imitation

Imitation suggests some awareness that other minds exist. The ability to imitate develops early in both humans and chimpanzees. Thus, both a very young home-raised chimpanzee and human infants have mimicked adults sticking out their tongue.[144] Humans show a higher ability to imitate than do apes, though some speculate that chimpanzees and bonobos have the same ability to imitate as humans but just don't use it.[145] Whether the abilities are the same or not, numerous apparent examples of imitation by enculturated chimpanzees, and a much smaller number of examples by unenculturated chimpanzees, in which they apparently assume the visual perspective of another, have been reported.[146]

For example, after a fight at Arnhem, Luit, the alpha male, injured a finger. He did not walk on that hand for days after, but used his wrist. To Frans de Waal's amazement, "all the young apes imitate him and suddenly begin stumbling around on their wrist."[147] In the late 1940s, Keith and Cathy Hayes played a "Simon Says" game with the chimpanzee Viki by asking her to perform an act such as clapping her hands, then doing it themselves. She learned to follow it immediately.[148] Roger Fouts later played an ASL version with Washoe. He would sign *"Do this"* and put his hands on his head or over his eyes and Washoe would do the same thing. Washoe also bathed her dolls in the same dishpan in which she was bathed.[149]

Both chimpanzees and bonobos have learned lexigrams on a keyboard through imitation at Savage-Rumbaugh's Language Research Center.[150] Indeed, every symbol-using ape has learned his or her symbols through imitation.[151] Nim learned many of his ASL signs by imitating his trainers, who signed primarily with their right hands.[152] In a twist, he demonstrated a more sophisticated ability to imitate by signing not just with his left hand, but with his feet![153] Ironically, even though Terrace and his teachers failed miserably at the task of teaching American Sign Language to Nim, he still managed to imitate his teachers so well that he duped them into *thinking* he had learned it.[154]

Finally, Professor Andrew Whiten of the University of St. Andrews recently devised an artificial fruit that contains a real fruit reward once it is opened. To open the fruit, two bolts must be removed and passed through rings, a pin has to be removed from a handle so it can be manipulated, and the handle has to be shifted either by pulling it from a barrel or turning it to an appropriate position. One way to open this artificial fruit was demonstrated to each of four chimpanzees from Savage-Rumbaugh's Language Research Center: Panzee, who was enculturated; Sherman and Lana, who were not enculturated but who had substantial contact with humans and had been taught how to use lexigrams; and Mercury, a mother-raised unenculturated chimpanzee with substantial human contact. Each chimpanzee imitated the human demonstration of the sequence required to open the "fruit."[155] Whiten, the lead author of the "Cultures in Chimpanzees" article that highlighted dozens of cultural variations in chimpanzee behaviors, and his colleagues thought it "difficult to see" how the chimpanzees' cultural behaviors "could be perpetuated by social learning processes simpler than imitation."[156]

Teaching

Active teaching is a type of cultural transmission that requires a teacher to understand at least some of the student's mental states.[157] Children generally don't actively teach until their fourth year, and formal teaching, like counting, is relatively unusual in some human hunter-gatherer societies.[158] Wild chimpanzees have only twice been observed actively teaching. Active human teaching only may therefore account for much of the explosion in the cognition of enculturated apes.

The two known examples of wild chimpanzee teaching emerged from the Tai National Park. Christophe Boesche and Michael Tomasello consider them "very important because they seem to be of the type characteristic of all human cultures as they instruct their youngsters in at least some important cultural activities."[159] Chimpanzee mothers were seen influencing their infants' attempts to crack nuts in three ways. They stimulated the infants to crack nuts by leaving hammers or nuts near anvils. They facilitated their infants' nutcracking by providing them with much better hammers or by collecting nuts for them. And they actually demonstrated nutcracking.[160]

David Premack argues that the chimpanzees' failure to engage in "a full-fledged pedagogy" cannot be explained by lack of cognition, as they have "all the prerequisites for pedagogy, including a theory of mind," but should be ascribed to lack of motivation.[161] Chimpanzees, he thinks, lack the drive and sense of excellence that impels humans relentlessly to hone their own skills and engage in self-training, activities that go along with training others.[162]

Enculturated apes, however, appear to teach more often than do their wild cousins. Roger Fouts and his colleagues sought to learn if the chimpanzees to whom they had taught sign language would in turn teach it to ten-month-old Loulis. No human was permitted to sign in Loulis's presence. Eight days after Loulis arrived, his first sign appeared. Washoe actively taught Loulis one of his earliest signs, "*Come*," by signing "*Come*" and retrieving and approaching her son until he would come when she signed "*Come*."[163] At the age of twenty-nine months, Loulis knew seventeen signs. At the age of sixty-three months, he knew forty-seven. Sometimes Washoe, who had learned some of her signs through active molding, actively molded Loulis's hands to form the sign in a manner familiar to the parents of deaf children. For example, while waiting for a candy bar, Washoe repeated the sign for "*food*." She molded Loulis's hand to make the sign for "*food*" and moved his hand in the motion associated with making that sign. She repeated this process for many signs, "*gum*," "*drink*," "*what*," "*brush*," and "*chair*."[164] In another example, Fouts found the irrepressible Lucy trying to teach ASL to her cat. Sitting on the floor, Lucy "placed the cat between her legs, and held up a book so that the cat could see it. Then, while pointing to the book, Lucy gave the cat a lesson in the sign for '*book*.'"[165] The bonobo Panbanisha is today teaching language skills to her son, Nyota, whose vocabulary approximates that of a one-year-old human child.[166]

Intentional Deception

More than fifty types of deceptive behavior have been reported in chimpanzees.[167] Intentional deception occurs when one animal intends to deceive another animal in order to alter her beliefs.[168] This occurs, say, when I tell my wife I am going to the grocery when I actually intend to go to a bookstore. Constant practice from a young age and the fact that our deceptions are often so very hard to detect that we get away with them has made most humans virtuoso deceivers.

Psychology professor Peter Mitchell thinks that it is "virtually impossible" to investigate the intentional deception of apes in the laboratory. Why? Because "[h]ow do we request or encourage an ape to engage in deception? How could we ask the ape to deceive, and in any case what would be the ape's incentive?"[169] If he is correct, we must rely upon careful field studies of ape behavior, that we can reasonably conclude is very difficult to interpret in any way other than as an act of intentional deception. But intentional deception can involve complicated behavior, and it is always hard to tell whether one animal's action is based on another animal's state of mind or on that other animal's behavior.[170] The primatologists Andrew Whiten and Richard Byrne have proposed an interesting test for detecting the intentional deception of apes in the wild: Look for one ape to employ a novel or unusual behavior to counter another ape's apparent deception.[171] Here is perhaps the best example.

Outside the Tai National Park, a lone wild chimpanzee was about to be fed bananas contained in a metal box that could be opened from afar. Just as the box opened, another chimpanzee ambled toward the clearing. The first chimpanzee immediately shut the box, then nonchalantly walked about ten feet away. The second chimpanzee left the scene. But as soon as he was out of sight, the second chimpanzee hid behind a tree and watched the first chimpanzee return to the box and open it. The second chimpanzee then sprang forward and seized the bananas.[172] Because it is so unusual for a grown chimpanzee to hide behind trees and spy on another chimpanzee, Byrne and Whiten propose this as one example of an intentional deception that is also the sort of *fourth-order* mental state that humans attain only in middle childhood or later: "[T]he chimp who hid *believes* that the other will *think* he does not *guess* that he really *knows* about hidden bananas."[173]

Jane Goodall reported numerous instances of apparently intentional deception at Gombe that are difficult to explain in any way other than as novel counterdeceptive behavior. In one, after Melissa gave birth to twins, with whom she was very protective, her juvenile daughter would touch the infants with her hands while looking away from them or would gaze into the sky while touching them with her toes.[174] In another, Goodall designed boxes into which she could place bananas and open by remote control.

Evered and Figan learned how to unscrew these boxes to get at the bananas that Goodall had placed in them. Evered always did so openly. But time and again, higher-ranking males patiently waited for him to complete his work, then stripped him of his hard-won bananas. After this also happened to him several times, Figan changed the tactics.

> Very nonchalantly, and with an apparent lack of purpose, he wandered to a handle. There he sat and performed the entire unscrewing operation with one hand, never so much as glancing at what he was doing. Thereafter he simply sat, gazing anywhere but at the box, one hand or one foot resting on the handle. There he outwaited the big males, sometimes for as long as thirty minutes. Only when the last one had gone did he release the handle and (silently) run to claim his well-earned reward.[175]

Figan used a different mode of apparently intentional deception to obtain bananas. Goodall watched him stride into the forest "with a brisk and purposeful walk," so purposeful that it apparently "indicated to (higher-ranking males) that he was headed for a good food source." But when they followed him, Figan would vanish, then circle back to eat the remaining bananas.[176]

The semiwild Arnhem Zoo chimpanzees also appeared intentionally to conceal. Here are four of Frans de Waal's observations. Two centered on Dandy. A subordinate male, he apparently pined for the sexual intercourse that more dominant males refused to tolerate. Thus, occasionally Dandy made a "date" with a female for surreptitious sex.

> Male chimpanzees start their advances by sitting with their legs wide apart revealing their erection. Precisely at the point when Dandy was exhibiting his sexual urge in this way, Luit, one of the older males, unexpectedly came around the corner. Dandy immediately dropped his hands over his penis concealing it from view.[177]

Dandy also once searched for buried grapefruits alongside some higher-ranking chimpanzees. None were found. But Dandy apparently located them during the search, though he kept that fact secret. When the others were asleep, he rose and made straight for the hidden grapefruits, immediately found and ate them.[178]

Luit was once challenged by Nikki, the subordinate male soon to overthrow him. Luit sat beneath a tree from which Nikki was hooting provocations. At each provocation, a nervous grin appeared on Luit's face, which he physically suppressed by pressing his lips together with his fingers.[179] In the final example, a male, Yeroen, was injured in a fight with Nikki. For a week after, he "hobbles pitifully (when Nikki can see him), but once he has passed Nikki his behavior changes and he walks normally again."[180]

Emil Menzel's experiments with captive chimpanzees produced a second strong candidate for Whiten and Byrne's intentional deception test. Belle was repeatedly shown hidden food. Rock, a dominant male, repeatedly seized it. Belle began sitting on her food when Rock was about. Rock began to shove her aside when she sat in any one place for too long. So Belle stopped approaching the food. Like a destroyer determined to sink a submarine, Rock began to search in wider and wider areas around Belle. Belle began to sit farther and farther from the food and would wait until Rock looked away before breaking for it. Rock began looking away and sometimes seemed to wander off. But at the moment that Belle was about to uncover the hidden food, Rock would suddenly wheel about.[181] Byrne and Whiten saw Rock's counterdeceptive action in waiting until Belle was about to eat the food, then suddenly turning, as strong evidence that Rock knew that Belle was trying to deceive him, because it was such an unusual way for a chimpanzee to act, yet it was entirely appropriate to the situation. Again, Byrne and Whiten found this persuasive evidence of a fourth-order intention: "Rock *believes* that Belle will *think* that he no longer *wants* to discover what she *knows* about hidden food."[182]

Numerous examples of apparently intentional deception have been reported in enculturated chimpanzees. Roger Fouts tells how Bruno was playing with a hose when a larger chimpanzee, Booee, seized it. Bruno went to the door and gave a false alarm call. Booee responded by dropping the hose and racing outside, leaving Bruno to play with the hose unmolested.[183] Austin, who was subordinate to Sherman, once walked outside in the dark, created strange noises that frightened Sherman, then rushed back into the lab with his hair erect, as if he, too, were terrified. Sherman, who is afraid to go outside in the dark, would come to Austin for reassurance, instead of dominating him.[184]

Kanzi is a master of deception. Finally, Kanzi may drop a ball and feign disinterest in it while keeping it within reach. Simultaneously, Savage-Rumbaugh may also feign disinterest.

> We are then in a sort of stalemate, me attempting to look disinterested until Kanzi moves far enough away for me to grab the ball, and Kanzi trying to look disinterested, though he is really watching my every move. Such stalemates may last 3–10 minutes during which both of us continue to feign disinterest in the ball, then suddenly we will both dive for the ball. Before that time, every glance and slight body movement is weighed by both parties, and both attempt by every move and glance to deceive the other in a game where deceit is the ground rule.[185]

Once he pulled his sleeping blankets over his body while he lay in bed fairly still for twenty minutes as a caretaker "searched" for him. When finally "discovered," Kanzi emerged with a grin.[186] After being refused permission to go outside, Kanzi once hid the small tool used to open an outside gate. Sue Savage-Rumbaugh searched thoroughly, but unsuccessfully, for the tool. She asked Kanzi to help her search, and he did so, with apparent diligence. No luck. But thirty seconds after Savage-Rumbaugh turned away, Kanzi produced the tool, unlocked the gate, and escaped.[187] Maurice Temerlin relates a similar escapade of Lucy's. While the Temerlins were at work, Lucy would escape from her locked spacious bedroom and raid the kitchen. They caught her letting herself out of her bedroom with a stolen key that she concealed in her mouth.[188]

Of course, Mitchell's rhetorical questions ("How do we request or encourage an ape to engage in deception? How could we ask the ape to deceive, and in any case what would be the ape's incentive?") assume that the ape lacks language. Temerlin lifted the following instance of deception from a signed conversation between Roger Fouts and Lucy after she defecated on the living-room floor. Every parent will recognize the attempted intentional deception.

ROGER: *What's that?*
LUCY: *Lucy not know.*
ROGER: *You do know. What's that?*

LUCY: *Dirty, dirty.*
ROGER: *Whose dirty, dirty?*
LUCY: *Sue's.*
ROGER: *It's not Sue's. Whose is it?*
LUCY: *Roger's!*
ROGER: *No. It's not Roger's. Whose is it?*
LUCY: *Lucy, dirty dirty. Sorry Lucy.*[189]

I end this section with a 1998 example of deception by captive un-enculturated chimpanzees described to me by Tetsuro Matsuzawa. Over the course of a month, Matsuzawa repeatedly showed one of two chimpanzees where food had been hidden in a container in an outdoor enclosure. The second chimpanzee could see the knowledgeable chimpanzee, but not where the food was hidden. Sometimes he reversed the roles of the chimpanzees, while other times neither chimpanzee could see where the food had been hidden. Both chimpanzees developed deceptive strategies and counter-strategies to mislead the other and obtain the food.[190]

Empathy

Frans de Waal defines "empathy" as "the ability to picture oneself in the position of another individual."[191] Savage-Rumbaugh believes that this suggests that apes "can understand that different people experience the world in different ways."[192] Here are nine strongly suggestive candidates for empathy.

The first is the only one observed in a controlled laboratory experiment. At Boysen's Primate Cognition Project, researchers constructed a two-sided apparatus with an "informant" side and an "operator" side. Sheba, Sarah, Darrell, and Kermit were taught that the operator could control four paired food trays using four handles but couldn't see what was in the food trays. When the operator pulled a handle, each tray in a pair of trays moved in opposite directions; one tray moved within reach of the operator while the other tray moved within reach of the informant. The informant could see what was in the food trays but couldn't reach the handles.

Each chimpanzee was then paired with a human researcher. Sheba and Kermit were the first chimpanzee informants. Each watched a researcher place food in one pair of the four pairs of food trays. The human operator, who couldn't see the food, waited for the chimpanzee to indicate one of the pair of trays. If the chimpanzee informant didn't manually indicate soon, the human asked the chimpanzee in English which tray contained the food. When the chimpanzee indicated, the human pulled the corresponding handle and both human operator and chimpanzee informant either obtained food or didn't.

Darrell and Sarah were the first chimpanzee operators. Each of their human informants hid food in one of the four pairs of food trays that the chimpanzee operator could not see, then pointed to the pair of food trays in which the food was hidden. The chimpanzee operators then learned that when they pulled the corresponding lever, both they and the human informants obtained food. One day the researchers suddenly switched roles. Sheba and Kermit were thrust into the role of operators, while Darrell and Sarah became informants. Would they understand what they and their human partners were supposed to do after this role reversal? Most did. When food was placed in one of the four pairs of food trays, Sarah and Darrell immediately began pointing to the tray in which food had been placed, so that when their human operator pulled the corresponding handle, both human and chimpanzee received food. Similarly, Sheba began to act as an operator, pulling the corresponding handle when her human informant pointed to a pair of food trays (though Kermit did not appear to get it as well).[193] Boysen and her colleagues argued that the chimpanzees' ability to switch roles strongly suggested that they were capable of empathy.

The other eight examples come from uncontrolled observations. The most recent occurred on April 10, 1998, in the Chimpanzees of Mahale Mountains exhibit at the Los Angeles Zoo. Here is how it was related to me by one eyewitness, Dr. Charles Sedgwick, the zoo's highly experienced director of Animal Health Services.

Jamal, a three-year-old, was playing with a twist-strand nylon rope. He separated a strand and placed it around his neck. Then he began to choke. A band of chimpanzees rushed to his aid. But their desperate attempts to help actually hastened, and perhaps even caused, his death. "It was the adult females that tugged at the little chimp, tugged at the hawser, bit at the rope and gave forth their loudest distress cries," Dr. Sedgwick wrote.

They drove the males away, though. The keepers and veterinarians tried to clear the chimps out of the exhibit, into their secure holding area with water hoses and they resisted valiantly for over thirty minutes (which didn't help Jamal of course). The last chimps to resist the hosing (all were furious) were Toto and Bonnie, allies, our oldest members. The chimps grieved. They sat, without activity for several days, heads down, solitary.[194]

Washoe once rescued a drowning chimpanzee. Double fencing normally kept the chimpanzees who lived on her Oklahoma island from the surrounding pond. A new chimpanzee, Penny, arrived. Like all chimpanzees, she could neither float nor swim. When she vaulted the fencing, she splashed into the pond. Despite her chimpanzee fear of water, Washoe also hurdled the fences, then anchored one arm to a fence pole, extended her other arm toward the thrashing Penny, and pulled her to safety.[195]

One of Washoe's caretakers, Kat, was pregnant. Washoe, who had lost two babies, was very interested in her swelling abdomen and would ask about "*baby*." When Kat miscarried and returned to the lab after an absence of several days, she signed to Washoe that she had lost her baby. Roger Fouts relates that "Washoe looked down to the ground. Then she looked into Kat's eyes and signed '*cry*,' touching her cheek just below her eye."[196]

At the Arnhem Zoo, Krom, a thirsty female chimpanzee, eyed a tire filled with water. But the tire was blocked by half a dozen heavy tires. In ten minutes of struggle, she only wedged the water-filled tire more securely. Finally she gave up. Jakie, a young male to whom Krom had acted as an "aunt," began pushing the heavy tires away. When he reached the water-filled tire, "he carefully removed it so that no water was lost and carried the tire straight to his aunt, where he placed it upright in front of her."[197] De Waal found it "hard to account for Jakie's behavior without assuming that he understood what Krom was after and wished to help her by fetching the tire. Perspective-taking revolutionizes helpful behavior, turning it into cognitive empathy, that is, altruism with the other's interests explicitly in mind."[198]

Jakie's acts recall those of Linda, a bonobo in the colony at the San Diego Zoo. Her two-year-old daughter whimpered and looked at her in the way that bonobo children look when they want to be nursed. But Linda had stopped lactating. She walked to a fountain and sucked up some water. Then she returned and puckered her lips so her daughter could drink. She repeated this four times.[199]

When Kuni, a female bonobo, captured a starling at the Twycross Zoo in England, the caretaker urged her to free him. Kuni brought the starling outside and set him upright. When he didn't move, Kuni threw him a short distance. She then carried him to the top of the highest tree and wrapped her legs around the trunk while she unfolded the bird's wings, spread them open, placed one in each hand and tossed him into the air.[200]

In 1998, Sally Boysen asked a man to dress as a veterinarian, then armed the impostor with the dart gun used to anesthetize the chimpanzees for their physical examinations, and placed him near Darrell's unoccupied cage. The chimpanzees don't like veterinarians. As Darrell began to enter his cage, Kermit, in a position to see the "veterinarian" lying in wait, loudly and repeatedly hooted. Darrell instantly froze. Later, when Darrell had obviously seen the "veterinarian," Kermit did not "warn" him again.[201]

Finally, Professor Michael Huffman of Kyoto University has spent years studying apparent examples of chimpanzee self-medication. He has seen healthy young chimpanzees watch their mothers, ill with the diarrhea, constipation, and malaise that come from nematode infections, engage in a rare behavior: The mothers may detour from a group's normal travel route to seek out and ingest some of the thirty or more species of bitter-tasting plants that can enable them to recover from illnesses within twenty-four hours. The youngsters then follow suit when they became ill. Empathy, Huffman says—the idea that the youngsters place themselves in their mother's position—"seems to be a perfectly parsimonious explanation."[202]

NONLANGUAGE SYMBOLIC COMMUNICATION

Convincing evidence of theory of mind lies in observations of ape communication with symbols. I'll save a discussion of symbolic language until the last section and give three examples of nonlanguage symbolic communication here.

After Sherman and Austin were taught the critical components of communication—requesting, naming, and comprehending—they spontaneously began to monitor and engage each other much more closely, gesture to each other, take turns, and announce what they intended to do. In *Kanzi*, Savage-Rumbaugh presents a series of photographs that demonstrate their progress toward a full-blown theory of mind.

Sherman can be seen watching bananas being placed into a box on his side of a divided room. Austin, on the other side of the divider, can see Sherman's side of the room through a window. But he cannot see the box of bananas and has no clue as to what has happened on Sherman's side. The banana-filled box is locked, but only Austin has the key. Sherman goes to his computer keyboard and presses the lexigrams for "*Give key*." Austin sees the message and removes the key from a tool kit on his side of the room. Sherman comes to the window. Austin hands him the key. With Austin watching intently, Sherman goes to the box, turns the key in the lock, and removes the bananas. Sherman then takes half of the bananas to the window (tasting Austin's share on the way) and hands Austin his share of the loot through the window.[203]

Austin and Sherman's theory of mind was not lashed to their computer keyboard. One day, Savage-Rumbaugh shut Austin's keyboard off to see how he would respond and if he would be able to communicate with Sherman without it. She then removed labels from foods they relished and scattered them about Austin's side of the room. With Sherman in a separate room, she slopped Peter Pan peanut butter into the box before Austin's eyes and locked it. Austin rushed to his keyboard where he learned it had been turned off. What to do? Spying the food labels, he picked up the Peter Pan peanut butter label and looked to Sherman. Sherman pounded his keyboard to ask for peanut butter. When Savage-Rumbaugh reversed the situation and placed Welch's grape jelly in Sherman's box, Austin requested it after Sherman held up the Welch's label. The test results held for M&Ms and Doritos corn chips and twenty-five other brand names.[204] Professor Richard Byrne states flatly that Savage-Rumbaugh's work here

> settle[d] two of the most hotly contested issues that were raised in the early days of "ape language" studies. There is now *no doubt* that these apes can understand and use the concept of *reference*; that is, the fact that one thing, a word, can be made to stand for and refer to another thing, an object or a class of objects. *Nor is there doubt* that they are capable of using words for real communication, requesting and offering new information when it is appropriate, as well as demanding things and commenting on performance [205]

The neuroscientist Terrence Deacon has cited the accomplishments of Sherman and Austin as "one of the most insightful
demonstrations" of how *any* creature makes the extraordinary
mental leap from simple associative learning to the understanding
of symbols.[206]

An extraordinary hypothesis is being investigated near the
Congo village of Wamba, where scientists have been studying
wild bonobos for more than a quarter century. They noticed that
parties of bonobos traveled on the ground, often along paths and
through dense forests, for much of the day. Although they would
split up, the parties always reunited at distant sites and seemed
to know where the bonobos ahead of them were going. Large
trampled plants and branches four to six feet long often appeared
where trails crossed, forked, or faded into unclarity. The scientists had often tracked the bonobos using those very signs of
progress. Now they began to wonder if it was possible that the
bonobos were intentionally marking the trail, not for the benefit
of the human trackers but for their companions coming after
them.

They began to scrutinize the trail "markers." Over a couple of
days, they found 110 possible "markers." Each appeared to show
the actual direction that the bonobos had taken, because when
they followed the markers, they located the apes. Some markers
seemed just to confirm that they were on the right trail. But most
cleared up a question of proper direction, such as which of two
crossing trails should be taken. More work needs to be done. But if
further research confirms the intentional character of the trail
markings, it will be evidence that bonobos are intentionally signal-
ing other bonobos, who then use the signaled information—a sophisticated theory of mind indeed and one attained without
human enculturation.[207]

Finally, the Tai National Park is thickly blanketed with rain forest that causes visibility to be less than twenty meters. Like the
bonobos at Wamba, the Tai chimpanzees split into daily foraging
parties. Yet as at Wamba, each party remains in contact throughout
the day with the rest of its community. Early on, Christophe and
Hedwige Boesch saw and heard Brutus, the alpha male, frequently
vocalize and drum upon buttressed trees with both his hands and
feet. Finally they realized that he was sending messages to the rest
of the group.

The Boeschs identified three messages. Brutus might instruct the others to change their direction of travel by drumming first on one tree, then on another within about two minutes. Christophe Boesch believed that the chimpanzees "inferred the new direction by mentally visualizing Brutus' displacement between the two trees and then transposing it to their own direction of travel."[208] Or Brutus might instruct the others to rest by drumming twice on the same tree within the two minutes.[209] Or Brutus might instruct the others both to change direction and rest by drumming once on a tree in the direction being traveled, then twice on another tree in the new direction he was instructing them to take, again within two minutes.[210] Boesch said,

> [M]aintenance of this system is dependent upon the comprehension of it by other group members. Comprehension here seems more demanding than production. The receiver, often out of visual contact with Brutus, has to mentally visualize Brutus' location from his drumming, and infer his movement between them, to understand the proposed direction. Tai chimpanzees may be silent for long periods of time and normally the group would follow Brutus' proposals without any vocalisation, with no sound being made for the next one or two kilometers . . . Therefore the risk of losing contact with other community members is real if receivers are unable to make such an inference.[211]

LANGUAGE

The writings of scientists are often cold, colorless, dispassionate, and dull. Not Steven Pinker's. In *The Language Instinct,* Pinker, perhaps the most prominent critic of ape language, employs an acerbic and colorful writing style to blast Kanzi as the last example of the "highly trained animal acts" that pass for ape language research. Entertaining though Pinker may be, when he ventures from his narrow field of expertise, he's wrong, and wrong, and wrong again. The following misinformation appears in just six pages of Pinker's book.

Pinker claims that the scientists who work with Kanzi and Company are just "chimpanzee trainers." Wrong. We will see that Kanzi, Panbanisha, Panzee, and others were never trained in language. They are utterly immersed in language. But never trained. Pinker says that chimpanzees are "vicious" (in my line of work, them's fightin' words, for vicious animals are frequently marked for execution). Wrong. No chimpanzee can hold a rifle to the likes of Hitler, Stalin, Mao Tse-tung, or Pol Pot, while bonobos are notorious for making love, not war. Ape language researchers work only with babies, says Pinker, because adults are so vicious. Wrong. Kanzi, at nineteen, is no baby. Neither is Ai, now in her early twenties. Tatu, Dar, and Moja are in their mid-to-late twenties. Sarah and Washoe are pushing forty. Tetsuro Matsuzawa reports that he is often alone in a small room with one of the nine chimpanzees with whom he works at the Primate Research Institute and feels perfectly safe.[212] "The stronger the claims about the animal's abilities, the skimpier the data made available to the scientific community for evaluation," Pinker says. Wrong. More videotapes, still photographs, written observations, scientific articles, and books publicly document Kanzi's life than all the lives of the Pinker family down to Adam and Eve. Pinker's bibliography reveals he just never looked at them. A chimpanzee responds correctly to a request like *"Would you please carry the cooler to Penny,"* Pinker says, because "all the chimp had to notice was the word order of the two nouns (and in most of the tests, not even that, because it is more natural to carry a cooler to a person than a person to a cooler)." Wrong. Pinker's lack of basic research is also responsible for this misstatement. With one short exception, Savage-Rumbaugh's very extensive writings are missing from Pinker's bibliography. The words "Fouts" and "Washoe" are absent from Pinker's book altogether. As we will see, Kanzi can comply with some distinctly strange and unnatural requests, such as "Put the toothbrush in the lemonade," "Hide the toy gorilla," "Put on the monster mask and scare Linda," "See if you can make the doggie bite the ball," and "Feed your ball some tomato," as well as more complicated requests that contain an embedded phrase such as "Get the ball that's in the group room" and "Get the noodles that are in the bedroom," where "group room" refers back to "ball" and "bedroom" refers back to "noodles."[213] Something far more cognitively complex is occurring when a ball lies before an ape and another ball sits in a bedroom, and the ape can differentiate a request to "go get the ball that's in the bedroom" from "take the ball to the bedroom."[214]

Meanwhile, Pinker's support for his claim for dramatic differences in the minds of language-using chimpanzees and two-year-old children is limited to this story about Pinker's two-year-old niece.

One night the family was driving on an expressway, and when the adult conversation died down, a tiny voice from the back seat said, "Pink." I followed her gaze, and on the horizon several miles away I could make out a pink neon sign. She was commenting on its color, just for the sake of commenting on its color.[215]

One can only imagine what Pinker would have said if Kanzi or Washoe had been sitting in the backseat and Kanzi had pointed to "pink" on his lexigram board or Washoe had signed "pink." Chimpanzees and bonobos do comment, just for the sake of commenting. Savage-Rumbaugh reports that Kanzi carries his portable lexigram keyboard away from humans and, once alone, touches lexigrams as if "talking" to himself.[216] Washoe's first teachers, Allen and Beatrix Gardner, wrote that from an early age, "Washoe used ASL to comment on things around her, or just to start a conversation."[217] After they left the Gardners and moved to the University of Oklahoma, Roger Fouts and Washoe sometimes rambled together through the nearby woods. During one walk, she pointed to a tree, bird, and cow and signed their names as she did.[218]

When Tatu arrived at the Fouts's Chimpanzee and Human Communications Institute in Ellensburg, Washington, she was temporarily assigned a windowless room and separated from the other chimpanzees. When freed to enter Washoe's room, which had a window, she "pointed to the asphalt parking lot and signed "black."[219] Through remote videotaping over a period of fifty-six weeks, Deborah Fouts captured twelve minutes a day of the chimpanzees signing to themselves when no humans were present. As children do, the chimpanzees referred to objects or events that did not involve them ("my book" or "cow"), to what they wanted ("more," "chase"); to an act that another might perform ("clean," "out"); to an ongoing sensory event ("listen," "listen there"), a description of their own activity ("peekaboo," "drink"), or an "imaginary object ("cat," "flower"). Both Moja and Tatu were seen to comment "red."[220]

Nim lived with the family of Stephanie LeFarge in a townhouse on New York City's West Side for eighteen months before he was moved to the Bronx. Terrace relates how nine months after Nim's relocation, he drove toward Stephanie's house. "A few blocks away, Nim began to hoot. When we turned into Stephanie's block, Nim's excitement intensified further. As we drove past Stephanie's house, Nim stood up in his seat and pointed to the house."[221] Nim also would thumb through magazines and books and sign about what he saw.[222] According to Terrace, Nim would sign spontaneously when he recognized a person, an object, or an attribute. One day, as Terrace and Nim were stopped at a traffic light, Terrace noticed that Nim was signing "*drink.*" A few seconds later, Nim pointed to a bus and Terrace could see that the bus driver had just poured himself a cup of coffee from a thermos and was drinking.[223] Terrace reported that Nim signed "*hat*" while passing a billboard of a cowboy wearing a hat and "*red*" at the sight of a red flower.[224] No wonder Savage-Rumbaugh recently entitled one of her book chapters, "Why Are We Afraid of Apes with Language?"[225]

To an unfrightened observer, chimpanzees and bonobos, often enculturated, wield powerful language abilities. The chimpanzees at Boysen's Primate Cognition Project use Arabic numbers. Washoe and the other chimpanzees at the Fouts's Chimpanzee and Human Communication Institute use more than one hundred ASL signs. At Matsuzawa's Primate Research Institute, Ai has learned to use more than eighty symbols, including all twenty-six letters of the English alphabet, forty-one of the many Japanese Kanji characters, and the Arabic numbers from 0 to 9.[226] Ai also learned to use letters of the English alphabet to stand for humans, herself, other chimpanzees, and an orangutan. Thus "L" stands for Ai; "A" stands for Akira, another chimpanzee; "U" stands for Dodou, an orangutan; and "Z" stands for Tetsuro Matsuzawa.[227] Nim sometimes used the ASL words for "*bite*" and "*angry*" to express his anger and desire to bite instead of actually biting.[228]

Upon what grounds do critics such as Pinker claim that apes lack any language abilities at all? I can think of two. First, Pinker and his mentor, Noam Chomsky, think that a "language acquisition device," or LAD, kind of a natural cross between *The Oxford Esperanto Dictionary* and Strunk and White's *Universal Elements of Style*, has been soldered into every one of our brains, but only ours.[229] But no one has ever seen a LAD, and most scientists doubt that anyone ever will. Language is more likely a product of activity spread across many parts of the brain. But *if* such a thing does exist and, especially, if it's a language *comprehension* device, then it's not likely to have been soldered just into our brains: We might have the silver Year 2000 Rolls Royce version, but bonobos and chimpanzees can hum along nicely with their basic black 1920 Model T. We may have the nifty Pentium III processor, while theirs lumbers on like ENIAC. But theirs works.

Second, Pinker claims that Kanzi can't *produce* any language and therefore doesn't *have* any language.[230] Kanzi can't speak. But the vocal tracts of bonobos, like those of human infants, chimpanzees, and the *Australopithecines*, are constructed so they can't.[231] Apes have tried valiantly to speak. In the mid-1950s, after intensive training, Keith and Cathy Hayes taught Viki to say, "papa," "mama," "cup," and "up."[232] I understand that watching Viki's struggle to overcome her physical limitations on videotape can be painful to watch.

No one has ever claimed that Kanzi or Panbanisha or Washoe or Loulis ever was or ever will be a great language producer. Or even a good one. Or even a fair one. Okay. Chimpanzees and bonobos have been, are, and probably always will be really lousy, lousy language producers. At least compared to a human adult. But given adequate conditions, they can *understand* it—and in the way of a human child, and with no training whatsoever! Thus today Panbanisha can produce about 250 words on a voice synthesizer but can comprehend about 3,000 words.[233] Some may quibble that they don't have any language. But if it's not language, it's some extraordinary mental feat for ape and human alike.

Savage-Rumbaugh argues that the core of language is *comprehension*, not production. Comprehension requires

> an active intellectual process of listening to another party while trying to figure out, from a short burst of sound, the other's meaning and intent—both of which are always imperfectly conveyed. Production, by contrast, is simple. We know what we think and what we wish to mean.[234]

Recall Christophe Boesch's thought that comprehension seemed more demanding than production when Brutus was drumming his messages to the Tai community. At age five, Washoe could reliably use more than one hundred ASL signs and comprehended hundreds more.[235] Until 1976, when Roger Fouts reported that Washoe comprehended at least ten English words, few thought that any nonhuman animal could possibly understand any human language.[236] Few doubt it any more.

Comprehension also demands some theory of mind, for a language comprehender must assume the perspective of a language producer in order to understand what he is telling her.[237] Every child language researcher and every parent knows that an infant in her second year who can only produce sonic booms at dinnertime still understands dozens of words and even some sentences.[238] This is why the child psychologist Elizabeth Bates thinks that "[i]f we want to understand what an organism knows about language, isn't comprehension the best place to start?"[239] It may turn out that comprehension and production even involve different brain processes altogether.[240]

Several early attempts were made to teach chimpanzees ASL.[241] The oldest, and longest ongoing, project began with Washoe in 1966 in the Gardners' backyard. They pioneered the idea of inducing language in an ape by creating a very rich social and linguistic environment, then communicating with her as they would with a human child. Washoe learned American Sign Language primarily because she learned it in the same natural way in which deaf human children do.[242] The Fouts have now been engaged in a thirty-four-year (and counting) ASL project with Washoe, and later with four other chimpanzees—her adopted son, Loulis; Dar; Moja; and Tatu—at Central Washington University. All have learned rudimentary ASL and routinely sign not only to humans but to each other.[243]

"We Don't Teach It, We Just Use It."

Early in her career, Sue Savage-Rumbaugh realized that "[k]nowing how to use the symbol 'banana' as a way of getting someone to give you a banana is not equivalent to knowing that 'banana' represents banana."[244] To learn that the word "banana" means the fruit requires that the word be separated from the fruit.[245] Comprehension meant being able to "respond appropriately when someone else indicated that *she* wanted to be tickled or that *he* wanted a banana."[246] We have seen that Sherman and Austin took giant chimpanzee steps, if small human steps, toward language through the use of lexigrams. But their progress encouraged Savage-Rumbaugh in her belief that comprehension held the key to language's door.[247] She had to admit that their comprehension was severely limited. True, they often appeared to understand spoken English. Many visitors thought they could. But Savage-Rumbaugh is neither one of Barnum's suckers nor one of Dennett's gullible souls but a tough and savvy investigator. She realized that like many apes, Sherman and Austin were masters at discerning what others wanted by analyzing their nonverbal cues. When Savage-Rumbaugh rigorously eliminated these nonverbal cues, they understood as much English as your average Outer Mongolian.[248]

She began teaching the lexigram system that Sherman and Austin had mastered to Kanzi's stepmother, Matata, a ten-year-old. Kanzi just hung out with his mother and played around her during her intensive lexigram training sessions. At the age of two and a half, he gave a hint of what was to come when he spontaneously begun to use Matata's lexigrams.[249] But Matata herself was a painfully slow student. In two years and 30,000 trials, she learned just six symbols—and those not very well. Eventually, Savage-Rumbaugh placed Matata on maternity leave and separated her from Kanzi.

The day his mother left, Kanzi approached the computer keyboard before which his mother had so diligently sat for two years. He used it once; he used it twice; he used it 120 times that first day. He seemed to know what the lexigrams meant, and he could use them both to request and to state his intentions.[250] This was further evidence that language was first a matter of comprehension and not production. Savage-Rumbaugh junked her plans to teach Kanzi anything. As Roger Fouts and the Gardners had done with Washoe and other chimpanzees, she set out to create as rich a social and linguistic bonobo environment as she could in which Kanzi could learn language as naturally as a human child does.[251] Speaking of language, Savage-Rumbaugh said, "We don't teach it, we just use it."[252] Recent stories show that the Language Research Center bonobos learn a word for a new object after listening to dialogues between their caretakers in which the new word is pronounced between 6 and 60 times. Half this time the object is present when its name is mentioned and half the time it is absent. It makes no difference in the bonobos' learning of the new word.[253]

Kanzi, not Savage-Rumbaugh, decided what words he would learn and in what order. Every available day, they roamed the surrounding fifty-five acres of forest, using the natural world, any toys they brought, and the games and foods that characterized each forest location to catalyze the daily language agenda. Savage-Rumbaugh developed a portable board upon which photographs of the lexigrams were affixed so that everyone could chat as they moved from one of the seventeen locations to another. Eventually these lexigrams totaled 256. Thinking back on this decision a decade later, Savage-Rumbaugh wrote, "It now seems odd that anyone had ever decided which words an ape was to learn—certainly no one does that with children, who elect to learn very different words, apparently focusing on the aspects of language that fascinate them."[254]

Because they believed that "[c]hildren do not learn language by talking; they learn by listening," Savage-Rumbaugh and her colleagues spoke English to Kanzi as often and as much as they could.[255] He began to show signs that he understood them at the age of two.[256] By the end of seventeen months, he understood at least 150 English words produced by a range of human speakers, even synthesized speakers, and Savage-Rumbaugh was writing that "[e]ven if the ape was unable to speak, an ability to comprehend language would be the cognitive equivalent of having acquired language."[257]

In 1972, Roger Fouts had demonstrated that Ally, a three-year-old chimpanzee, could apply a rule of grammar and comprehend novel ASL sentences like *"Put ball in purse"* or *"Give Bill toothbrush."*[258] Testing Ai's comprehension of individual spoken Japanese words, Tetsuro Matsuzawa realized that she could comprehend many words, and some gestural signs, too.[259] How much language did Kanzi comprehend, and how did it compare to the comprehension abilities of a human child?

For nine months in 1988 and 1989, Savage-Rumbaugh and her colleagues compared the English comprehension abilities of Kanzi, then almost eight years old, with a two-year-old human girl, Alia. So that there could be no possible cueing, the person asking the question was out of Kanzi's and Alia's sight. Sentences were spoken at a normal rate. Kanzi and Alia were asked to take 660 different actions such as to "pour the milk on the cereal" and "tickle Rose with the bunny." But as I mentioned, they were also asked to perform more unusual tasks, such as to "put the toothbrush in the lemonade," "hide the toy gorilla," "put on the monster mask and scare Linda," "see if you can make the doggie bite the ball," "feed your ball some tomato," "get the ball that's in the group room," and "get the noodles that are in the bedroom." Kanzi was generally able to respond correctly when the same word was used both as a noun and a verb. For example, with a play syringe before him, he was able to differentiate between the requests to "give the shot to Liz" and "give Liz a shot" and the requests to "knife the orange" and "put the knife in the hat."[260] And he knew that a request to "give the pillow to Sue" and "give Sue the pillow" meant the same thing.[261]

Sometimes Kanzi responded in an unexpected, but correct, way that showed how hard he struggled to *make sense of the mind of the speaker* and understand the often strange things he was asked to do. When asked to "Put some water on the carrot," he threw a carrot outside—into the rain. When asked to "Put some water on the vacuum cleaner," Kanzi drank from a glass of water, walked to the vacuum cleaner, and dribbled the water onto it. When requested to "Wash the hot dogs," neither Kanzi nor Alia responded as expected. Savage-Rumbaugh thought they would bring the hot dogs to the sink then run water over them. Instead, Kanzi turned a hose on them, while Alia wiped them. One of the most extraordinary examples of this "sense-making" occurred when Kanzi was asked to "feed your ball some tomato." But he had no "ball" before him. The closest thing was a soft sponge Halloween pumpkin with a jack-o'-lantern face. Kanzi picked it up, turned the face toward him, and pretended to feed the mouth some tomato.[262]

Given the large number of objects, places, and people arrayed before them, the chance that either Alia or Kanzi would randomly perform the requested action was exceptionally low. Yet Kanzi ended up responding correctly to 74 percent of the questions, Alia to 66 percent, though often both responded partially correctly or mixed up words that sounded similar, such as potato and tomato. On the more complicated embedded sentences, Kanzi responded correctly 77 percent of the time, Alia 52 percent.[263] Kanzi had the most trouble with sentences in which he was asked to produce two objects, such as to "show me the doggie and the milk." He would offer one, but not the other, an error that Savage-Rumbaugh chalked up to a problem in short-term memory. She relates what happened when Kanzi clearly misunderstood one sentence in which she asked him to "put the paint in the potty." Kanzi

promptly picked up some clay (a similar play object to paint), and put it in the potty. I said "What about the paint?" Kanzi put more clay in the potty. I said "thank you," but "now put the paint in the *potty.*" Kanzi clearly thought me a little dumb, and so brought me the potty and placed it right in front of my face so I could see that he had done what I was so persistent in asking for.[264]

In psychologist Peter Mitchell's opinion, "Kanzi's ability to comprehend novel and even bizarre sentences is nothing short of remarkable."[265]

If Alia understood syntax, then Kanzi did, too. But could he produce it? In a five-month period during 1986, Savage-Rumbaugh and her colleagues recorded almost 14,000 of Kanzi's keyboard utterances and analyzed them to see how they compared with criteria for syntax. Because his word combinations tended to follow rules of English word order, they concluded that he was using a primitive syntax or grammar, in other words, a "protogrammar."[266] He had begun to produce the words and short sentences characteristic of a human eighteen-month-old. But unlike Nim, more than 90 percent of his multiword productions were spontaneous and imitated nothing.[267]

Matsuzawa has also concluded that Ai uses a protogrammar. In a less complicated example, Ai tended to name herself first when presented with a picture of herself and another chimpanzee. When shown a picture of a chimpanzee and a human, she tended to name the chimpanzee first.[268] In a more complicated example, Ai learned the symbols for number, color, and object. When presented with five objects (pencil, paper, brick, spoon, and toothbrush) in five colors (red, yellow, green, blue, and black), in arrays of one, two, three, four, or five, she could combine these symbols to identify the number and color of each object. While naming them, she spontaneously developed a "word order" in which she almost exclusively favored two of the six possible alternatives: color-object-number and object-color-number.

Matsuzawa applied the same four-level analysis to symbol use that he developed to describe tool use in wild chimpanzees. A conditioned response was assigned Level 0. The first level at which a chimpanzee or bonobo understands that an ASL sign or a lexigram symbolizes a word or an Arabic numeral symbolizes a number is Level 1. At Level 2, she can grasp the relationship between two symbols. Level 3 reveals her understanding of the relationship among three symbols. This is the level of Ai's understanding of the relationship among the symbols for number, color, and object.[269]

When Mulika, Kanzi's half-sister, arrived at the Language Research Center, she was immediately immersed in a language-rich and extremely social environment. She responded by spontaneously using her first lexigram more than a year and a half before Kanzi had. The same thing occurred with another of Kanzi's half-sisters, Panbanisha. She also used symbols at the age of one.[270] The process was repeated with Panzee, a chimpanzee. Although she did not produce her first spontaneous lexigram until the age of eighteen months, still a year before Kanzi, Panzee almost caught up to the bonobos.[271] All learned the words they wanted to learn, which is why there is scarcely any overlap among the first ten words that each of the bonobo siblings learned.[272] In a test of comprehension similar to that given Kanzi and Alia, Panbanisha, at the age of three, responded appropriately to 92 percent of 483 sentence requests.[273]

There seems little doubt that Kanzi, Panzee, Mulika, and Panbanisha are intentionally *producing* language as well as comprehending it. It's not Shakespeare; heck, it's barely Dr. Seuss. But it's real language. Savage-Rumbaugh wearily demands that critics of ape language production play fair and use the same basic assumptions in analyzing how apes might produce language as they do when analyzing how children might acquire it. Just as human parents do when rearing their children, scientists routinely assume that children are intending to convey some information to a listener when they are studying child communication, that their intentions can be encoded in complicated and differing ways that often depend upon context, and that the information the listener receives may not cause him to do anything with it immediately, or ever.[274] There is a reason they make these assumptions: They can't prove them. But if they don't make these assumptions, they can't study the language of children! In a manner reminiscent of the two ways in which Skinner and I might have described a baseball game, Savage-Rumbaugh shows us what a report on child language might look like if a scientist did *not* make these assumptions.

Children produce a wide array of graded grunt-like sounds, as well as graded clicks and click-vowel combinations that appear to be somewhat more variable than those of other nonhuman primates and . . . these vocalizations are under considerable respiratory control. Little, if any, meaning could be assigned to the sound, since each sound . . . occur[s] in a setting that [i]s unique to that sound. The most likely interpretation [is] that the sounds reflect the emotional state of the child.[275]

Requiring Savage-Rumbaugh to *prove* that Kanzi, Mulika, Panbanisha, or Panzee are intentionally conveying information is the same thing as demanding that she prove that they're conscious or that *she's* conscious, and just about as useful. People who still want you to believe that these apes are not intentionally communicating must think that they are the only conscious creatures in the universe. And if you believe that, I have a bridge I'd like to offer you.

Savage-Rumbaugh has shown that the Language Research Center apes can easily comprehend even bizarre and embedded sentences based solely upon what they learned by observation and can produce simple sentences, too. "Observation" is the key word. No training is involved. Instead, they engage in "sense-making."[276] Savage-Rumbaugh emphasizes the large difference between associative learning by training and learning by observation. If I want to train my dog, Marbury, to sit, I do not call him into the family room and say, "See here, Marbury, I would like you to sit. Here is how it is done," then sit, then sit again, then sit again until he gets it.[277] Long after the couch legs buckled, Marbury would be standing. Instead, Marbury learned to sit after I said "sit," and pushed his butt to the floor, then repeated it over and over again—and pushed his butt down and down again—until he understood that when I said "Sit!" he should sit. How well he learned this was driven home the other week when Christopher climbed onto the family room couch and unsteadily swayed to his feet. Thinking he was about to topple to the hardwood floor, I rushed toward him yelling "Sit!" Christopher just laughed and started to dive off the couch. But in the kitchen, Marbury promptly sat.

Journalism professor Deborah Blum, who won a Pulitzer Prize for writing about primate research and researchers, has written that the work at the Language Research Center is "considered by some to be revolutionary."[278] Today Kanzi, Panbanisha, Mulika, and Panzee, all graduates of Savage-Rumbaugh's School of Language Comprehension (a new student, Nyota, was born of Pan-

banisha in 1998), all comprehend more than 500 words and pro-
nounce more than 150 words[279], and know that word order is vital
to sentence meaning: "Feed your ball some tomato" is not the
same thing as "Feed your tomato some ball." They understand
past and future verb tenses. They understand pronouns of possession such as "mine" and "yours." (Remember Ai and her green
feeding bowl.) They understand expressions relating to time, such
as "now" or "later." They understand qualifications of state such
as "hot" and "cold." They understand that one clause within a sentence can modify another portion of the same sentence, for example, "Get the ball that is outdoors, not the one that is here."[280] They
can eavesdrop on conversations in which they are not involved,
and understand them. Savage-Rumbaugh reports that they "easily
participate in three- and four-way conversations that deal with the
intentions, actions, and knowledge states" of others and understand narrative dialogue.[281]

There can be no doubt that if these apes were children, no one
would dare claim that they didn't comprehend language. And
since Wise's Rule Three for assessing the minds of nonhuman animals requires us to play fair and ignore special pleading, different
rules should not apply to human and nonhuman animals. So at the
least, these enculturated chimpanzees and bonobos comprehend
human language, if children do.[282] Nor should their abilities to produce language be unfairly diminished. Savage-Rumbaugh argues
that comparisons of ape and child utterances typically overlook
the great disadvantage that an ape must encounter because its input modality (spoken English) differs from its output modality
(printed symbols). It is not easy to locate symbols, even when they
are well known, on a board of hundreds of symbols. It is bother-
some to have to find a keyboard each time one wishes to communicate. And it is time consuming and laborious to produce
elaborate, highly structured sentences, especially if a word or two
plus a gesture and a vocalization, in context, gets the message
across. The enormity of this accomplishment is often overlooked
because the difficulty of the task is not widely understood. It is the
equivalent of learning to read, in order to be able to talk.[283]

But Savage-Rumbaugh goes further and hypothesizes that the
fact that several apes—bonobos and chimpanzees—have shown
that they can "make sense" of language without any training
strongly suggests that enculturated wild apes, meaning those

treated by other apes as intentional communicators, have the same capacities.[284]

Like humans over the age of four or five years old, Kanzi, Mulika, and Panbanisha appear to have acquired both halves of a "belief/desire psychology." They use a folk psychology to "constantly impute states of minds to others and recognize the value of communication in altering the perceptions of others."[285] In one formal theory of mind test, a researcher, Liz, asks Savage-Rumbaugh for some M&Ms in Panbanisha's presence. Savage-Rumbaugh stashes them in a box. Liz leaves, saying she will soon return. With Liz gone, Savage-Rumbaugh takes out the M&Ms and inserts a bug. Liz returns and Savage-Rumbaugh hands Liz the box. While Liz is pretending to struggle to open the box, Savage-Rumbaugh asks Panbanisha what Liz is looking for. Panbanisha hits the lexigram for "candy," and not "bug" or any other object, showing that she could differentiate between what she knew (that the bug was in the box) and what she thought Liz knew (that candy was in the box, even when she knew that what Liz thought was false).[286]

LOOKING BACKWARD

I have mentioned just a small faction of the large and rapidly growing number of research articles and books that swarm the libraries, detailing the astonishing cognitive capacities of chimpanzees and bonobos. But almost 150 years after Darwin reversed Aristotle, a tiny band of scientists stubbornly fights a rear-guard action aimed at turning the clock back to ancient Greece by persuading you that even chimpanzees and bonobos are just the "automatic puppets" that Aristotle said they were, the unreasoning, undesiring, perhaps even unconscious creatures the Stoics described.

The most prominent example today is probably Daniel J. Povinelli, whose experiments with the New Iberia Seven I discussed in Chapter 9. Povinelli claims that chimpanzees have no idea of cause and effect, cannot point to show, and don't understand such basic building blocks of theory of mind as seeing and joint attention.[287] Recently, he speculated that chimpanzees may not even be conscious, which is an extraordinary stance for any ape researcher.[288] Remember that Povinelli based his conclusions primarily on a series of experiments he began in 1992 in which the New Iberia Seven were just as likely to gesture for food to a human who could not see them, as she was blindfolded, had her eyes closed, had a bucket on her head, or was turned away, as they were to gesture to a human who could see. From this he concluded that chimpanzees do not understand that eyes connect internal states of attention to the world.

But that same year, a team led by Juan Carlos Gómez placed humans in front of six chimpanzees, three of whom were hand-reared and three of whom had been raised in a zoo. Sometimes these humans were visually attentive and responsive to the apes, sometimes they were visually attentive to them but not responsive, and sometimes they paid no visual attention to them at all, either because their eyes were closed, they were looking to one side or over the chimpanzees' heads, or they were sitting with their back to them. The critical test was whether the chimpanzees would try to get a human's attention through touching and vocalizing when she was not paying attention to them but not try to get her attention when she was paying attention to them. The three hand-reared chimpanzees passed this test. They actively searched out eye contact when a human would fail to respond to a request, and then the chimpanzees repeated the request. If the human was looking away, they would engage in some attention-getting behavior such as touching and repeat the request only after they were able to establish eye contact with her. The three zoo-raised chimpanzees failed.[289]

Gómez concluded that Povinelli's data "seems to be the result of faulty experimental procedure, perhaps combined with [the chimpanzees'] lack of appropriate experience with humans."[290] Aside from the chimpanzee's lack of human experience, he saw two, and perhaps, three flaws in Povinelli's methodology. The first was that Povinelli assumed that the New Iberia Seven would request food only from humans who could see them and would not request food from humans who could not see them because their eyes were covered or they were looking away or were turned away. This, in turn, assumed that the chimpanzees understood that their gesture of reaching toward a human through one of the two holes that Povinelli cut through a Plexiglas window was intended to *communicate* that they wanted food. The problem was that the humans who were supposed to be visually attentive to the chimpanzees did not actually *look* at them. Instead, they stared at an imaginary point midway between the two holes or at one of the two holes in the plexiglass, but never at the chimpanzees. The chimpanzees were actually choosing between two humans, *neither* of whom was actually looking at them!

There was a second methodology flaw. Povinelli systematically trained the seven chimpanzees to gesture toward a researcher based not on whether she was looking at them but upon her physical stance in relation to the holes cut into the plexiglass. Unfortunately, this may also have trained them to think that the human's visual attention was irrelevant. Gómez thinks that training chimpanzees at all to make the critical communicative response—here, begging for food through the holes in the plexiglass—is worse than unnecessary. It is outright counterproductive, because the chimpanzees may actually learn associations that the experimenter never foresaw or intended.[291]

Gómez detected a third possible flaw because of his experience with his own attention experiments with chimpanzees. He noticed that the six chimpanzees with whom he worked only made eye contact with the human half of the time before they gestured. Apparently Povinelli did not ascertain whether the chimpanzees were actually checking the human's eyes before they gestured toward them.[292]

Sue Savage-Rumbaugh has spotted a fourth and a fifth flaw in Povinelli's methods. The more important was that he never asked the New Iberia Seven to *communicate* anything. Instead, they responded to the discrimination learning problem that Povinelli had unwittingly trained them to solve—"Do I stick my hand through the right or the left hole?" Sometimes the chimpanzees received food when they stuck their hands through the left hole and sometimes when they stuck their hands through the right hole. Because they got a food reward based on which hole they stuck their hand through, they probably didn't treat the problem as one in which they were to gesture to a human but rather as one in which they were to choose a hole. Whether the humans were looking at them or not was irrelevant to them "and thus one would expect apes to behave just as Povinelli reports that they do."[293] She completely agrees with Gómez that chimpanzees should not need any training at all for the critical communicative response, which was begging in the case of the New Iberia Seven: "[I]f they can tell another person cannot see, they should demonstrate this at once—wouldn't you?"[294]

Savage-Rumbaugh has begun a series of as-yet-unreported experiments based upon Povinelli's work. She reports that in preliminary tests, by reconfiguring the experiment so that the chimpanzees have one hole or no holes to reach through, setting the humans farther apart, and taking their glances and facial orientation into account, two unenculturated chimpanzees at the Language Research Center easily passed the tests that the New Iberia Seven failed, just as the three hand-raised chimpanzees in Gomez's experiment passed.[295]

Another team led by graduate students Dan Shillito and Rob Shumaker and also including the highly respected Gordon Gallup and Ben Beck ran a series of trials testing the visual attention of Indah, an orangutan at the National Zoo in Washington, D.C., in 1995.[296] According to Shillito, although Indah was not enculturated, she had experienced more human contact than the average zoo orangutan.

One day, Shillito noticed that Indah could see food near her cage that Rob Shumaker, who was standing in front of her cage, could not see. To Shillito's surprise, Indah reached through the bars and maneuvered Shumaker into a new visual position: Now he faced the food and could give it to Indah. Intrigued, Shillito and Shumaker constructed a series of experiments to determine whether Indah understood that eyes connect internal states of attention to the world. When confronted with one human whose eyes were covered and another who could see, would Indah manually move the sighted human, but not the blinded one, toward the food?

Indah easily understood that humans with buckets over their heads could not see to help her obtain the food. Nearly every time, she reached out and manually moved the sighted researcher into a position to see the food. She would even move the sighted person *around* the researcher standing with a bucket over his head. She often removed the bucket from a person's head and then moved him into a position to see the food, especially if he was closer to the food than a sighted person was. When the researchers sawed off the bottom of a bucket so that when put over the head, it did not block the eyes, Indah treated the researcher as if he were sighted. In short, Indah, a member of a species that split from the common ancestor of humans, chimpanzees, and bonobos around 14 million years ago (the ancestor of chimpanzees and bonobos split from the human line about 5 or 6 million years ago) easily demonstrated at least the Level 1 visual perception that children develop around the age of three.

Harvard psychology and anthropology professor Marc Hauser has pointed out a sixth flaw in Povinelli's methodology. In the training phase of the experiment, Povinelli taught the chimpanzees to beg from one trainer by reinforcing them 100 percent of the time for begging for food. Hauser explains that scientists who do this wipe out any ability that the chimpanzees might have to search for the key discriminative cue—their visual attention—when two trainers are present.[297] Strikingly, soon-to-be-published data from Hauser strongly suggests that cotton-top tamarins, New World monkeys whose common ancestor split from the human-ape lineage approximately 35 million years ago, have passed a nonverbal version of the Sally-Anne theory of mind test.[298]

Povinelli has conceded that his results are controversial and "not the final word of whether chimpanzees have a theory of mind. Indeed, it is not even the first sentence of the first chapter of the future volume, *Principles of Chimpanzees' Theory of Mind*."[299] He also admits that the ways that chimpanzees and humans connect behaviors and mental states may not be the same.[300]

These are significant concessions. Both the "theory theorists" of mind (who look to an abstract rule to understand the minds of others) and the "simulation theorists" of mind (who think we try on others' "mental shoes") accept that socialization plays an important part either in developing or triggering our ability to understand the minds of others.[301] Professor Angeline Lillard has pointed out that "[t]hroughout much of the literature on children's theories of mind, and a parallel literature in philosophy concerning the process by which we read minds, runs an assumption that 'our' theory is universal."[302] But as I mentioned in Chapter 8, Western researchers don't yet appreciate how even humans in many non-Western cultures understand the minds of other humans, much less how captive apes do. The Quechua children of Peru and the Tainae children of Papua New Guinea, for example, appear to Western psychologists not to have developed a theory of mind when they are on the cusp of adolescence.[303] But Western psychologists do not therefore assume they lack one. Lillard believes it more likely that these cultures simply emphasize and discuss minds to a smaller degree than Westerners do. Unable to build on the knowledge of others about mental states that active teaching and even discussion would impart, she speculates that these cultures may produce humans who must rediscover the mental states of others for themselves and therefore have "much simpler and more individualized understandings of those states."[304] Along similar lines, child psychology professor Janet Astington suggests that "[p]erhaps Western theory-of-mind tasks, or even testing itself, are not an appropriate way to assess their understanding . . . All people have some way of interpreting social action, but it may not be *our* way."[305] There may not even be a "Western" theory of mind. Lillard and her colleagues recently discovered that dramatic differences in folk psychology exist even between urban and rural American children, and she believes that differences might well turn on regional, class, or other characteristics.[306]

As children do, chimpanzees and bonobos probably require socialization in order to learn to "mind read" efficiently, and the more they are socialized by Westerners the more their mind reading may resemble Western theory of mind.[307] Gómez believes that a theory of mind may develop not just in response to active human teaching but in response to the social behaviors of the members of one's own species, which scientists call "conspecifics." If so, Gómez thinks that the theories of mind they develop "must be theories of conspecifics' minds."[308] A natural chimpanzee or bonobo theory of mind might be more simple than a natural human theory of mind for two reasons. First, the chimpanzee or bonobo minds that they normally "read" may not be as complicated as human minds; certainly, they are somewhat different. Second, because chimpanzees and bonobos do not discuss mental states, each ape must rediscover them for herself. As with humans who live in cultures in which mental states are not discussed, this could lead to much simpler and more individualized understandings of them.

Chimpanzees may thus develop a theory of the *chimpanzee* mind, bonobos a theory of the *bonobo* mind, human children a theory of the *human* mind. Urban Western children might even develop a theory of the urban *Western human* mind, Quechua children of the *Quechua* mind. Human enculturation of apes may then allow apes to develop a theory of the human (or Western or perhaps even middle-class Western academic) mind that they would not ordinarily develop. But without enculturation or at least substantial experience with humans, an ability to develop a theory of the *human* mind may be as difficult for chimpanzees and bonobos to achieve as a theory of the ape mind apparently is for humans to achieve.[309]

GETTING READY FOR JUSTICE

Critics of animal cognition offer occasional bromides intended to communicate just how wonderful animals are. For instance, the journalist Stephen Budiansky says that "an honest view of animal minds ought to lead us to a more profound respect for animals as unique beings in nature, worthy in their own right" and that "[t]he intelligence that every species displays is wonderful enough in itself; it is folly and anthropomorphism of the worst kind to insist that to be truly wonderful it must be the same as ours."[310] Steven Pinker asks,

> Why should language be such a big deal? It has allowed humans to spread out over the planet and wreak large changes, but is that any more extraordinary than coral that build islands, earthworms that shape the landscape by building soil, or the photosynthesizing bacteria that first released corrosive oxygen into the atmosphere, an ecological catastrophe of its time?[311]

Writing about issues of "human uniqueness, animal welfare, and the dignity—and even rights—of other species," Daniel Povinelli and his colleagues have said that "separation is in the interest of all parties" and claim that "their research will actually uncover the unique qualities of chimps."[312] To what end? To gain "insight about the type of animals that are [an] appropriate model for specific research."[313]

But cognition is a very big deal because the fundamental legal rights of animals, the least porous barrier against oppression and abuse that humans have ever devised, depend on it. We are well within what *National Geographic* recently called the "Sixth Extinction" of life on earth. The first five were probably caused by climate changes or comets smacking into the planet. But the Sixth Extinction is being caused by what one scientist called "the exterminator species."[314] Us. My grandchildren may never see a chimpanzee or bonobo, for they are severely endangered. They are shot, eaten, kidnapped, jailed, vivisected, and deprived of habitat—not because we "respect" them or see them as "unique" or because we believe that they are "truly wonderful" but because we don't. No one but a professor or a deep ecologist thinks that a language-using animal is not a bigger deal than island-building coral, soil-building earthworms, or photosynthesizing-bacteria.

Whatever legal rights these apes may be entitled to spring from the complexities of their minds. Today they are legal Harveys, invisible to law, without personhood, lacking rights. We know what happens when humans are stripped of their legal personhood and dignity-rights. Australian aborigines, African slaves, Turkish Armenians, German Jews, Rwandan Tutsis and Burundian Hutus, and Kosovar Albanians fall victim to genocide. It can be no surprise that not only bonobos and chimpanzees but also thousands of other species of animals have been pushed into extinction or teeter on its brink. We do what we do to a Jerom because he can't stop us. None of them can. We can only stop ourselves. Or some of us can try to stop the others. The entitlement of chimpanzees and bonobos to fundamental legal rights will mark a huge step toward stopping our unfettered abuse of them, just as human rights marked a milepost in stopping our abuse of each other.

We have learned the sorry history of law toward nonhuman animals. We realize what legal rights are, what they do, and what entitles someone to them. We understand that the common law provides a flexible framework upon which principle and policy can operate. We have deciphered the nature of the minds of chimpanzees and bonobos as well as we could. Now we're ready to bend toward justice.

11

Bending Toward Justice

[T]he moral arc of the universe is long, but it bends towards justice.
—Henry Mayer, quoting Martin Luther King Jr., 1998[1]

What if Sue Savage-Rumbaugh called me one day to say that five of the Language Research Center's apes—Kanzi, Panbanisha, Panbanisha's son, Nyota, Sherman, and the mother-reared bonobo, Tamuli—had been shipped to a biomedical research laboratory, there to be subjected to invasive brain studies that will kill them? Savage-Rumbaugh doesn't own them, the way that Roger Fouts owns Washoe and Loulis. Terrible things have happened even to language-using chimpanzees. Remember Lucy, the ASL-using sexually aware tea hostess? The Temerlins could no longer care for her and, at the age of twelve, she was flown to a chimpanzee rehabilitation center in Senegal. After a rocky adjustment, she was moved to Gambia. There poachers shot and skinned her. They hacked off her feet and hands to be sold as trophies.[2] Language-using chimpanzees have never even been exempt from lethal biomedical research and testing. In 1982, Ally, who used ASL, was sold to the White Sands Research Center, which Roger Fouts described as "a private laboratory that tests drugs, cosmetics, and insecticides on animals."[3] Officials never acknowledged receiving him. But Fouts says he was told that Ally had died after being injected with insecticide.[4] That same year, Booee and Bruno, both ASL signers, were trucked to the Laboratory for Experimental Medicine and Surgery in Primates (LEMSIP) in New York. Six years later, Fouts dispatched a student there. When he signed to Booee, now infected with Hepatitis C, he received this response: "*Key out.*" Fouts could not find the key to Booee's cell for seven more years. In the meantime, Bruno died inside LEMSIP.[5]

In a 1776 letter, Abigail Adams pleaded with her husband, John, to ensure that the new Continental Congress not place "unlimited power into the hands of Husbands. Remember that all Men would be tyrants if they could."[6] Gender isn't the issue; power is. As Lord Acton recognized, "absolute power corrupts absolutely."[7] Our absolute power over chimpanzees and bonobos has corrupted us absolutely. But immunities blunt power and forestall tyranny. Without these barriers, liberty is destroyed. Chapter by chapter, I have constructed the individual building blocks of the legal personhood of chimpanzees and bonobos: the anachronism of their present legal thinghood, the meaning of legal rights, the critical importance of liberty and equality to justice, how the common law changes, and the natures of these apes. Now it is time to do what lawyers do: get to work assembling these blocks into the dignity-rights that will convince fair-minded judges that the time has come to end our tyranny over these apes.

Writing Horses and Talking Dogs

On Thursday, February 6, 1837, Abigail's son, John Quincy, rose in the House of Representatives. The former president asked the Speaker to rule on whether House consideration of a petition originating from twenty-two black slaves was forbidden by the gag rule enacted to forbid the submission of just such antislavery petitions. Cries for Adams's expulsion rang from the House floor. The patience many Southern congressmen had shown with Adams, who could produce an unending stream of abolitionist petitions, had expired. Waddy Thompson, Jr., a combustible South Carolina upcountryman, thundered on the House floor: "Slaves have no right to petition. They are property, not persons; they have no political rights, and even their civil rights must be claimed through their masters. Having no political rights, Congress has no power in relation to them, and therefore no right to receive their petitions." His fellow South Carolinian Henry Laurens Pinckney, ever so slightly more moderate and a soon-to-be ex-Congressman because of it, growled that "he would just as soon have supposed that the gentleman from Massachusetts would have offered a memorial from a cow or horse—for he might as well be the organ of one species of property as another." To this Adams would reply on the Tuesday next: "Sir . . . if a horse or a dog had the power of speech and of writing, and he should send [me] a petition, [I] would present it to the House."[8]

Lloyd Weinreb, a Harvard Law School professor, says there exists "a single uniform rule that the category of persons is coextensive with the class of human beings: All human beings are persons, and all persons are human beings."[9] Thompson, Pinckney, and a majority of the U.S. Supreme Court justices of their day would have thought this "Human Rule" too broad: Blacks were both human beings and legal things.[10] Adams implied that this "Human Rule" was too narrow if it excluded writing horses and talking dogs. Today, Professor L. H. Sumner, a philosopher at the University of Toronto, thinks it "quite inconceivable that the extension of any right should coincide exactly with the boundary of our species. It is thus quite inconceivable that we have any rights simply because we are human."[11]

Both the law professor and the philosopher are correct. Weinreb accurately describes what the law *is*. But Sumner tells us what the law *should be*, for the "Human Rule" *is* inconceivable, in the sense that it is irrational or incredible.[12] A species, after all, is just a population of genetically similar individuals naturally able to interbreed, and no one suggests that it is the human ability to interbreed that justifies our legal personhood.[13]

Rights for Neandertals and *Homo erectus*?

Although chimpanzees and bonobos have sophisticated minds, the differences among our three species may remain too large for many readily to appreciate our similarities. Perhaps we should try to begin by imagining species whose minds are even more similar to ours. But once we didn't have to imagine them. They lived just over the mountain, down the river, and across the valley.

The ape/human split from our common ancestor occurred about 5 or 6 million years ago. One branch led to chimpanzees and bonobos.[14] The other led to the hominids, those erect bipedal primates that include you and me and the other species of the genera *Homo*, *Australopithecus*, *Ardipithecus*, and *Paranthropus*. Some of them are our ancestors. Most were not. The general outline of the main branches are widely agreed upon, though new twigs are added and old twigs rearranged with each new major archaeological find. With one exception, those details need not concern us.

The earliest known hominid, who appears in the fossil record almost four and one-half million years ago, has been placed in both the genus *Australopithecus* and the genus *Ardipithecus*. A few hundred thousand years later, undisputed Australopithecines arrived, alongside two hominids who are sometimes classified under the genus *Paranthropus*. *Homo* emerged perhaps 2.5 million years ago with the trio of *Homo habilis*, *Homo ergaster*, and *Homo rudolfensis*. Then *Homo erectus* appeared about 1.8 million years ago, followed by Archaic *Homo sapiens* and *Homo heidelbergensis* maybe 1 million years later, followed by us, *Homo sapiens sapiens*, and the Neandertals. As the hominids evolved, so did brain size (expanding from 400–500 cc to 1200–1700 cc) and brain structure (including the expansion of the prefrontal cortex). More advanced cognitive abilities, such as the sophisticated use of a complex language and mathematics, resulted.

All the many hominid species have gone extinct, except for us. But for perhaps 100,000 years, we may have lived alongside both Neandertals and the older *Homo erectus*. *Homo erectus* walked upright, had a brain two-thirds to three-fourths the size of ours, probably possessed a degree of self-consciousness greater than modern chimpanzees and bonobos and a well-developed Broca's area and a Wernicke's area (brain areas often associated with speech), and crafted technically complex hand axes and other tools with exceedingly even, regular, and sharp edges that were often beautiful and symmetrical. They probably used fire and possessed a primitive, maybe even a complex, culture. They colonized several continents, piloted crafts over hundreds of miles of deep, fast-moving water, used primitive symbols, and almost certainly employed a range of sounds to communicate what may, or may not, have been too simple to be classified as language.[15] Neandertals were even larger-brained than we. They were, quite possibly, our neurological and mental equals. They probably possessed an even more-advanced degree of self-consciousness than *Homo erectus*, crafted even more sophisticated tools, organized group hunts, regularly used fire, possibly, or even probably, spoke a rudimentary language, and developed a complex culture that probably included taking care of their sick and weak, burying their dead, and creating primitive art.[16]

We need not become entangled in the controversy over the precise relationship between modern humans and the Neandertals other than to note it. Not long ago, our physical similarities caused many scientists to believe that Neandertals were merely a subspecies of *Homo sapiens* and that we had evolved from them. Indeed a 24,500-year-old boy with our face, but seemingly with a Neandertal's body and legs, has recently been unearthed in Portugal.[17] But in the mid-1990s, molecular and other evidence began to strongly suggest that they were a species apart, *Homo neandertalis*, from whom we did not evolve.[18] The situation is fluid and not likely to be soon resolved. Because I think that the evidence today weighs more heavily toward Neandertals as a species apart, when I refer to "humans," I will just mean us—*Homo sapiens sapiens*.

Imagine that a hardy band of Neandertals tomorrow descends from the mountains of Spanish Andalusia or a tribe of *Homo erectus* emerges from the mists of Java, where they have lived isolated for three thousand generations. May we without hesitation capture and exhibit them, breed and eat them, and force them into biomedical research? If their minds make us hesitate and in that moment, if we open our minds to the possibility that they might be eligible for dignity-rights, shouldn't the minds of chimpanzees and bonobos make us hesitate as well?

Autonomy, Dignity, and "Human Rights"

In the *New York Times Magazine*, Nobel Prize–winning Nigerian playwright, Wole Soyinka recently wrote that one of the best ideas of the last thousand years was that "certain fundamental rights are inherent in all humans."[19] In Soyinka's view, such "human rights" as bodily integrity and bodily liberty, which we have called "fundamental liberty rights," spring from "intuition."[20] But "human rights" are bottomed on firmer stuff. Much of the world, certainly the Judeo-Christian West, links them to a dignity produced by autonomy.[21] This is why I call these fundamental immunities "dignity-rights." Modern courts simply assume that "a right of personal autonomy cannot exist independent of a recognition of human dignity" and that "Anglo-American law starts with the premise of thorough-going self-determination."[22]

Having dignity-rights without autonomy is a little like being a bird without feathers or a Buddhist pope. But a succession of post–World War II treaties, constitutions, and judicial decisions have created just such a category of rights-holders. "Can it be doubted," many American high courts rhetorically ask about competent and the most incompetent human beings, "that the value of human dignity extends to both?"[23] The U.S. Supreme Court, which often refers to the "dignity" or "worth" of all humans, has recognized, for example, the inextinguishable fundamental right to personal security and bodily liberty of Duane Youngberg, a thirty-three-year-old man with an IQ below ten, a tendency to violence, and the mental capacity of an eighteen-month-old child.[24] Although permanently and involuntarily committed to an institution for the retarded, Youngberg was entitled to be kept in conditions of reasonable care and safety and in a reasonably nonrestrictive confinement.[25] One must understand that the courts are not merely erecting barriers of fundamental immunities, such as bodily liberty and bodily integrity, about our bodies and personalities. They are giving humans who lack consciousness, sometimes even brains, the power to choose to exercise rights that they don't even know they have, with judges sometimes confusing immunities and powers by saying that to deny the exercise of a right is to deny the right.[26] Thus, the Michigan Supreme Court says that "[i]t would violate the concept of human dignity to measure the value of a person's life by that person's mental or physical condition,"[27] and the Delaware Supreme Court says that even an incompetent human must receive "the benefit of that very precious and protected right to choose and [is] accord[ed] . . . the dignity as a human being of making [a] choice."[28]

In one 1992 Massachusetts case, the court-appointed guardian for Beth, a ten-month-old in a permanent vegetative state, objected to a "do-not-resuscitate" order, arguing that she "had no dignity interest in being free from bodily invasions," as she "has *no* cognitive ability." The high court retorted that "[c]ognitive ability is not a prerequisite for enjoying basic liberties." Bereft of all cognition, even consciousness and sentience, Beth was nevertheless "entitled to the same respect, *dignity*, and *freedom of choice* as competent people."[29] The court affirmed the "do-not-resuscitate" order as *"the child* would refuse resuscitative measures in the event of a cardiac or respiratory arrest."[30] "Any other view," said the New Jersey Supreme Court, discussing the right of Claire Conroy, an eighty-four-year-old woman suffering from severe and irreversible mental and physical problems that were on the brink of killing her, to remove the nasogastric tube that supplied her food, "would permit obliteration of an incompetent's panoply of rights merely because the patient could no longer sense the violation of those rights."[31]

Fifteen years before Beth's case, the Massachusetts high court had pondered the fate of sixty-seven-year-old Joseph Saikewicz, who had an IQ of ten and a mental age of thirty-two months and suffered from an acute myeblastic monocytic leukemia that was rapidly killing him. Chemotherapy might kill him, too; it was certain to inflict painful side effects. Obviously, he could not decide treatment for himself. The high court said that respect for his autonomy required that those making the treatment decision place themselves in his shoes. The correct decision would then be "that which would be made by the incompetent person, if that person were competent, but taking into account the present and future incompetency of the individual as one of the factors which would necessarily enter into the decision-making process of the competent person."[32] (No matter how many times you read that sentence, it does not get any clearer.) Massachusetts judges are hardly the only Alices in this legal Wonderland.[33] The New Jersey Supreme Court ruled that Ann Grady, a severely retarded woman, "has the same constitutional right of privacy as anyone else to choose whether or not to undergo sterilization. Unfortunately, she lacks the ability to make that choice for herself. We do not pretend that the choice of [others] is her own choice. But it is a *genuine choice* nevertheless."[34] "Allowing *someone* to choose," the California Court of Appeals has said, "is more respectful of an incompetent person than simply declaring that such a person has no more rights."[35]

It is not just nonsense to permit one person to waive the dignity-rights of another who lacks the autonomy needed to create them, but dangerous nonsense. Simply assigning "human rights" to every human may accord with Soyinka's intuition, the intuitions of many modern judges and legislators, and even mine. Many came to regret that their "human rights" had been entrusted to the intuitions of Hitler, Stalin, Pol Pot, Mao Tse-tung, Idi Amin, and Slobodan Milosevic. Better we tether "human rights" to a dignity tied to some more objective property, such as autonomy. Otherwise, dignity and "human rights" inevitably degenerate into articles of faith, fashions that move in and out of favor, and values that compete with genocide in the political marketplace.

A large portion—though not all—of the paradox of awarding "human rights" to humans who lack autonomy can be resolved by examining what courts mean by autonomy. Philosophers often understand it to mean what the German philosopher Immanuel Kant intended it to mean two hundred years ago. We'll call Kant's notion of autonomy "full autonomy." Although whole books have been written about what Kant meant, I will try to catch much of its core meaning in a single sentence. I have autonomy if, in determining what I ought to do in any situation, I have the ability to understand what others can and ought to do, I can rationally analyze whether it would be right for me to act in some way or another, keeping in mind that I should act only as I would want others to act and as they can act, and then I can do what I have decided is right. My ability to perform something like this calculus is what makes me autonomous, gives me dignity, and requires that I be treated as a person. But if I can't do this, I lack autonomy and dignity and can justly be treated as a thing.

Whether I have summarized Kant's ideas perfectly or not is irrelevant. What is important is that anything that resembles this analysis demands an ability to reason at an inhumanly high level. Perhaps our Aristotles, Kants, Freuds, and Einsteins achieve it some of the time. But it's only a glimmering possibility for infants and children, most normal adults never reach it, and the severely mentally limited and the permanently vegetative don't even start. How did Kant deal with them? Well, he didn't, and his "deep silence" on the moral status of children and nonrational adults has not gone unnoticed.[36] Even Aristotle and Company pass significant portions of their lives on "automatic pilot" or often act out of desire, and not reason, which is precisely how Kant thought nonhuman animals act.[37] Were judges to demand full autonomy as a prerequisite for dignity, they would exclude most of us, themselves included, from eligibility for dignity-rights.

Moral and legal philosophers often emphasize the importance of less-complex forms of autonomy. Some argue that a being is autonomous if she has preferences and the ability to act to satisfy them; or if she can cope with changed circumstances; or if she has desires and beliefs and can make at least some sound and appropriate inferences from them; or if she can make choices, even if she can't evaluate their merits.[38] These "realistic" autonomies more accurately depict what humans actually have and what judges actually try to protect. While the most complex self-consciousnesses of the most rational adult humans lie at the high end of a full autonomy, the more complex of the simpler consciousnesses, those of young children or the adults of many species of mammals, probably approximate the lower end of a realistic autonomy.

Because dignity, like pregnancy, is all or nothing, even a flickering autonomy produces the same dignity that full autonomy does. Judges then respect almost any choice, regardless of whether it is rational, reasonable, or even inimical to the chooser's interests. Consider the California gentleman who fired a gun and was charged with assault with a dangerous weapon and assorted related crimes. Wanting no part of lawyers, he demanded the opportunity to defend himself in court. When the trial judge turned him down, the California Court of Appeals reversed. "[R]espect for the dignity and autonomy of the individual is a value universally celebrated in free societies," it said. "Out of fidelity to that value, [the] defendant's choice must be honored even if he opts foolishly to go to hell in a handbasket."[39] One Massachusetts defendant, having been granted his wish to defend himself against serious criminal charges, had regrets after conviction and argued that the judge should have appointed him a lawyer anyway. The Massachusetts high court demurred: "[E]ven in cases where the accused is harming himself by insisting on conducting his own defense, respect for individual autonomy requires that he be allowed to go to jail under his own banner if he so desires."[40]

Although this explains why judges give "human rights" to humans with a low-level realistic autonomy, it doesn't explain why judges give them to humans who have no autonomy at all. We find that explanation embedded in a debate on the propriety of terminating the feeding and hydrating of Jane Doe, a profoundly retarded woman in a permanent vegetative state, between the four-judge majority of the Massachusetts high court and the three dissenters. All agree that Jane Doe's autonomy is a "legal fiction."[41]

Legal fictions have allowed judges to attribute legal personhood not just to nonautonomous, nonconscious, nonsentient humans like Jane Doe but to ships, trusts, corporations, even religious idols.[42] Judges are perfectly capable of pretending that an animal without feathers is a bird or that the pope is a Buddhist. They have pretended even stranger things, that slaves are real estate and that Paris is located in England.[43] Indeed—shades of George Orwell—legal fictions always demand that judges pretend that what they know to be false is true. Because a legal fiction may cloak abuses of power, Jeremy Bentham characterized it as a *"syphilis . . . [that] carries into every part of the system the principle of rottenness."*[44] Perhaps he exaggerated. But the danger of injustice is greatest when a legal fiction contradicts bedrock values. It happens. For hundreds of years, the common law conclusively presumed that husband and wife were one—the husband—to the wife's frequent sorrow.[45]

Giving Beths and Jane Does the dignity-right to bodily integrity through the legal fiction that "all humans are autonomous" appears benign, as it places them off-limits to exploiters, treats them as important, and violates no one's dignity-rights. In contrast, judges deny rights to all animals by employing a second legal fiction—"no nonhuman animals are autonomous"—and then wink at the autonomy that an animal may have and arbitrarily withhold the dignity-rights their autonomy creates. This leaves demonstrably autonomous beings acutely vulnerable to the exploitation and abuse that the unhappy Jerom in Chapter 1 suffered at the hands of the Yerkes AIDS researchers.

Two hundred years ago, "dignity" signified one's elevated rank in a social hierarchy, such as the dignity of the king or a noble or a bishop. This notion collapsed beneath the weight of the French and American Revolutions.[46] The legal fiction that "no nonhuman animals are autonomous" revives this discarded idea of dignity as an upper rank that humans arbitrarily reserve for themselves, erodes the modern claim that dignity arises from autonomy, and saps the strength from "human rights."[47]

The Liberty Rights of
Chimpanzees and Bonobos

I will not argue that any chimpanzee or bonobo has full autonomy. But no bright line divides full autonomy from realistic autonomy or realistic autonomy from the legal fiction that "all humans are autonomous." However, a little mountain geography might help us understand the relationships among them a little better. We'll start at the top of the world. The summit of 29,038-foot Mount Everest in the Himalayas will represent those few humans who may have attained full autonomy. A few more occupy the apex of K-2 in the Karakoram Range in northern Kashmir, which at 28,250 feet is the second-highest mountain in the world. Millions cluster atop the highest mountain in the Hindu Kush Range, 25,260-foot Tirich Mir, located in Pakistan along the Afghan border. The nearby Pamir Range in Tajikistan is filled with peaks above 20,000 feet. The autonomies of most adult *Homo erectus*, Neandertals, and *Homo sapiens* can be found among those peaks.

Sea level represents autonomy's absence. The top of Cadillac Mountain, jutting 1,530 feet from the Atlantic on the eastern side of Mt. Desert Island, Maine, stands for the minimum autonomy that judges require to trigger human dignity and entitle one to dignity-rights, unaided by any legal fiction. We'll let 250-foot Flying Mountain, on Mt. Desert Island's west side, stand for the highest autonomy that Aristotle and the ancient Stoics believed nonhuman animals, whom they thought could only perceive and act on impulse, ever attained.

As we saw in the last three chapters, the autonomies of the Language Research Center's four adult apes tower above Cadillac Mountain. If a phalanx of primatologists is right, we can place the autonomy of mother-raised Tamuli, who has a complex cognition but lacks the ability to use symbols, at the tip of Mount Etna, which rises 10,902 feet above Sicily. If Juan Carlos Gómez is correct, hand-reared but languageless Sherman, who has all Tamuli's cognitive abilities and uses symbols to boot and has an implicit theory of mind and implicit self-consciousness, can be symbolized by Mt. Blanc, which at 15,771 feet is the highest Alp. If Sue Savage-Rumbaugh, Michael Tomasello, Roger Fouts, and others are correct, human-enculturated Kanzi and Panbanisha have all Sherman's cognitive abilities but also possess at least some, and perhaps all, the elements needed for an explicit theory of mind, comprehend human language at the level of at least a three-year-old human, produce language like a human two-year-old, are explicitly self-conscious, and demonstrate a raft of other complicated cognitive abilities. Their autonomies may not top the 20,000-foot level, as do most human adults with advanced language abilities, complex consciousnesses and theories of mind, and numerous other highly advanced cognitive abilities. But they reach the pinnacle of Mt. Kilimanjaro, which at 19,340 feet is the tallest African mountain.

Today, the autonomy of Nyota, at just one year old, probably lies just shy of Cadillac Mountain's peak. But early reports suggest that his cognition is advancing so rapidly that his may be the first nonhuman autonomy to reach the height of Mt. McKinley, which at 20,320 feet is the highest peak in North America. At birth, my daughter Siena's autonomy floated just about at sea level. So did those of Kanzi, Panbanisha, Sherman, Tamuli, and Nyota. As did they, Siena vaulted past 250-foot Flying Mountain early in her first year. Kanzi, Panbanisha, Sherman, and Tamuli all topped 1,530-foot Cadillac Mountain late in their first year or perhaps early in their second, and Siena and Nyota probably will, too, and on just about the same schedule. But now Siena will begin to accelerate. Sometime in the second half of her second year or perhaps early in her third year, she will race past Tamuli and overtake Sherman. If she continues to develop normally, by age four, or by age five at the latest, she will overhaul Kanzi and Panbanisha, then spend the rest of her life slowly ascending toward Everest, probably never reaching it. Nyota's probable course of development is less well-known.

Notice how I used "probably," "likely," "if she continues to develop normally," and "probable course of development" in describing Nyota's and Siena's future autonomies. They may not advance an inch—who can know? In determining entitlement to dignity-rights, should one look to a being's *present* or *potential* autonomy? At one extreme, H. Tristram Englehardt Jr. argues that the nature of rights demands that entitlement be determined at the time the question "Do you have a right?" is asked. He says flatly that "if X is a potential Y, it follows that X is not a Y. If fetuses are potential persons, it follows clearly that fetuses are not persons. As a consequence, X does not have the actual rights of Y, but only potentially the rights of Y."[48] At the other extreme, the philosopher John Rawls argues that one with a potential should have full rights, whether or not one's capacities have been developed.[49] Between these poles lie a host of other arguments.[50]

We need not decide the question. Under the common law, autonomy determines entitlement to dignity and legal personhood. The evidence is strong that normal humans, chimpanzees, and bonobos all reach Cadillac Mountain's height (the minimum necessary autonomy to trigger dignity) by the end of their first or the beginning of their second year. If we accept the argument for potential autonomy, then both bonobo and child are entitled to dignity-rights. If we reject it, then neither is *entitled* to dignity-rights. Whether one or both gets them will turn on the willingness of judges to use a legal fiction that one or both is autonomous until they actually become so. If judges choose the usual "all humans are autonomous" fiction to extend dignity-rights to Siena alone, Nyota will have no argument that as a matter of *liberty*, he is being treated unjustly. As we will see, equality will be another matter altogether.

EQUALITY

Similar Autonomies

In 1988, Jerom, infected with two strains of HIV, was languishing in the small, windowless, cinder-block cell within the Yerkes Regional Primate Research Center. He had eight more pain-wracked years to live. That year, Yerkes' director, Frederick King, coauthored an article, "Primates," with three Yerkes colleagues in the prestigious journal *Science*. His justification of routine invasions of the bodily liberties and bodily integrities of primates on the ground of their *similarity to humans* is shocking stuff to those who value equality.

> [S]imilarities in the biological mechanisms of humans and primates underlie the value of these animals for research in a broad range of disciplines . . .
>
> Primates have played a major role in increasing our knowledge of the structure, organization, chemistry, and physiology of the human brain. The complexity of the primate brain and its similarity to that of humans makes [sic] primates excellent subjects for the study of motivational states such as hunger, thirst, and emotion. . . .
>
> The large, convoluted cerebral cortex, with great areas devoted to associational activities, is almost certainly responsible for the primate's ability to learn highly complex cognitive tasks beyond the capacities of species other than humans.
>
> Because the primate brain shares with humans a high degree of plasticity, their cognitive and social behaviors are heavily dependent

on learning and the environment, as is the human behavioral reper-toire. Hence in studies of the relationship of neural plasticity and the emergence of behaviors dependent upon social learning primates are often the subject of choice. . . .

[P]rimates in general develop socially and relate to each other and their environments in ways that are more similar to humans than to other animals.

King concluded by noting that the animal-rights activists' "cam-paign against primate research is actually based on the scientists' rationale for studying primates: the biological and behavioral sim-ilarities of primates to humans."[51]

We saw in Chapter 6 that like Creon and Jefferson Davis and the killing doctors of Hadamar, King believes that all rights are be-stowed by human beings. Here he fails to grasp that equality de-stroyed anywhere, even for chimpanzees, threatens the destruction of equality everywhere. That is why, near the onset of the American Civil War, Abraham Lincoln told Congress that "[i]n giving freedom to the slave, we assure freedom to the free."[52] To deny freedom to the slave, the Confederacy had to shackle its white citizens. Had Pickett's Charge split the Union lines at Get-tysburg, the American South might today be dotted with biomed-ical research laboratories using not just slaves, instead of nonhuman primates, but anyone that the government, like King, thought most useful.

King's unembarrassed advocacy, in one of the world's most re-spected scientific journals, of using raw power to exploit nonhu-man primates *because they are like us* rests on an argument that is arbitrary, unprincipled, and corrosive to equality, which at bottom demands that likes be treated alike. The equality rights of chim-panzees and bonobos must be determined by comparing them to others who already have rights. If alikes are treated differently or if unalikes are treated the same, for no good and sufficient reason, equality is violated. The following three equality arguments all lead to the same conclusion: King was wrong, at least with respect to chimpanzees and bonobos.

The first of three equality arguments that we can present in court will be this: Kanzi, Panbanisha, Sherman, Tamuli, and Nyota are entitled to the rights to bodily integrity and bodily liberty *if* hu-mans with similar autonomies are entitled to them. The second argument will be: Kanzi, Panbanisha, Sherman, Tamuli, and Nyota's autonomies entitle them to bodily integrity and bodily liberty *if* humans who completely lack autonomy have their rights. The third argument: If the autonomies of any of the five chimpanzees and bonobos are insufficient to entitle them to the rights to bodily integrity and bodily liberty in full, they are still entitled to them in proportion to the degree to which their autonomies approach the necessary minimum *if* humans with similar autonomies are entitled to these rights in proportion to the degree to which *their* autonomies approach the necessary minimum. In each argument, we will compare chimpanzees and bonobos to a different class of humans who possess the fundamental rights to bodily integrity and bodily liberty.

Few would argue that the complex autonomies of chimpanzees and bonobos would not entitle them to bodily liberty and bodily integrity if they were human. We have seen that they are not disqualified from these rights as *liberties* because they are not human. The strongest assault against ape equality has been mounted by the philosopher Carl Cohen. "The issue," Cohen says, "is one of kind."[53] Sure, Cohen says, some humans lack autonomy. But others are as fully autonomous as Kant could ever have dreamed. Cohen argues that their full autonomies should therefore be imputed to every human, regardless of actual ability. But Cohen's argument is illogical. The species *Homo sapiens* cannot rationally be designated as the boundary of any relevant "kind" that includes every fully autonomous human. Other "kinds" exist. Some categories are broader, for example, those of animals, vertebrates, mammals, primates, and apes. At least one "kind" is narrower, that of normal adult humans, which also contains every fully autonomous human.

Cohen's argument for group benefits is not just logically but morally flawed. The philosopher James Rachels has pointed out how it "assumes that we should determine how an individual is to be treated, not on the basis of *its* qualities but on the basis of *other* individual's qualities."[54] You get straight As; I go to Harvard. I jump thirty feet; you go to the Olympics. You look like Elle MacPherson; I get a modeling contract. Unsurprisingly, many people are bothered by this kind of argument. Surprisingly, Cohen is one of them. He has attacked racial affirmative-action policies as "illegal and immoral" because they favor blacks over whites in precisely the way he favors humans over apes.[55] But even Harvard Law School professor Laurence Tribe, who supports racial and sexual affirmative action, has acknowledged the "tenaciously held principle . . . with undeniable constitutional roots . . . that each person should be treated as an individual rather than as a statistic or as a member of a group—particularly of a group the individual did not knowingly choose to join."[56]

This does not mean that racial or sexual affirmative action is always wrong or should be illegal. Something like Cohen's notion of group benefits is occasionally used to *correct* the effects of prior discrimination in the United States. But even then, judges often manifest what Tribe called a "considerable unease" in sharply divided endorsements of affirmative action plans.[57] Judges are not alone. In a New York Times–CBS poll taken at the end of 1997, 25 percent of Americans said they wished to abolish affirmative-action programs outright, 43 percent wanted them changed, and just 24 percent would leave them as they are.[58]

A 1996 case, *Hopwood v. State of Texas*, exemplifies the rising judicial resistance to the use of affirmative action for reasons other than to correct past discrimination. There, the Fifth Circuit Court of Appeals flatly barred the use of race as a factor in university admissions to achieve the goal of a diverse student body. Affirmative action, it said:

> treats minorities as a group, rather than as individuals. . . . The assumption is that a certain individual possesses characteristics by virtue of being a member of a certain racial group. This assumption, however, does not withstand scrutiny. The use of a racial characteristic to establish a presumption that the individual also possesses other, and socially relevant, characteristics exemplifies, encourages, and legitimizes the mode of thought and behavior that underlies most prejudice and bigotry in modern America.[59]

Group benefits stir intense controversy even in racial affirmative-action plans that are anchored in the laudable desire to correct ancient discrimination and achieve racial equality for a long-oppressed minority. But outside of apartheid-era South Africa or Nazi Germany, group benefits have never been used in the way that Cohen and others would use it against apes, as a sword instead of a shield.

Greater Autonomies

Some humans—infants, young children, the anencephalic (who suffer from the congenital absence of major portions of the skull, scalp, and brain, never attain consciousness, can neither feel nor suffer, and usually die within a few months of birth), the severely mentally retarded, and those in persistent vegetative states—either lack autonomy or have autonomies too "low" to be called "realistic" (falling below the peak of Cadillac Mountain). Judges routinely award them dignity-rights anyway by using the "all humans are autonomous" legal fiction. But if judges recognize the liberties of these humans but reject the liberties of apes with *greater* autonomy, they act perversely, and their decisions cannot be explained except as acts of naked prejudice. Of course, it is always open to judges to sever autonomy from dignity. But only a foolish hydroelectric-plant manager would stem flooding in the control room by diverting the river that generates the electricity.

The lowest autonomy of any adult ape, Tamuli's, lies at the level of 10,902-feet Mount Etna. I won't argue, here, that equality entitles nonhuman animals whose autonomies exist at the level of 1,530-foot Cadillac Mountain or even at the height of 3,491-foot Mt. Greylock, highest of the Berkshires, to any fundamental right, though compelling arguments can be made. I will leave it at this: At some point the disparity between the autonomies of nonhuman animals with *no legal rights* and the virtual sea-level autonomies of humans *with* dignity-rights becomes completely indefensible.

I said that if *potential* autonomy was insufficient to justify an entitlement to dignity-rights, the immunities of both my infant daughter, Siena, and Nyota would turn on the sufferance of judges until their autonomies rose to the level of Cadillac Mountain. I also conceded that if judges extended these rights to bodily integrity and bodily liberty just to Siena, not because they were her due but as a gift, Nyota would have no claim that as a matter of liberty, he was being treated unjustly. But the situation is different with equality. Neither Siena nor Nyota are autonomous now; both have only the potential for autonomy. If Siena's potential for achieving an autonomy peak in the mid-to-high 20,000-foot range is ignored, Nyota has no equality claim to the recognition of his potential for achieving an autonomy peak in the low 20,000-foot range. But if Siena's potentiality is sufficient for giving her fundamental rights, Nyota's potential must equally be recognized.

Proportional Autonomies

Humans infants, young children, and the severely mentally retarded or autistic who lack the autonomy necessary to entitle them to full liberty rights are not totally denied them. Judges give them dignity-rights, and even the right to choose, by using the legal fiction that "all humans are autonomous." However, as their autonomies *approach* the minimum, the *scope* of their fundamental rights may be varied *proportionately*. If so, equality demands that the rights of animals who possess the same degree of autonomy as these humans possess vary proportionately, too.

We saw in Chapter 6 that "proportionality rights" can vary in three dimensions. First, humans who lack sufficient autonomy may be allowed *fewer* legal rights than autonomous humans. But judges do not characterize them as legal things. For example, their right to engage in a political process that they cannot understand might be restricted by refusing them the right to vote.[60] But we cannot vivisect them. Severely developmentally disabled humans cannot testify in a court, but we do not enslave them. To the contrary, courts strain to give even the most severely disabled humans "some measure of personal control over their lives" in an effort to "maximize the personal autonomy and dignity" along with conditions of reasonable care and safety and a reasonably nonrestrictive environment.[61] To the extent that any of the Language Research Center apes that Savage-Rumbaugh wants me to help her return home lack the autonomy necessary to entitle them to liberty rights in full, they should be given *fewer* rights than one whose autonomy is sufficient, *if* humans who lack minimum autonomy are given these rights. They are not to be reduced to legal things, enslaved, or vivisected. Equality requires that they be given the same rights to "maximize the personal autonomy and dignity" that they possess. If the rights of apes must be restricted—and in America they must—they are also entitled to live in similar conditions of reasonable care and safety and in a reasonable nonrestrictive environment.

Second, judges may *narrow* the liberty rights of humans who lack minimum autonomy. Their right to bodily liberty may be restricted so that they can exercise it only in a manner consistent with both public and their own safety. That is why state institutions for the mentally retarded may physically restrain the movements of their dangerous residents and parents can commit children to a state institution without an adversary proceeding.[62] But these patients are not reduced to legal things then enslaved or vivisected. To the extent that any of the Language Research Center apes lack the autonomy necessary to entitle them to liberty rights in full, they should be given *narrower* rights than one whose autonomy is sufficient, *if* humans who lack minimum autonomy are given these rights. Their right to bodily liberty might be restricted in a manner consistent with the public safety and their own safety. But they are not to be reduced to legal things, enslaved, or vivisected.

Third, courts and legislatures might give humans who lack minimum autonomy *partial elements* of a complex right. Remember that most of what we normally think of as legal rights are actually a bundle of rights. My legal right to bodily integrity means that I have the negative liberty-right that you not hit me, an immunity-right that disables you from legally hitting me, a claim-right against you if you do hit me, a power-right to sue you in a court, and more. Humans whose autonomies fall below the necessary minimum are sometimes given all these rights, but not the power-right to waive them, because they can't understand what it means to waive their rights. Thus, an adult woman in a permanent vegetative state might lack the right to refuse life-saving medical treatment and a twelve-year-old boy might lack the right to waive his right to remain silent when accused of crime.[63] But neither the woman nor the boy is reduced to a legal thing and enslaved or vivisected. To the extent that any of the Language Research Center apes lack the autonomy necessary to entitle them to liberty rights in full, they are entitled to partial elements of a complex right, *if* humans who lack minimum autonomy are given them. They are not to be reduced to legal things, enslaved, or vivisected.

COMMON LAW RIGHTS OF APES

A judge of the Virginia Supreme Court in 1827 described the legal metamorphosis of a slave to free woman, legal thing to legal person, in almost religious terms. She "becomes a new creature, receives a new existence."[64] Legal personhood is the frame upon which we stretch fundamental immunities that block abuses of power, whether that power is rooted in precedent, policy, principle, or prejudice. Governments, legislatures, and judges can blind themselves to legal personhood and violate dignity-rights. But the lesson of the killing doctors of Hadamar, of the murderers of the White Rose, of the Nuremberg trials, of Eichmann in Jerusalem, and of the killers of Bosnia, Rwanda, and Kosovo is that even governments cannot make dignity-rights disappear. That is why no lawyer will argue that any human being could be subjected to the torments that the biomedical researchers plan for Kanzi, Panbanisha, Sherman, Tamuli, and Nyota.

Now that we understand why liberty and equality entitle them to fundamental rights, there remains the problem of how to hurdle the barrier of the common law's characterization of them as rightless things for so many centuries. Whether history or taxonomy will foreclose their "new existence" in law will depend largely upon our imaginary judge—we'll call her Judge Juno. What is her vision of law? And do her beliefs disable her from recognizing the possibility that nonhuman animals might be entitled to legal rights? Remember what Judge Athena had to consider when she analyzed how a sixty-five-year-old decision of the House of Lords might influence the outcome of an imaginary lawsuit I brought in London (after my friend Sarah offered me a cold beer in a clear bottle, which had a mouse rear end bobbing about in the bottom, that she had purchased from the microbrewery owned by the lovely man around the corner from her London condominium)? Judge Juno will have to consider very much the same things.

We know that nearly every common law judge at least begins to decide a question by reviewing what earlier judges have said. If Judge Juno is a Precedent (Rules) Judge, the apes will go under the knife pronto. She doesn't care if circumstances, facts, morality, or values have shifted. She doesn't think she makes law; she doesn't think she *can*. She cares about precedents and will rule on a legal question the way other judges have ruled on it, not because they were right, but because that's how they ruled. She will confine her search for justice to the law library. There she will learn that a common law rule has made things of all nonhumans for centuries. QED. But Precedent (Rules) Judges are uncommon today. A few are scattered throughout the United States, and it will be our bad luck if Judge Juno is one of them.

Far more likely, she will not believe that law is just a system of rules. Instead, she will feel it represents the community's present sense of right and justice and not what it felt a hundred years ago, or a thousand. She will hesitate to perpetrate ancient injustices blindly. More numerous than Precedent (Rules) Judges today are Policy Judges. If Judge Juno is a Policy Judge, she thinks that good policy is good justice and that it promotes society's important values. She cares mostly about the effects of their rulings. The biomedical research laboratory will bombard her with policy arguments. Consider the benefits that could accrue from using the apes in biomedical research. Aphasia might be conquered. Children will learn how to read and speak better. If we can't open ape skulls for language investigations, what will happen when we need them for cancer research? What of the economic consequences? Biomedical research centers will have to close their doors. Breadwinners will lose their jobs. American research will fail to compete with countries whose laws are not so foolish.

But when economic rights clash with dignity-rights, the policy of modern Western law is to choose liberty at nearly every juncture. This was not always so. The policy of promoting economics fortified the determination of some of the finest Northern judges to uphold American slavery in the years before the Civil War. Today this policy is widely conceived as the foulest blot on American justice. No economic argument can justify human slavery.

Another policy-based destruction of dignity-rights, this time not based on economics, was inflicted upon 120,000 Japanese Americans living on the West Coast during World War II. Because some thought Japanese Americans were disloyal and we were at war, the Supreme Court upheld their imprisonment.[65] Fifty-nine years later, the U.S. Justice Department has just finished paying $1.6 billion in reparations to the prisoners.[66] We come to rue the day we subordinate dignity-rights to policy.

The plea that "if it is language research today, it will be cancer tomorrow" distills to "necessity." We must harm these apes if we are to help ourselves. But it is such a thoroughly immoral argument, resting upon "might makes right," that only a "pure" Policy Judge, blind to principle, could ever accept it. Nazi doctors invoked "necessity" to justify immersing Jews, Russians, and convicted criminals in freezing water, forcing them to drink seawater, infecting them with typhus and gangrene, trying to regenerate and transplant their bones, and exposing them to mustard gas in the concentration camps.[67] "Necessity" justified the Japanese 731st

Regiment's vivisecting living Chinese, Koreans, Russians, and Mongolians without anesthetics, replacing their blood with horse's blood, infecting them with syphilis, bubonic plague, anthrax, and cholera, immersing them in cold water and throwing them into the winter, exposing them to high doses of X rays, and systematically starving them in vitamin and nutrition research in Harbin, China, during World War II.[68] Yet the two most famous legal "necessity" cases in the English-speaking world denied its power to justify the taking of innocent life. In an American case, sailors who threw passengers from a leaky life raft were convicted of manslaughter, though they had saved other passengers from drowning and all would otherwise have died.[69] In an English case, two drifting sailors were convicted of killing a dying boy, then eating his body when food and water ran out.[70]

At least on the issue of fundamental rights, most judges today are neither predominantly Precedent (Rules) Judges nor Policy Judges. Instead, they look to principle. If this is Judge Juno, she respects the common law's impulse to "work itself pure" and thinks that law should promote a society's fundamental principles. Remember that Precedent (Principles) Judges believe they are *bound* to follow legal principles laid down in earlier decisions but need not blindly hew to narrow legal rules. Both Precedent (Principles) Judges and Principle Judges want to do the "right" thing. When equality and liberty are threatened, they stand shoulder-to-shoulder, because these principles are both once and future. The dignity-rights of chimpanzees and bonobos will be pronounced by these judges.

The 1954 school desegregation case of *Bolling v. Sharpe*, in which the U.S. Supreme Court overruled the fifty-eight-year-old precedent, *Plessy v. Ferguson*,[71] exemplifies how a sure grasp of principle, history, scientific fact, and the evolution of public morality can lead judges to a volte-face from a disgraceful series of rulings that conflict with a more principled tradition.[72] *Bolling* was a companion case to the more famous *Brown v. Board of Education*.[73] Between them, the justices unanimously held that segregated schools violated both the liberty rights and the equality rights of black children.

By 1954, black chattel slavery, the *Dred Scott* case, slave codes, Jim Crow laws, and *Plessy v. Ferguson*'s "separate but equal" doctrine had been undermining fundamental principles of American liberty and equality for three hundred and fifty years. As Lord Mansfield finally branded English slavery too "odious" for the common law, no matter what "inconveniences" might follow, "inconsistent [in Professor Dworkin's words] with more fundamental principles necessary to justify law as a whole," in *Bolling*, the justices unanimously struck down invidious racial classifications because they were *"contrary to our traditions."*[74]

Justice Jackson, the Nuremberg prosecutor, and two of his law clerks illustrate the competing legal currents. One clerk told Jackson that *"Plessy* was wrong" but he thought that "separate but equal" had burrowed deeply into American society and "where a whole way of life has grown up around such a prior error, then I say we are stuck with it."[75] Jackson's other clerk, the future chief justice of the United States, William Rehnquist, warned not only that the black plaintiff's argument was "palpably at variance with precedent, but that *Plessy* was right and should be re-affirmed."[76] If Jackson had been a Precedent (Rules) Judge on civil rights, he wouldn't have cared about any "separate but equal" effect or whether it was right. If Jackson had been a Principles Judge on civil rights who, like Rehnquist, embraced "separate but equal," he would have ruled against the black children. But their days had passed with respect to "separate but equal." Although the justices advanced a policy reason for their ruling— "separate educational facilities are inherently unequal" and generated a permanent feeling of inferiority[77]—its core was that racial classifications were "contrary to our traditions." Not the baser traditions of chattel slavery, slave codes, Jim Crow laws, and "separate but equal," but the grand tradition of equality. Decisions like *Bolling* both honor and invigorate fundamental principles.[78] That is why they may strike us as fairly reached, even when we disagree with the outcome. And that's why, though we may agree with an outcome that conflicts with our fundamental principles, our guts churn when the lights go out. Loose cannons fire in all directions.

Long ago we learned how to domesticate wild animals. We applied that learning to enslave each other.[79] But as our domestication of wild animals served as an unprincipled model for our enslavement of human beings, so the destruction of human slavery and all its badges can model the principled destruction of chimpanzee and bonobo slavery. Determining the dignity-rights of chimpanzees and bonobos in accordance with fundamental principles of Western law—equality, liberty, and reasoned judicial decisionmaking—reemphasizes and reinvigorates these principles just as the abolition of slavery and its badges did.

Beliefs and Genes

A judge's deepest cultural and religious beliefs may disable even a Principle Judge from deciding according to principle. Think of a Fundamentalist Protestant faced with a decision about teaching evolution in the public schools or a Roman Catholic deciding a question of abortion rights. Is it surprising that Nazi judges dispensed Nazi justice and that racist judges dispensed racist justice? Religious and cultural beliefs can, at least theoretically, change. But consider U.S. Court of Appeals Judge Richard Posner's mother of all beliefs:

> The main "reason" why the "philosophical" idea that . . . talking apes might have more rights than newborn or profoundly retarded children seems outlandish and repulsive may simply be that our genes force us to distinguish between our own and other species and that in this instance disembodied rational reflection will not overcome feelings rooted in our biology.[80]

When I asked him, Judge Posner said that his point had been that "it would be bizarre to think that apes should have greater rights than profoundly retarded human beings; that leaves open the question whether apes should have any rights."[81] When I countered that I had never heard of anyone who claimed that apes should have *more* rights than profoundly retarded humans, Posner briskly pointed me to Professor Bruce Ackerman, who had.[82]

Well, not quite. It turns out that Ackerman thinks that if an ape can use language enough to stake a moral claim to the right to be treated as a fellow member of the political community, then the ape should be a citizen.[83] Ackerman thinks that any being, of whatever species, who can stake a moral claim is entitled to citizenship. Any being, of whatever species, who cannot is not so entitled. That was Ackerman's point.

But what could be outlandish, repulsive, or bizarre about giving a "talking ape," along with an anencephalic human infant, a profoundly retarded child, or a "human vegetable," the same dignity-rights, then giving the ape the additional rights to vote, to run for political office, and any other right that might be appropriate to such an ape, but not to an anencephalic infant, a profoundly retarded child, or a "human vegetable?" If an ape, a Neandertal, or a

Homo erectus argues like Socrates, we violate our most fundamental principles of liberty and equality to deprive him of those legal rights. If Judge Posner thinks this is outlandish, repulsive, and bizarre, it is not because his genes "force" him to.

CHILDISH THINGS

But perhaps Judge Posner is partially correct. Both normally developing, but very young, human children and most autistic humans of every age are unable to recognize that other humans have minds and so relate to them as if they were machines. But since other humans have minds, autistic "mindblindness" evidences severe pathology. Perhaps we are an autistic species, biologically incapable of recognizing that nonhumans have minds. Pathologically self-absorbed, we relate to them as if they were machines.

Certainly any such deficit, if it exists, is not universal within our own species. From ancient times, some humans have believed that at least some nonhuman animals have sophisticated minds.[84] Chapter 10 was one long catalogue of scientific evidence for the complex minds of chimpanzees and bonobos. The U. S. Congress, by enacting a 1985 amendment to the Animal Welfare Act that required "a physical environment adequate to promote [their] psychological well-being," implicitly recognized that primates have minds.[85]

But "animal mindblindness" is only pathological if we cannot help but believe it, as an autistic child cannot help but believe that no one else has a mind. Some cognitive scientists argue that we humans have a tendency to view objects as designed for a purpose because our minds naturally bend in that direction.[86] Thus children and adults with limited understandings of scientific fact may follow their natural predispositions to think in just that way.[87] Thus young children may assert that they can predict in what direction an object that produces random results such as a spinner will end up.[88] Meanwhile their parents spend billions of dollars betting that they can predict when a lottery number will turn up or how a pair of dice will land. Numerous ancient religions teach that things and people and events are designed for some purpose. Our forebears, intelligent, but scientifically naive or drenched in religious belief, could straight-facedly conclude that rain fell for corn, that oceans tided for ships, that slaves breathed for their masters, and that women existed for men. This press towards purpose can be so powerful that even today some who understand and accept the overwhelming evidence for evolution by natural selection, which describes a process without a design or a Designer, begin to think of it in terms of designing organisms so that they will be better able to adapt to their environments.[89]

Deborah Kelemen writes that:

> For all the scientific sophistication we acquire, people are built in a particular way—intentional explanations are a powerful way of making sense of the world and consequently [it] remains a part of most of us ... It does not take much for it to reassert itself into domains where it should not be applied when explanations seem inadequate or difficult to grasp. For example, it is revealing that a widespread response to first reports of the AIDS epidemic in both the US and Europe was that the disease was an intentionally contrived global punishment. Such ideas are easy for people to latch onto.[90]

All this suggests that while our animal mindblindness may be a tendency, even a strong one, it is not irresistible. Certainly it has resisted change, but strongly-held beliefs always do. Even beliefs about how the physical world operates, which can be falsified, can be exceedingly difficult to alter. Pope Paul V and Cardinal Bellarmine went to their graves convinced that the earth was the center of the universe, though with the telescope this belief could be falsified as easily as the claim that gravity makes objects fall up. Today many outright disbelieve an evolution by natural selection that every competent biologist and even Pope John Paul II take as granted.[91] How much harder it is to change moral, religious, or jurisprudential beliefs.

Neither the ancient Greeks nor Hebrews nor Christians had difficulty accepting that the end of everything in the universe was themselves. But no scientific evidence today exists that other animals, or anything, were made for us.[92] While few educated men and women today still think that nonhuman animals were made for humankind, they may believe it, perhaps because their natural tendencies to accept it are fanned by their religions, which arose at a time before science when everyone believed it and which command them to believe it, and commandments are etched in stone. But as children mature, they can learn. Saint Paul said, "When I was a child, I spoke as a child, I understood as a child, I thought as a child: but when I became a man I put away childish things."[93] Our religions, our laws, our parents, even our own minds may incline us towards embrace of the childish notions that rain falls for corn, oceans tide for ships, slaves breathe for their masters, women exist for men, and that Kanzi, Panbanisha, Sherman, Tamuli, and Nyota are mindless automatons that were made for us. But when we become educated men and women, we should put these childish things away.

GENOCIDE

Chapter 1 concluded with the statement about chimpanzees and bonobos that "their abuse and their murders must be forbidden for what they are—genocide." This was not intended as metaphor. The word "genocide" emerged from the Holocaust. The *Oxford English Dictionary* defines it is as "the deliberate and systematic destruction of an ethnic or national group."[94] *Merriam-Webster's Collegiate Dictionary* explains it as "the deliberate and systematic destruction of a racial, political, or cultural group."[95] The Convention on the Prevention and Punishment of the Crime of Genocide says it means

> any of the following acts committed with intent to destroy, in whole or in part, a national, ethnic, racial or religious group, as such: (a) killing members of the group; (b) causing serious bodily or mental harm to members of the group; (c) deliberately inflicting on the group conditions of life calculated to bring about its physical destruction in whole or in part; (d) imposing measures designed to prevent births within the group; or (e) forcibly transferring children of the group to another group.[96]

Genocide, in short, is the deliberate and systematic destruction, or attempted destruction, of any group that shares a nation, a politics, a culture, a race, a language, a religion, a tribe, or a history. Genocide is a crime whenever and by whomever committed. Genocide "shocks the conscience."[97] Genocide need not spring from hatred. Had Westerners worked every African slave to death merely for profit, there would have been genocide. Nazis who murdered Jews to wash out racial impurities committed genocide. American settlers who wiped coveted land clean of whole Indian tribes were genocidal.

The Latin roots of "genocide" are "*genus*" and "*caedere*." *Caedere* means "to kill." *Genus* generally means a class or kind that share common attributes. So genocide carries not just an explicit sense of destroying or trying to destroy a discrete group but also the implicit sense that the destroyed and the destroyer share membership in some larger group. Morris Goodman showed us that chimpanzees, humans, and bonobos are literally all members of the genus *Homo*, or should be. Chimpanzees and bonobos share not just our taxonomic trunk but our bough, our branch, our twig. If we don't all share a language, then we share "language" or something remarkably like it. If we don't all share a common culture, we share "culture."[98] If human politics and chimpanzee politics are not the same thing, both are still "politics."[99] Perhaps in the end, we simply need to convince Judge Juno that we may not be an autistic species, just a narcissistic one, transfixed by our own reflection and that she needs to put aside childish things and allow her mature reason, her passion for liberty and equality, and her sense of fair play to open her eyes so that when she gazes into the mirror of justice, she sees Jerom.

12

Epilogue:
Other Cages, Other Peaks

I hope I have convinced you that we must replace the legal *thing-hood* of chimpanzees and bonobos with a legal *personhood* that immunizes them from serious infringements upon their bodily integrity and bodily liberty—and that we must do it now. But I never meant to imply that these two dignity-rights are the *only* legal rights to which they might be entitled. Should they, for example, have the legal rights to reproduce, to keep their offspring, or to have sufficient and proper habitat?

One particularly thorny question arises around human enculturation. An increasing body of evidence strongly suggests that as with the enculturation of human children, the human enculturation of chimpanzees, bonobos, and perhaps other nonhuman animals, enhances, activates, perhaps even creates such advanced cognitive abilities as language, mathematics, other forms of symbolic representation, and explicit theory of mind. When a human "wild child" appears, every effort is made to socialize him. But is socialization his legal right? If human enculturation can flower the minds of chimpanzees and bonobos should they have a positive liberty-right to it? Or should they have the negative liberty-right to be left alone? Alas, our struggle with all these questions will have to be deferred. When we take them up again, we must determine the answers for nonhuman animals in the same careful and logical way that we determine them for human beings.

I also never meant to imply that chimpanzees and bonobos are the only nonhuman animals who might be entitled to the fundamental legal rights to bodily integrity and bodily liberty. Judges must determine the entitlement to dignity-rights of any nonhuman animal the same ways they determine the entitlements of chimpanzees, bonobos, and human beings—according to autonomy.[1] Autonomy, of course, arises from minds. Nonhuman animals who lack minds are little more than animate versions of "the MIT 3"— COG, YUPPEE, and KISMET—and their entitlement to legal rights should be seriously doubted.

Remember, because no one can ever know for certain that anyone, human or nonhuman, has a mind, all we can do is apply our reasoned judgment to the facts that science reveals. In order to keep our judgments about "who's in and who's out" both accurate and fair, we must keep Wise's usual three rules in mind. We cannot trust ourselves to judge ourselves or place ourselves fairly in nature. We must be at our most skeptical when we evaluate arguments that confirm the very high opinion that we have of ourselves. We must play fair and ignore special pleading that different rules apply when we assess the mental abilities of human and nonhuman animals.

Through careful analyses similar to those I have done for chimpanzees and bonobos, we can determine the next best candidates to whom the dignity-rights to bodily integrity and bodily liberty might be extended. Where might we start? Arguably the best place might be those species of animals evolutionarily closest to the three species that are most clearly entitled to fundamental legal rights: chimpanzees, bonobos, and ourselves. Tantalizing clues exist that such extensions might be warranted, or even required.

Orangutans, a third great ape, have been able to follow gazes.[2] Indah demonstrated that she could assume the visual perception that children develop around the age of three. Another orangutan's performance on a rough version of a false-belief test suggests that she might have elements of a theory of mind.[3] Orangutans and the fourth great ape, gorillas, have passed the mirror self-recognition test.[4] One zoo-raised gorilla's performance on another test has suggested her ability to share joint attention.[5] A heavily human-enculturated gorilla trained in sign language has performed well above the level of chance on the standard Assessment of Children's Language Comprehension test both in sign language and spoken English and tested at an average IQ level of 80 on a battery of standard tests.[6]

Although most monkeys fail, some rhesus monkeys have been able to follow the gazes of other monkeys and at least one Japanese macaque and one pigtailed macaque have passed a mirror self-recognition test.[7] Some capuchins (New World monkeys from Central and South America who branched from our common ancestor about 35 million years ago) have passed Stage 6 tests.[8] (Remember that passing Piaget's Stage 6 test is generally understood to be strong evidence that a subject can mentally represent.) Other monkeys communicate by pointing and deceive by refusing to point.[9] Some capuchins make stone tools that resemble those made by early *Homo* and in a similar fashion.[10] Vervet monkeys may understand analogies that allow them to think, "You hurt my sister, now I'm going to hurt yours."[11] Cotton-top tamarins, a New World monkey, appear to have passed a version of the Sally-Anne false-belief test.[12] Evidence increasingly suggests that some members of still other monkey species, such as the rhesus, an Old World monkey who separated from our common ancestor more than 20 million years ago, can count and add low numbers.[13]

But the capacity for complex mental representations may not be limited to primates or even to species evolutionarily close to humans. Bottlenose dolphins have passed the mirror self-recognition test.[14] The songs of humpback whales may be constructed from a complicated syntax.[15] Both elephants and African gray parrots use mirrors to help them search for objects.[16] Dogs have demonstrated Stage 6 object task permanence.[17] (Daniel Dennett, for one, would not be surprised if thousands of generations of human enculturation have not caused the canine brain to reorganize and produce a more advanced state of consciousness.)[18] New Caledonia crows regularly use hooks and tools that they manufacture to a high degree of standardization to aid in the capture of prey.[19] Common ravens size up a complex problem of thirty steps or more and solve it with no training and on the first try.[20] Scrub jays display "episodic recall." They remember where they stored food, what they stored, and when they stored it. If enough time has elapsed for the food they stored to spoil, they will ignore it. This ability in humans has been said to involve "the conscious experience of self."[21] Alex, an enculturated African gray parrot, not only understands shapes, colors, and materials, but can tell you—in English—how objects that he has never seen differ in shape, color, or material and whether they are made of cork, wood, paper, or rawhide.[22]

Because we have no time to investigate any of these claims here, they must remain tantalizing clues. But it should now be obvious that the ancient Great Wall that has for so long divided humans from every other animal is biased, irrational, unfair, and unjust. It is time to knock it down. The decision to extend common law personhood to chimpanzees and bonobos will arise from a great common law case. Great common law cases are produced when great common law judges radically restructure existing precedent in ways that reaffirm bedrock principles and policies. All the tools for deciding such a case exist. They await a great common law judge, a Mansfield, a Cardozo, a Holmes, to take them up and set to work. Until that day arrives, the legal thinghood of chimpanzees and bonobos will gnaw at the heart of what we believe, what we stand for, and who we are.

NOTES

Chapter One

1. Frederick Law Olmsted, *A Journey in the Back Country* 64 (1860).
2. Personal communication from Rachel I. Weiss, dated March 3, 1999; Rachel I. Weiss, "Yerkes: On the Inside" (Unpublished manuscript 1998); Rachel I. Weiss, "For Jerom," (Unpublished manuscript 1997); Frank J. Novembre, 71 *Journal of Virology* 4086 (1997).
3. Roger Fouts and Stephen Tukel Mills, *Next of Kin: What Chimpanzees Have Taught Me About Who We Are* 54 (William Morrow and Company, 1997), quoting Vincent Sarich, "Immunological Evidence on Primates," in *The Cambridge Encyclopedia of Evolution* (Cambridge University Press 1994).
4. Roger Fouts and Stephen Tukel Mills, *supra* note 3, at 239-40.
5. Roger Fouts and Stephen Tukel Mills, *supra* note 3, at 342-3.
6. Michael Bogdan, "Article 6," in *The Universal Declaration of Human Rights: A Commentary* 111 (Scandinavian Press 1992).
7. *International Covenant on Civil and Political Rights*, adopted Dec. 16, 1966, 99 U.N.T.S. 171 (entered into force Mar. 23, 1976)(reprinted in Richard B. Lillich, *International Human Rights Instruments* 170.7 [2nd ed. 1990])(Article 16); *American Covenant on Human Rights*, opened for signature Nov. 22, 1969, O.A.S/T/S. No. 36 (entered into force July 18, 1978)(reprinted in Richard B. Lillich, *supra* at 190.2)(Article 3)(leaves out the word "everywhere"); *Universal Declaration of Human Rights*, adopted Dec. 10, 1948, G.A. Resolution 217A (III), 3 GAOR (Resolutions, part 1) at 71, U.N. Doc. A/810 (1948)(reprinted in Richard B. Lillich, *supra* at 440.2)(Article 6).
8. Michael Bogdan, *supra* note 6, at 111.
9. Donald G. McNeil Jr., "The Great Ape Massacre," *The New York Times Magazine* 54, 55 (May 9, 1999).
10. Roger Fouts and Stephen Tukel Mills, *supra* note 3, at 329.
11. Donald G. McNeil Jr., *supra* note 9, at 54; Tom Butynski, "Africa's great apes: An overview of current taxonomy, distribution, numbers, conservation status, and threats," unpublished manuscript presented at a conference, "Great Apes and Humans at an Ethical Frontier," held at the Disney Institute, Orlando, Florida, June 21-24, 1998.
12. Frans de Waal and Frans Lanting, *Bonobos: The Forgotten Ape* 61-2 (University of California Press 1997).

13. Robin A. Weiss and Richard W. Wrangham, "From Pan to pandemic," 397 *Nature* 385, 385 (1999).

14. Deborah Blum, *The Monkey Wars* 23–4 (Oxford University Press 1994); Dale Petersen and Jane Goodall, *Visions of Caliban: On Chimpanzees and People* 266–67, 277–79 (Houghton-Mifflin 1993).

Chapter Two

1. Aristotle, "Politics," in 2 *The Complete Works of Aristotle* 1990–1991 (Jonathan Barnes, ed., Princeton University Press 1984).

2. Porphyry, *On Abstinence*, Book 3.20.1

3. Keith Thomas, *Man and the Natural World* 20 (Pantheon Books 1993).

4. Peter J. Bowler, *Evolution: The History of an Idea* 53 (rev. ed. University of California Press 1989).

5. *People v. Hall*, 4 Cal. 399, 405 (1854).

6. *Dred Scott v. Sandford*, 60 U.S. (19 How.) 393, 403 (1856).

7. *Loving v. Virginia*, 388 U.S. 1, 3 (1967).

8. *The Illiad of Homer* 3 (Ennis Rees trans. Modern Library 1963).

9. E.R. Dodds, *The Greeks and the Irrational* 6 (University of California Press 1951).

10. Hesiod, "Works and Days," in *Hesiod and Theognis* 67 (Dorthea Wender, trans. Penguin Books 1981).

11. Michael Grant, *The Rise of the Greeks* 160–163 (Charles Scribner's Son 1988); Drew A. Hyland, *The Origins of Philosophy* 96–126 (G.P. Putnam 1973); Reginald E. Allen, *Greek Philosophy: Thales to Aristotle* 1–5 (The Free Press 1966).

12. The most famous double-play combination known to baseball history.

13. Henry Fairfield Osborne, *From the Greeks to Darwin: An Outline of the Development of the Evolution Idea 2nd ed.* 39–40, 42 (MacMillan and Co. 1894); Aristotle, *Physics*, Book 2.4, 196a, in Jonathan Barnes, ed., *supra* note 1, at 334–335; Aristotle, "On the Parts of Animals," Book 1, 640a, in *id.* at 995–996.

14. Xenophon, *Recollections of Socrates*, 4.3.9–10, at 115 (Anna S. Benjamin, trans. Bobbs Merrill Co., Inc 1965).

15. Arthur O. Lovejoy, *The Great Chain of Being: A Study of the History of an Idea* 52, 59, 242 (Harvard University Press 1960).

16. *Id.* at viii (emphasis in the original).

17. Matt Cartmill, *A View to a Death in the Morning* 99 (Harvard University Press 1993).

18. Aristotle, "On the Generation of Animals," Book 2.1, 726b, 732a–33b, in Jonathan Barnes, ed., *supra* note 1, at 1128, 1136–7; D. R. Oldroyd, *Darwinian Impacts: An Introduction to the Darwinian Evolution* 5–7 (Humanities Press 1980); Charles Singer, *A Short History of Anatomy From the Greeks to Harvey 2nd ed.* 26 (Dover Press 1957).

19. Plato, *Timaeus*, secs 69D–73A, at 97–103 (John Warrington, ed. and trans. J.M. Dent & Sons Ltd. 1965). *See* Richard Sorabji, *Animal Minds and Human Morals* 10–1, 37 (Cornell University Press 1993).

20. Richard Sorabji, *supra* note 19, at 70.

21. Aristotle, "On the Soul," Book 2.2–2.3, 413a–415a, in Jonathan Barnes, ed., *supra* note 1, at 657–660; Aristotle, "On the Parts of Animals," Book 1.1, 641b, in *id.* at 998; Arthur O. Lovejoy, *supra* note 15, at 58–59.

22. Nancy Tuana, *The Less Noble Sex* 3, 18–21 (Indiana Universtiy Press, 1993).

23. Aristotle, "Politics," Book 1.13, 1260a, in Jonathan Barnes, ed., *supra* note 1, at 1999–2000.

24. Aristotle, *On the Soul*, Book 2.2–2.3, 414a–414b, in Jonathan Barnes, ed., *supra* note 1, at 659–660; Aristotle, "Nichomachean Ethics," Book 1.7, 1098a, in *id.* at 1735; Book 1.13, 1102b–1103a, in *id.* at 1741, 1742; Aristotle, *Eudemian Ethics*, Book 2.1, 1219b, in *id.* at 1931.

25. Aristotle, "Politics," Book 1.5, 1254b, in Jonathan Barnes, *supra* note 1, at 1990; Aristotle, "Eudemian Ethics," Book 2.1, 1219b, in *id.* at 1930–1931; Aristotle, "Nichomachean Ethics," Book 1.7, 1098a, in *id.* at 1735; Book 1.13, 1102b–1103a, in *id.* at 1735, 1741–1742. *See* Richard Sorabji, *supra* note 19, at 70, 135.

26. Aristotle, "Nichomachean Ethics," Book 8.11, 1161a–1161b, in Jonathan Barnes, *supra* note 1, at 1835; Peter Green, *Alexander of Macedon, 356–323 B.C.: A Historical Biography* 58 (University of California Press 1993).

27. Aristotle, *Politics*, Book 1.8, 1256b, in Jonathan Barnes, ed., *supra* note 1, at 1993–1994; Aristotle, *Physics*, Book 2.8, 198b, in *id.* at 339. *E.g.*, Aristotle, *On The Parts of Animals*, Book 1.1, 639b–640a, in *id..* at 994–996; Aristotle, *On the Generation of Animals*, Book 1.1, 715a, in *id.* at 1111. *See* James Rachels, *Created From Animals* 112–114 (Oxford University Press 1990); Stephen Jay Gould, "Hutton's Purpose," in *Hens' Teeth and Horses' Toes* 79–81 (W.W. Norton & Co. 1983); E.E. Spicer, *Aristotle's Conception of the Soul* 45–47, 118–124 (University of London Press 1934).

28. Robert S. Brumbaugh, "Of Man, Animals, and Morals: A Brief History," in *On the Fifth Day: Animal Rights and Human Ethics* 8 (Acropolis Books Ltd. 1978). Lovejoy used the synonym, "anthropocentric teleology," Arthur O. Lovejoy, *supra* note 15, at 188.

29. *New Developments in Biotechnology: Patenting Life, Special Report*, OTA-BA 98 (U.S. Printing Office, April, 1989); Peter J. Bowler, *supra* note 4, at 5, 59; Oliver L. Reiser, "The Concept of Evolution in Philosophy," in *A Book That Shook the World: Anniversary Essays on Charles Darwin's Origin of Species* 38–41 (University of Pittsburgh Press 1967); David Brion Davis, *The Problem of Slavery in Western Culture* 68, 152 (Cornell University Press 1966); John Dewey, *The Influence of Darwin on Philosophy* 5–7, 9–11 (Indiana University Press 1965); Arthur O. Lovejoy, *supra* note 15, at 52, 57, 59, 200–7.

30. Stephen Jay Gould, "Bound by the Great Chain," in *The Flamingo's Smile* 282 (W.W. Norton & Co. 1985).

31. Aristotle, "Movement of Animals," Book 1.7, 701b, in Jonathan Barnes, ed., *supra* note 2, at 1092; Martha Craven Nussbaum, *Aristotle's De Motu Animalium*, 60 (Princeton University Press 1978).

32. Aristotle, "On the Soul," Book 3.3, 428b, in Jonathan Barnes, ed., *supra* note 2, at 681; Aristotle, "Metaphysics," Book 1.1, 980a–980b, in *id.* at 1554; Aristotle, "Posterior Analytics," Book 2.19, 100a, in *id.* at 165; Aristotle, "On The Parts of Animals," Books 2.3, 415a; 3.3, 428a, 433b–434a; 3.10–3.11, 433b–434a, in *id.* at 660, 680–681, 689–690.

33. Richard Sorabji, *supra* note 19, at 12.

34. Aristotle, "Politics," Book 1.2, 1253a, in *id*. at 1988; Aristotle, "On the Soul," Book 3.7, 431a, in *id*, at 685; *See* Richard Sorabji, *supra* note 19, at 55–8.

35. Aristotle, *Politics*, Book 1.8, 1256b, in Jonathan Barnes, ed., *supra* note 1, at 1993–1994; Stephen R.L. Clark, *The Moral Status of Animals* 163 (Oxford University Press 1984); Richard Sorabji, *supra* note 19, at 110, 199.

36. M.L. Clarke, *The Roman Mind* 133 (Harvard University Press 1960); William C. Morey, *Outlines of Roman Law 2nd ed.* 108 (G.P. Putnam's Son 1914).

37. Ludwig Edelstein, *The Meaning of Stoicism* 73–74, 83–84 (Harvard University Press 1966). *See* Richard Sorabji, *supra* note 19, at 134; Robert J. Harris, *The Quest for Equality: The Constitution, Congress and the Supreme Court* 4–8 (Louisiana State University Press 1960).

38. Eduard Zeller, *The Stoics, Epicureans and Sceptics* 185–186 (Oscar J. Reichel, trans. Russell & Russell, Inc. 1962); Gerard Verbeke, "Ethics and Logic in Stoicism," in *Atoms, Pneuma, and Tranquility* 4, 24 (Margaret J. Oscer, ed. Cambridge University Press 1991).

39. Richard Sorabji, *supra* note 19, at 134–5.

40. Cicero, *De Naturum Deorum,* 277 (H. Rackhman, trans. William Heineman Ltd. 1933); *id*. at 159.

41. Eduard Zeller, *supra* note 38, at 186 note 2; E. V. Arnold, *Roman Stoicism* 205 note 40 (1911); Stephen R.L. Clark, *supra* note 35, at 15. *See* Cicero, *supra* note 40, at 275–276; John Passmore, *Man's Responsibility for Nature* 14 (Charles Scribner's Sons 1974); Porphyry, *supra* note 2.

42. Richard Sorabji, *supra* note 19, at 125; Eduard Zeller, *supra* note 38, at 208; Daniel A. Dombrowski, *The Philosophy of Vegetarianism* 76 (University of Massachusetts Press, 1984); John Passmore, *supra* note 41, at 15.

43. Richard Sorabji, *supra* note 19, at 52.

44. *Id*. at 20–21.

45. *Id*. at 40–2; *id*. at 52 (quoting the Stoic, Seneca, *Epistles* 124, 16); *id*. at 60, (discussing Seneca, *On Anger* 1.3); *id*. at 86 (discussing Seneca, *Epistles* 121, 19–23). *See* Cicero, *De Officiis, Book* 1.11, at 13 (Walter Miller, trans. 1968).

46. Richard Sorabji, *supra* note 19, at 40–42, 59–60; A.A. Long, "Language and Thought in Stoicism," in *The Problems of Stoicism* 86–9 (A.A. Long, ed. The Athene Press 1971); Ludwig Edelstein, *supra* note 37, at 35–36.

47. Richard Sorabji, *supra* note 19, at 42.

48. *Id*. at 114.

49. *Id*. at 81; A.A. Long, in A.A. Long, ed., *supra* note 46, at 87.

50. Cicero, *De Finibus Bonorum et Malorum* 287 (H. Rackham, trans. William Heineman Ltd. 1967).

51. Richard Sorabji, *supra* note 19, at 122–124; Stephen R.L. Clark, *supra* note 35, at 23; Daniel A. Dombrowski, *supra*, note 42, at 75–6 155, note 3; S. G. Pembroke, "Oikeiosis," in A.A. Long, ed., *supra* note 46, at 121; Eduard Zeller, *supra* note 38, at 313; A.B. Bruce, *The Moral Order of the World in Ancient and Modern Thought* 387 (Hodder & Stoughton 1899).

52. Lloyd L. Weinreb, *Natural Law and Justice* 1, 15 (Harvard University Press 1987). *See* J. T. Dobbs, "Stoic and Epicurean doctrines in Newton's system of the world," in Margaret J. Oscer, ed., *supra* note 38; Ludwig Edelstein, *supra* note 37, at 32; Charles H. Kahn, *Anaximander and the Origins of Greek Cosmology* 191 (Columbia University Press 1960).

53. A.A. Long, *Hellenistic Philosophy: Stoics, Epicureans, Skeptics* 169 (Duckworth 1974); 1 *The Hellenistic Philosophers* 331 (A.A. Long and D. N. Sedley, eds. Cambridge University Press 1987).

54. George H. Sabine, *A History of Political Theory* 4th ed. 161 (rev. Dryden Press 1973).

55. Daniel Dombrowski, *supra* note 41, at 76–77; Richard Sorabji, *supra* note 19, at 199; M.L. Clarke, *supra* note 36, at 148; Lloyd L. Weinreb, *supra* note 52, at 36; J.R. Pole, *The Pursuit of Equality in American History* 5, 156 (University of California Press 1978).

56. Genesis 1.28 (Authorized King James Version). *See* Robin Lane Fox, *The Unauthorized Version: Truth and Fiction in the Bible* 21(Alfred A. Knopf 1992).

57. Genesis 1.29 (Authorized King James Version).

58. Marilyn A. Katz, "Ox-Slaughter and Goring Oxen: Homicide, Animal Sacrifice, and Judicial Process," 4 *Yale J. L. & Human* 249, 274 (1992).

59. Genesis 9:1–3 (Authorized King James Version). *See* Robin Lane Fox, *supra* note 56, at 177.

60. Mark 5:2–13 (Authorized King James Version).

61. I Corinthians 9:10–11 (Authorized King James Version).

62. Lloyd L. Weinreb, *supra* note 52, at 47, quoting Romans 2:14–15.

63. Jeffrey Burton Russell, *Satan: The Early Christian Tradition* 109–111 (University of California Press 1981).

64. Origen, *Contra Celsum*, Book IV:74–80, at 242–248 (Henry Chadwick, ed. and trans. Cambridge University Press 1953); Richard Sorabji, *supra* note 19, at 200.

65. Lloyd L. Weinreb, *supra* note 52, at 47; St. Augustine, *The Confessions of St. Augustine*, Book 8, Chap. 7.

66. Gerard Watson, "The Natural Law and Stoicism," in A.A. Long and D.N. Sedley, eds., *supra* note 53, at 235–236.

67. Augustine, *The City of God* 360 (Marcus Dods, trans. Modern Library 1950).

68. Supra note 67, at 26.

69. Richard Sorabji, *supra* note 19, at 196; Gerard O'Daly, *Augustine's Philosophy of Mind* 11 and note 32, 14 (University of California Press 1987); Augustine, *supra* note 67, at 158, 228.

70. Gerard O'Daly, *supra* note 69, at 12–14; Augustine, *supra* note 67, at 158.

71. Gerard O'Daly, *supra* note 69, at 97–98.

72. *Id.* at 99.

73. Augustine, *supra* note 67 at 372.

74. Gerard O'Daly, *supra* note 69, at 47, 89.

75. *Id.* at 90.

76. Augustine, *The Catholic and Manichaean Ways of Life* 102 (D. A. Gallagher and I. J. Gallagher, trans. The Catholic University Press 1966). *See* Matthew 21:19 and Mark 11:13–14 (stories of the fig tree).

77. William V. O'Brien, *The Conduct of Just and Limited War* 34 (Praeger Publishers 1981).

78. *Id.*

79. Augustine, *supra* note 67, at 1.20. *See* D. R. Oldroyd, *supra* note 18, at 2–3.

80. Robert S. Brumbaugh, *supra* note 28, at 11.

81. Arthur O. Lovejoy, *supra* note 15, at 186.

82. Keith Thomas, *supra* note 3, at 17–21; *e.g.*, Immanuel Kant, *Critique of Teleological Judgment*, part 2.22.431, at 93–94 (J.C. Meredith, trans. 1928).

83. St. Thomas Aquinas, "The Summa Theologica" Q.96, Art. 1, in 1 *Basic Writings of Saint Thomas Aquinas* 918 (Anton C. Pegis, ed. Random House 1945). *See* Thomas Aquinas, "The Summa Contra Gentiles," Chap. 112, in 2 *Basic Writings of Thomas Saint Thomas Aquinas* 122 (Anton C. Pegis, ed. Random House 1945); Arthur O. Lovejoy, *supra* note 15, at 186–187.

84. Arthur O. Lovejoy, *supra* note 15, at 188, quoting Galileo Galileo, *Dialogo di due massimi systemi*, III, 400.

85. Arthur O. Lovejoy, *supra* note 15, at 108–143.

86. Keith Thomas, *supra* note 3, at 168.

87. *Id.* at 167–168.

88. Arthur O. Lovejoy, *supra* note 15, at 188, quoting Rene Descartes, *Antidote Against Atheism*, II, ch. 9, 8.

89. James Rachels, *supra* note 27, at 113–115.

90. Peter J. Bowler, *supra* note 4, at 26–49; Keith Thomas, *supra* note 3, at 168–169; Francis C. Haber, "Fossils and the Idea of a Process of Time in Natural History," in *Forerunners of Darwin: 1745–1859* 222–61 (Bentley Glass, Oswei Tomkin, and William C. Strauss, Jr., eds. The Johns Hopkins Press 1959).

91. D. O. Oldroyd, *supra* note 18, at 20. *See id.* at 15; Peter J. Bowler, *supra* note 4, at 56, 64–65.

92. D. O. Oldroyd, *supra* note 18, at 21–22; Peter J. Bowler, *supra* note 4, at 56.

93. Arthur O. Lovejoy, *supra* note 15, at 242–287.

94. C.C. Gillespie, *The Edge of Objectivity: An Essay in the History of Scientific Ideas* 272 (Princeton University Press 1960).

95. Peter J. Bowler, *supra* note 4, at 165–167; D. O. Oldroyd, *supra* note 18, at 85–90; John Dewey, *supra* note 29, at 8.

96. Ernst Mayr, *Toward a New Philosophy of Biology* 196–212 (Harvard University Press (1988).

97. Ernst Mayr, "Introduction," in Charles Darwin, *On the Origin of Species* xv (Harvard University Press 1964)(A Facsimile of the First Edition).

98. James Rachels, *supra* note 27, at 115–126.

99. Ernst Mayr, *supra* note 96 at 3.

100. Edward J. Larson and Larry Witham, "Scientists and Religion in America," *Scientific American* 88, 90 (September 1999).

101. *Id.* at 21.

102. Daniel J. Boorstin, *The Discoverers* 457 (Vintage Book 1985).

103. Arthur O. Lovejoy, *supra* note 15, at 205.

104. *Id.*

105. Steven M. Wise, "Of Farm Animals and Justice," 3 *Pace Environmental Law Review* 191, 203 (1986); 4 *Am. Jur. 2d*, "Animals," secs. 1, 2, and 5, at 250–251, 253 (1962); 3A *C.J.S.* secs 3,4,6–10, at 474–475, 477–481 (1973).

CHAPTER THREE

1. J. J. Finkelstein, "The Ox That Gored," 71 *Am. Phil. Soc.* (part 2) 17–21 (1981); Michael Walzer, "The Legal Codes of Ancient Israel," 4 *Yale J. L. & Human* 335, 340 (1992); *The Ancient Near East: Supplementary Texts and Pictures Relating to the Old*

Testament 21 (James B. Pritchard, ed., J.J. Finkelstein, trans., Princeton University Press 1969)(a suit involving an ox from the Middle Babylonian period of about 1300 B.C.); Reuven Yaron, "The Goring Ox in Near Eastern Laws," 1 *Israel L. Rev.* 396, 398 (1966).

2. Dig. 1.5.2 (Hermogenianus, *Epitome of Law,* book 1)(Theodor Mommsen, Paul Krueger, and Alan Watson, eds., University of Pennsylvania Press 1985).

3. P.A. Fitzgerald, *Salmond on Jurisprudence* 300 (Sweet & Maxwell, 12th ed. 1966).

4. Sheldon M. Novick, *Honorable Justice: The Life of Oliver Wendell Holmes* 148–149 (Little, Brown & Company 1989).

5. Alan Watson, *Legal Transplants: An Approach to Comparative Law* 95 (University of Georgia Press, 2nd ed. 1993).

6. *Id.* at 7–8.

7. *Id.* at 93–94, 99, 112–114; Alan Watson, *Society and Legal Change* 104–105 (Scottish Academic Press Ltd. 1977); Alan Watson, "Legal Change: Sources of Law and Culture," in *Legal Origins and Legal Change* 73 (The Hambleton Press 1991).

8. George Santayana, *Life of Reason* 284 (Prometheus Books 1905).

9. Oliver Wendell Holmes, *The Common Law* 8 (Mark DeWolfe Howe, ed. Little, Brown and Co., 1963)(1881).

10. Alan Watson, *Legal Origins and Legal Change, supra* note 7, at 132–133; Oliver Wendell Holmes, Jr., "The Path of the Law," 10 *Harv. L. Rev.* 457, 469 (1897).

11. R.F. Willetts, *The Civilization of Ancient Crete* 91–92 (University of Chicago Press 1977); Shalom M. Paul, *Studies in the Book of the Covenant in the Light of Cuneiform and Biblical Law* 3 (E.J. Brill 1970).

12. Raphael Sealey, *The Justice of the Greeks* 23, 150, 151, 155 (University of Michigan Press 1994); Roscoe Pound, *An Introduction to the Philosophy of Law* 1 (Yale University Press 1954).

13. Shalom M. Paul, *supra* note 11, at 6.

14. *Id.* at 7, 8.

15. Shalom M. Paul, *supra* note 11, at 100. *Id.* at 36–37; Carl S. Ehrlich, "Israelite Law," in *The Oxford Companion to the Bible* 421 (Bruce M. Metzger and Michael D. Coogan, eds., Oxford University Press 1993); Michael Walzer, *supra* note 1, at 335, 336, 340–341.

16. The manner in which the various Near Eastern law codes recognized this is detailed in Steven M. Wise, "The Legal Thinghood of Nonhuman Animals, 23 *Boston Coll. Env. Aff. L. Rev.* 478–82 (1996).

17. James B. Pritchard, ed., supra note 1, at 87.

18. *Ancient Near Eastern Texts Relating to the Old Testament* 159 (2nd. ed., James B. Pritchard, ed., Princeton University Press 1955).

19. Johnson dates these Laws to about 1920 B.C., Paul Johnson, *A History of the Jews* 32 (1987). Another recent estimate is that they are of the nineteenth century, B.C., Carl S. Ehrlich, *supra* note 15, at 422.

20. James B. Pritchard, ed., *supra* note 18, at 163.

21. Carl S. Ehrlich, *supra* note 15, at 421; J.J. Finkelstein, "Ammisaduqua's Edict and the Babylonian 'Law Codes,'" *J. Cuneiform S.* 15: 102–103 (1961).

22. J.J. Finkelstein, *supra* note 1, at 15–16 and note 5.

23. Josephus, *Contra Apion* 2.154, at 353, 355 (H. St. J. Thackery, trans., William Heinman and G.P. Putnam's Sons 1926).

24. "The Hebrew word most often translated as 'law,' tora (Torah), actually means teaching or instruction. As such it expresses the morally and socially didactic nature of God's demands on the Israelite people," Carl S. Ehrlich, *supra* note 15, at 421. Michael Grant argues that "Torah originally carried the sense of "teaching conveyed by divine revelation," Michael Grant, *Saint Peter: A Biography* 17 (Charles Scribner's Sons 1995). In Mesopotamia "a technical term for law itself did not exist, nor was there an expression for 'by the application of the law' or 'in virtue of such and such a law," Shalom M. Paul, *supra* note 11, at 5.

25. Robert H. Pfeiffer, *Introduction to the Old Testament* 210 (Harper and Row 1948); Carl S. Ehrlich, *supra* note 15, at 422–423.

26. David J. A. Clines, in *The Oxford Companion to the Bible, supra* note 15, at 580; Robin Lane Fox, *The Unauthorized Version: Truth and Fiction in the Bible* 85 (Alfred A. Knopf, Inc. 1991); J.J. Finkelstein, *supra* note 1, at 17; Shalom M. Paul, *supra* note 11, at 104.

27. John I. Durham, in *The Oxford Companion to the Bible, supra* note 15, at 213; Paul Johnson, *supra* note 19, at 32.

28. J.J. Finkelstein, *supra* note 1, at 16–17.

29. The parallel between this provision and section 53 of the Laws of Eshunna is the closest known between a biblical law and a Near Eastern law code, Reuven Yaron, *supra* note 1, at 398.

30. J.J. Finkelstein, *supra* note 1, at 36.

31. The two omitted verses provide that when a an ox or ass falls into an open pit, the pit owner must pay the owner of the ox or ass, but may keep the dead body.

32. The Laws of Eshunna, secs. 56 and 57, treated a vicious dog who bit in the same way it treated a goring ox, Reuven Yaron, *supra* note 1, at 402, 404. In Yaron's opinion, the Laws of Hammurabi and the Covenant Code dispensed with the vicious dog only because he was seen as redundant to the goring ox, *id.* at 404.

33. Shalom M. Paul, *supra* note 11, at 82.

34. *Id.* at 8.

35. Henri Frankfort, *Before Philosophy: The Intellectual Adventure of Ancient Man* 152 (Pelican Books 1949).

36. J.J. Finkelstein, *supra* note 1, at 8–13, 39. Henri Frankfort, *supra* note 35, at 12.

37. J.J. Finkelstein, *supra* note 1, at 12.

38. Steven M. Wise, *supra* note 16, at 30–31; J.J. Finkelstein, *supra* note 1, at 7, 46, 71–2. *See* Genesis 9:5–6.

39. Moshe Greenberg, "Some Postulates of Biblical Criminal Law," *Yehezkel Kaufmann Jubilee Volume* 16 (Menaham Haran, ed., Magnes Press 1960).

40. Shalom M. Paul, *supra* note 11, at 37, 100.

41. *Id.* at 82.

42. Michael Walzer, *supra* note 1, at 340; Shalom M. Paul, *supra* note 11, at 61–62.

43. Genesis 9:5–6.

44. Shalom M. Paul, *supra* note 11, at 39.

45. J.J. Finkelstein, *supra* note 1, at 28, quoting Gen. 1:26, 28, 37. *E.g.,* Shalom M. Paul, *supra* note 11, at 79. *See United States v. Seifuddin,* 820 F. 2d 1074, 1076 (9th Cir. 1987).

46. Moishe Greenberg, *supra* note 39, at 15.

47. Exodus 21:32.

48. All three codes regulated only the ox who fatally gored a human being, but not less harmful human gorings, Reuven Yaron, *supra* note 1, at 404.

49. J.J. Finkelstein, *supra* note 1, at 32; the Tractate "Sanhedrin" 1.4, in *The Mishnah* 383 (Herbert Danby, trans., Oxford University Press 1964). The rabbis later reinterpreted the biblical rule to exonerate goring oxen if the human death had been accidental, J.J Finkelstein, *supra* note 1, at 32. While the later Talmudic rabbis allowed for the automatic death-by-stoning penalty to be modified in the case of ox goring, they stood fast on the automatic death penalty for nonhuman animals used in bestial acts, *id.* at 71.

50. *A Dictionary of the Bible* 481–2 (J.B. Burr & Hyde, William Smith, ed. 1873).

51. J.J. Finkelstein, *supra* note 1, at 57 and note 50.

52. *Id.* at 26–27.

53. *Id.* at 28. *E.g.,* 1 Emanuel B. Quint and Neil S. Hecht, *Jewish Jurisprudence: Its Sources and Modern Applications* 37–38 (Harwood Academic Publishers 1980).

54. Shalom M. Paul, *supra* note 11, at 83.

55. J.J. Finkelstein, *supra* note 1, at 28, 58.

56. Raphael Sealey, *supra* note 12, at 61–62.

57. Edith Hamilton, *The Echo of Greece* 23 (W.W. Norton & Company, Inc. 1957).

58. Douglas M. MacDowell, *The Law in Classical Athens* 138 (Cornell University Press 1978).

59. Paul Cartledge, "Fowl play: a curious lawsuit in classical Athens (Antiphon xvi, frr. 57–59 Thalheim)", in Paul Cartledge, Paul Millett, and Stephen Todd, eds., *supra* note 62, at 41–61; Douglas M. MacDowell, *supra* note 58, at 146, 233.

60. Stephen G. Miller, *The Pyrtaneion: Its Function and Structural Form* 18 (University of California Press 1978); Note, "Rats, Pigs, and Statues on Trial: The Creation of Cultural Narratives in the Prosecution of Animals and Inanimate Objects," 69 *N.Y.U. L. Rev.* 288, 288, 290, 294–296 (1994); Douglas M. McDowell, *supra* note 58, at 117; J.J. Finkelstein, *supra* note 1, at 58–60; Walter Woodburn Hyde, "The Prosecution and Punishment of Animals and Lifeless Things in the Middle Ages and Modern Times," 64 *U. Pa .L. Rev.* 696, 696, 696–700 (1916); E.P. Evans, *The Criminal Prosecution and Capital Punishment of Animals: The Lost History of Europe's Animal Trials* 9, 172–174 (Faber and Faber, Limited 1987 repr.)(1906).

61. Michael Gagarin, *Early Greek Law* 132–133 (University of California Press 1986).

62. *Id.* at 53, 54, 123–4, 154.

63. Alan Watson, *Legal Origins and Legal Change, supra* note 7, at 73, 79, 98; Alan Watson, *Legal Transplants, supra* note 5, at 90. *E.g.,* Donald R. Kelley, *The Human Measure: Social Thought in the Western Legal Tradition* 64–65 (Harvard University Press 1990); *Lane v. Cotton,* 88 Eng. Rep. 1458, 1463 (1701)(Holt, C.J.).

64. Dig. 1.5.3 (Gaius, *Institutes,* book 1).

65. Thomas Collett Sandars, *The Institutes of Justinian* 33 (William S. Hein & Co. 1984).

66. *Id.* at 26, 50.

67. Alan Watson, *Rome of the XII Tables: Persons and Property* 165 (Princeton University Press 1975); Roscoe Pound, *supra* note 12, at 110.

68. Theodor Mommsen, Paul Krueger, and Alan Watson, eds., *supra* note 2, at lv-lvi; Donald R. Kelley, *supra* note 63, at 53–54; Glanville Downey, *Constantinople in the Age of Justinian* 69, 72, 75 (Dorset Press 1991); John Julius Norwich, *Byzantium: The Early Centuries* 196 (Alfred A. Knopf 1988).

69. Dig. 1.1.1.3.

70. J. Inst. 1.2.1; J. Inst. 1.3.2; Dig. 1.1.4 (Ulpian, *Institutes*, book 1); Dig. 1.5.4.1 (Florentinus, *Institutes*, book 9); Lloyd L. Weinreb, *Natural Law and Justice* 45 (Harvard University Press 1987).

71. David Brion Davis, *Slavery and Human Progress* 20 (Oxford University Press 1984).

72. Thomas Collett Sandars, *supra* note 65, at 77; Alan Watson, *Roman Slave Law*, 7 (The Johns Hopkins Press 1987); David Brion Davis, *supra* note 71, at 20, 83; Thornton Stringfellow, "A Brief Examination of Scripture Testimony on the Institution of Slavery," in *The Ideology of Slavery* 136 (Drew Gilpin Faust, ed., Louisiana State University Press 1981).

73. Dig. 9.2.2 (Gaius, *Provincial Edict.*, book 7). *See* Alan Watson, *supra* note 72, at 46, 54–57.

74. Alan Watson, *supra* note 72, at 49–50. *See* Oliver Wendell Holmes, Jr., *supra* note 9, at 10–17.

75. Dig. 41.1.6 (Gaius, *Common Matters or Golden Things*, book 2). See J. Inst. 2.1.15, .16, citing Dig 41.1.5.6 (Gaius, *Common Matters or Golden Things*, book 2).

76. Thornton Stringfellow, *supra* note 72, at 136.

77. Alan Watson, *The Law of Property in the Later Roman Republic* 215–216 (Oxford University Press 1968).

78. Alan Watson, *Legal Origins and Legal Change, supra* note 7, at 215; A.P. d'Entreaves, *Natural Law: An Introduction to Legal Philosophy* 34 (2nd ed. Hutchinson Publishing Group 1971).

79. Michael Grant, *supra* note 24, at 4–10 (discussing how the Jews of Christ's time did not readily distinguish between the natural and miraculous); David R. Dow, "When Words Mean What We Believe They Say: The Case of Article V," 76 *Iowa L. Rev.* 1, 64–65 (1990)(discussing the apparent unconcern of the early Talmudic rabbis with the paradox of God determining all human action and humans using free will).

80. Donald R. Kelley, *supra* note 63, at 49; Roscoe Pound, *supra* note 12, at 110.

81. Dig. 41.1.3 (Gaius, *Common Matters or Golden Things*, book 2); Dig. 41.1.55 (Proculus, *Letters*, book 2); Dig. 41.1.44 (Ulpian, *Edict*, book 19).

82. Dig. 41.1.5 (Gaius, *Common Matters or Golden Things*, book 2); Dig. 41.2.3.14 (Paul, *Edict*, book 54.); J. Inst. 2.1.12.

83. Alan Watson, *supra* note 77, at 13; Joseph L. Sax, "The Public Trust Doctrine in Natural Resource Law: Effective Judicial Intervention," 68 *Mich. L. Rev.* 471, 475 (1970); R.W. Lee, *The Elements of Roman Law* 110 (4th ed., Sweet & Maxwell, Limited 1956); Patrick Mac Chombaich de Colquhoun, *A Summary of the Roman Civil Law Illustrated by Commentaries and Parallels from the Mosaic, Canon, Mohammedan, English, and Foreign Law* 4–5 (Fred B. Rothman & Co. 1988)(1851); W.W. Buckland, *supra* note 109, at 184–185; W. A. Hunter, *A Systematic and Historical Exposition of Roman Law in the Order of a Code* 255–257 (J. Ashton Cross, trans., 4th ed. 1903); Thomas Collett Sandars, *supra* note 75, at 160.

84. *Toomer v. Witsell*, 334 U.S. 385, 402 note 37 (1946); *Commonwealth v. Agway, Inc.*, 210 Pa. Super. 150, 154–155, 232 A. 2d. 69, 71 (1967); Roscoe Pound, *supra* note 12, at 111.

85. Donald R. Kelley, *supra* note 63, at 44, 47; 1 William E. H. Lecky, *History of European Morals* 314–315 (D. Appleton and Co. 1869); Edward D. Re, "The Roman Contribution to the Common Law," 29 *Fordham L. Rev.* 447, 452 (1960). *E.g.*, H. McCoubrey, *The Development of Naturalist Legal Theory* 31–36 (Croom Helm Ltd. 1987); A.P. D'Entreves, *supra* note 78, at 20, 23, 25, 27; C. H. Ziegler, *Roman Private Law* (R. W. Leage, 1st ed) 22 (2nd. ed., MacMillan and Co., Limited 1948); Charles Grove Haines, *The Revival of Natural Law Concepts: A Study of the Establishment and of the Interpretation of Limits on Legislatures with special reference to the Development of certain phases of American Constitutional Law* 8, 8 note 4, 9 (Harvard University Press 1930); William C. Morey, *Outlines of Roman Law Comprising Its Historical Growth and General Principles* 107, 109–10 (12th impression, G.P. Putnam's Sons 1902); Henry Sumner Maine, *Ancient Law* 46–7 (Dorset Press 1986)(1861).

CHAPTER FOUR

1. E.P. Evans, *The Criminal Prosecution and Capital Punishment of Animals: The Lost History of Europe's Animal Trials* 31 (Faber and Faber, Ltd., 1987)(1904).
2. *Id.* at 18–9; Walter Woodburn Hyde, "The Prosecution and Punishment of Animals and Lifeless Things in the Middle Ages and Modern Times," 64 *U. Pa. L. Rev.* 696, 706–7 (1916).
3. E.P. Evans, *supra* note 1, at 28, 41–9, 93–4, 115–6, 123–4, 51–2.
4. J.J. Finkelstein, "The Ox That Gored," 71 *Trans. A. Phil. Soc.*, Part 2 31 (1981).
5. E.P. Evans, *supra* note 1, at 37–50.
6. *Id.* at 31–2; J.J. Finkelstein, *supra* note 4, at 64–5.
7. William Shakespeare, *The Merchant of Venice*, act 4, scene 1, II.133–5.
8. E.P. Evans, *supra* note 1, at 140.
9. *Id.* at 16, 287.
10. *Id.* at 141, 288–9.
11. *Id.* at 290–1.
12. *Id.* at 143.
13. *Id.* at 304–5.
14. *Id.* at 160–1, 310.
15. *Id.* at 176.
16. *Id.* at 169–70.
17. *Id.* at 165.
18. J.J. Finkelstein, *supra* note 4, at 71; Esther Cohen, *The Crossroads of Justice: Law and Culture in Late Medieval France* 175 (E.J. Brill 1993); Piers Beirnes, "The Law is an Ass: Reading E.P. Evans's The Medieval Prosecution and Capital Punishment of Animals" in 2 *Society and Animals* 30, 40 (1994).
19. Coke, *Third Part of the Institutes of the Laws of England* 57 (Garland Publishing Co. 1979)(1644);
20. E.P. Evans, *supra* note 1, at 146–53. *See* Jonas Liliequist, "Peasants against Nature: Crossing the Boundaries between Man and Animal in Seventeenth and

Eighteenth-Century Sweden, 13 *Focaal: Tijdschrift voor Anthroplogie* 28, 50–1 note 5 (1990).

21. E.P. Evans, *supra* note 1, at 153; *id.* at 165.

22. Jonas Liliequist, *supra* note 20, at 29.

23. *Id.* at 147–53.

24. Jonas Liliequist, *supra* note 20, at 29.

25. *Id.* at 37.

26. E.P. Evans, *supra* note 1, at 150.

27. J.J. Finkelstein, *supra* note 4, at 64, 69.

28. Esther Cohen, *supra* note 18, at 83, 113; Jonas Liliequist, *supra* note 20, at 29, 31, 32, 33, 34; Esther Cohen, "Law, Folklore and Animal Lore," in *Past and Present* 11 (1986).

29. J.J. Finkelstein, *supra* note 4, at 73 (emphasis in the original). *E.g.*, Esther Cohen, *supra* note 18, at 110, 113.

30. J.J. Finkelstein, *supra* note 4, at 70, quoting E.P. Evans, *supra* note 1, at 171.

31. *Id.*

32. *Calero-Toledo v. Pearson Yacht Leasing Co.*, 416 U.S. 663, 681–682 (1974); *J.W. Goldsmith, Jr.-Grant Co. v. United States*, 254 U.S. 505, 511 (1921). *See* Jacob J. Finkelstein, "The Goring Ox: Some Historical Perspectives on Deodands, Forfeitures, Wrongful Death and the Western Notion of Sovereignty," 46 *Temple L. Q.* 169, 185 (1973).

33. 1 William Blackstone, *Commentaries on the Law of England* *301 (1766).

34. Jacob J. Finkelstein, *supra* note 32, at 181; J.J. Finkelstein, *supra* note 4, at 69, 75, 77, 80, 82; W.E. Lunt, *History of England* 45 (3rd. ed. 1949); Glanville Williams, *Liability for Animals: An Account of the Development and Present Law of Tortious Liability for Animals, Distress Damage Feasant and the Duty to Fence in Great Britain, Northern Ireland and the Common-Law Dominions* 267 and note 1 (Cambridge University Press 1939); *Calero-Toledo, supra* note 32, at 682 note 17; *J.W. Goldsmith, Jr.-Grant Co., supra* note 32, at 512.

35. Piers Beirnes, *supra* note 18, at 34; *Calero-Toledo, supra* note 32, at 681–682; *J.W. Goldsmith, Jr.-Grant Co., supra* note 32, at 510–511.

36. *E.g.*, Alan Watson, *Rome of the XII Tables: Persons and Property* 3 (Princeton University Press 1975); Oliver Wendell Holmes, *The Common Law* 10–1, 19 (Mark deWolfe Howe, ed. Little Brown & Co. 1963); Glanville Williams, *supra* note 34, at 267; Jacob J. Finkelstein, *supra* note 32, at 181; E.C. Clark, *History of Private Roman Law, Part III, Regal Period* 60 (Wm. W. Gaunt & Sons, Inc. 1990 repr.). *See Willis v. Schuster*, 28 So.2d 518, 521 (Ct. of App. of La. 1946), *cert. den.* (1947)(discussing Article 2321 of the Louisiana Civil Code); *Winter v. Mudianse*, 22 Ceylon L.R. 153 (1920).

37. *Id.* at 79; Jacob J. Finkelstein, supra note 32, at 170–171.

38. Coke, *supra* note 19, at 57; Walter Woodburne Hyde, *supra* note 2, at 726–7 note 104 (1916), quoting Coke, *supra* note 19, at 57; Jacob J. Finkelstein, *supra* note 32, at 182, 185.

39. Oliver Wendell Holmes, *supra* note 36, at 23–4; William Blackstone, *supra* note 33, at *301.

40. *Regina v. Eastern Counties Railway Co.*, 10 M. & W. 58 (Exch. 1845).

41. Henri de Bracton, *On the Laws and Customs of England* 39, 42, 43, 44 (Samuel E. Thorne, trans. Harvard University Press 1968).

42. *Id.* at 29, 30, 31, 36, 40, 43, 48.

43. *Id.* at 442–449.

44. *The Case of Swans*, 7 Coke Rep. 16 (K.B. 1592).

45. Thomas Hobbes, *Leviathan* 88–91 (Richard Tuck, ed., Cambridge University Press 1992) (*Leviathan*); Thomas Hobbes, "De Cive," in 2 *The English Works of Thomas Hobbes of Malmesbury* 85 (William Molesworth, ed., John Bohn 1841) (*De Cive*); Norberto Bobbio, *Thomas Hobbes and the Natural Law Tradition* 41 (Daniela Gobetti, trans., University of Chicago Press 1993); Johann P. Sommerville, *Thomas Hobbes: Political Ideas in Historical Context* 49 (St. Martin's Press 1992).

46. Noberto Bobbio, *supra* note 45, at 54–55; Howard Warrender, *The Political Theory of Thomas Hobbes: His Theory of Obligation* 189 (Oxford University Press 1957); *Leviathan, supra* note 45, at 151.

47. *Leviathan, supra* note 45, at 124; Noberto Bobbio, *supra* note 45, at 70–72, 138; Johann P. Sommerville, *supra* note 45, at 89.

48. *Leviathan, supra* note 45, at 97 (emphases in the original).

49. "De Cive" *supra* note 45, at 113–114.

50. John Locke, *Two Treatises of Government* sec. 1.39, at 168, sec.2.6, at 271 (Peter Laslett, ed., student ed., Cambridge University Press 1988); Ian Harris, *The Mind of John Locke* 153 (Cambridge University Press 1994); A. John Simmonds, *The Lockean Theory of Rights* 21 (Princeton University Press 1992); John Locke, *Essays on the Law of Nature* 157 (Wolfgang Von Leyden, ed., Oxford University Press 1965).

51. John Locke, *supra* note 50, sec. 2.6 at 271; Ian Harris, *supra* note 50, at 34, 153–154, 214–215. *See* Genesis 1:28 and 9:1–3.

52. Ian Harris, *supra* note 50, at 153, 154, 155, 214, 215; John Locke, *supra* note 50, sec. 1.86, at 204–206; sec. 2.25, at 285–286.

53. John Locke, *supra* note 50, sec. 1.24, at 157, sec. 27, at 160; sec. 29, at 161; sec. 39, at 168; sec. 140, at 169; sec. 43, at 171; sec. 67, at 190; sec. 87, at 206; secs. 2.25–26, at 285–286. *See* A. John Simmons, *supra* note 50, at 239.

54. John Locke, *supra* note 50, sec. 2.3, at 269, sec. 2.27, at 287–8.

55. John Locke, *supra* note 50, sec. 2.28, at 288; sec. 2.30, at 289; sec. 2.30, at 290; Sec. 2.37, at 294–295; H. McCoubrey, *The Development of Naturalist Legal Theory* 70 (Croom Helm Ltd. 1987).

56. 2 William Blackstone, *Commentaries on the Law of England* , at *2-*3, quoting Genesis 1:28.

57. *Id.* at 390, *400, *403, *404-*405, *411, quoting J. Inst. 2.1.12; *id.* at I,*258, quoting Dig. 41.1.3 (Gaius, *Common Matters or Golden Things, book* 2).*403, *411, Alan Watson, "The Impact of Justinian's *Institutes* on Academic Treatises: Blackstones' *Commentaries*," in *Roman Law and Comparative Law* 165–176 (University of Georgia Press 1991); Peter Stein, *The Character and Influence of the Roman Civil Law* 155–6, 173 (The Hambleton Press 1988).

58. Carl F. Stychin, "The Commentaries of Chancellor James Kent and the Development of an American Common Law," 37 *Am. J. Legal Hist.* 440, 440, 445, 447–451, 453–4, 462 (1993).

59. James Kent, 2 *Commentaries on American Law* *318–*319, *342-*343, *348. *350-*351, *360-*361 (1896).

60. *Geer v. Connecticut*, 161 U.S. 519, 522, 523, 529 (1896), *overruled on other grds.*, *Hughes v. Oklahoma*, 441 U.S. 322 (1979).

61. Oliver Wendell Holmes, *supra* note 36, at 187.

62. Steven M. Wise, "The Legal Thinghood of Nonhuman Animals," 23 *Boston Coll. Env. Aff. L. Rev.* 471, 535–6; Daniel Boorstin, *The Mysterious Science of the Law* 130–1 (Beacon Press 1958); *Blades v. Higgs*, 11 H.L. 621, 637 (1865). *See* Barry

Nicholas, *An Introduction to Roman Law* 131 (Clarendon Press 1962); Earl C. Arnold, "The Law of Possession Governing the Acquisition of Animals *Ferae Naturae*," 55 *Amer. L. Rev.* 395, 403–404 (1921).

63. 3A *C.J.S.* "Animals," sec. 4, at 475 (1973).

64. Joel Prentiss Bishop, *Bishop on Criminal Law*, sec. 594, at 434 (John M. Zane & Carl Zollman, eds. 9th ed. 1923). *E.g.*, David S. Favre and Murray Loring, *Animal Law* 122 (Quorum Books 1983); Charles E. Friend, "Animal Cruelty Laws: The Case for Reform," 8 *U. Rich. L.R.* 201, 201–202 (1974).

65. Daniel R. Coqillette, "Radical Lawmakers in Colonial Massachusetts: The 'Countenance of Authoritie' and The *Lawes* and *Libertyes*," *New Eng. Q.* 179, 189, 191, 192 (1994); Samuel Eliot Morrison, *Builders of the Bay Colony* 232 (Houghton-Mifflin 1930). *See* Joel Prentiss Bishop, *supra* note 64, at 192.

66. George Lee Haskins, *Law and Authority in Early Massachusetts: A Study in Tradition and Design* 141 (Macmillan and Co., Inc. 1960).

67. *See Grise v. State*, 37 Ark. 456, 458 (1881); *The Stage Horse Cases*, 15 Abb. Pr. (n. s.) 51, 77 (N.Y. Comm. Pleas 1873).

68. William H. Whitmore, *The Colonial Laws of Massachusetts. Reprinted from the Edition of 1672, with the Supplements Through 1686* 55 (Rockwell and Churchill 1890)(Liberty 94); *Grise, supra* note 67, at 458.

69. William Ewald, "Comparative Jurisprudence (I)- What was it Like to Try a Rat?," 143 *U. Pa. L. Rev.* 1889, 1905 note 5 (1995).

70. David Favre and Vivien Tsang, "The Development of Anti-Cruelty Laws During the 1800's," *Det. Coll. L.Rev.* 1, 3–4 (1993).

71. Richard D. Ryder, *Animal Revolution* 83 (Basil Blackwell 1989); Keith Thomas, *Man and the Natural World* 119–20 (Pantheon Books 1983).

72. Richard D. Ryder, *supra* note 71, at 83–4.

73. Lord Erskine, "Speech to the House of Lords," *Parliamentary Debates* columns 555, 557 (May 15, 1809).

74. David Favre and Vivien Tsang, *supra* note 70, at 4.

75. 3 Geo.4 c.71 (1822); Me. Laws ch.IV, sec.7 (1821). *See* David Favre and Vivien Tsang, *supra* note 70, at 8, 9.3 Geo.4 c.71 (1822).

76. *Stephens v. State*, 65 Miss. 329, 331–332, 3 So. 458, 458–459 (1888). See *Knox v. Massachusetts Society for the Prevention of Cruelty to Animals*, 12 Mass. App. 407, 408, 425 N. E.2d. 393, 396 (1981), quoting *Commonwealth v. Higgins*, 277 Mass. 191, 194, 178 N. E. 536, 538 (1931).

77. James Turner, *Reckoning with the Beast* 79–95 (Johns Hopkins University Press 1980).

78. *Commonwealth v. Lufkin*, 89 Mass. (7 Allen) 579, 581 (1863). *See* Annotation, "What Constitutes Statutory Offense of Cruelty to Animals." 82 *ALR 2d.* 794, 799 (1962).

79. *Hunt v. State*, 3 Ind. App. 382, 385 (1892).

80. *People* ex. rel *Freel v. Downs*, 136 N.Y.S. 440, 445 (Mag. Ct. 1911).

81. *Baker v. Bolton*, 1 Camp. 493, 170 Eng. Rep. 1033 (*Nisi Prius* 1808).

82. Jacob J. Finkelstein, *supra* note 32, at 178, 179–80 196–197; *e.g.*, Percy H. Winfield, "Death as Affecting Liability in Tort ," 29 *Colum. L. Rev.* 239, 253 (1929); *Admiralty Commissioners v. S.S. Amerika*, A.C. 38 (1916); *E.g.*, *Connecticut Mutual Life Insurance Company v. New York and New Haven R.R.*, 25 Conn. 265 (1856); *Worley v. Railroad Co.*, 1 Handy 481 (Ohio 1855).

83. *Hyatt v. Adams*, 16 Mich. 180, 191–192 (1868). *See* Jacob J. Finkelstein, *supra* note 32, at 173, 196, 253.

84. *J.W. Goldsmith, Jr.- Grant Co., supra* note 32, at 511. *E.g., United States v. One Black Horse*, 129 F. 167 (D. Me. 1904).

85. 135 Tenn. 509, 514–515, 188 S.W. 54, 55 (1916). *E.g., State v. Champagne*, 206 Conn. 421, 428–429, 538 A.2d 193, 197 (1988). *See Calero-Toledo, supra* note 32, at 683.

86. 9 &10 Vict. c. 62 (1846). *See* Steven M. Wise, "Recovery of Common Law Damages for Emotional Distress, Loss of Society, and Loss of Companionship for the Wrongful Death of a Companion Animal, 4 *Animal Law* 33, 58 (1998); Jacob J. Finkelstein, *supra* note 32, at 183.

87. Steven M. Wise, *supra* note 86, at 58–9.

88. J.J. Finkelstein, *supra* note 4, at 77, 81.

89. Harold J. Berman, "Toward an Integrative Jurisprudence: Politics, Morality, History," 76 *Cal. L. Rev.* 779, 782 (1988).

90. *Id.* at 41–42.

91. See, e.g., *New York Times*, at p. 4 (August 15, 1999); "Darwin and the Candidates," *Washington Post*, at p. 18 (August 30, 1999).

92. Stephen Jay Gould, "Spin Doctoring Darwin," 104 *Natural History* 6, 6 (July 1995). Believers that evolution by natural selection is an unproveable "theory" should examine Jonathan Weiner's, *The Beak of the Finch: A Story of Evolution in Our Time* (Alfred A. Knopf, Inc. 1995), a Pulitzer Prize–winning account of one painstaking, but by no means isolated, documentation of evolution by natural selection as it occurred on the island of Daphne Major in the Galapagos.

93. Anthony T. Kronman, "Precedent and Tradition," 99 *Yale L. Rev.* 1029, 1032–1033 (1990).

94. Moses I. Finley, *The Ancient Greeks: An Introduction to their Life and Thought* 36 (Viking Press 1963).

CHAPTER FIVE

1. 20 Howell's State Trials 1 (K.B. 1772).

2. 60 U.S. 393 (1857).

3. *Citizens to End Animal Suffering and Exploitation v. the New England Aquarium* 836 F. Supp. 45, 49 (D. Mass. 1993). *See* Joseph Mendelson, III, "Should Animals have Standing? A Review of Standing Under the Animal Welfare Act," 24 *Boston College Environmental Affairs Law Review* 795, 805 (1997).

4. William M. Wiecek, "Somerset: Lord Mansfield and the Legitimacy of Slavery in the Anglo-American World," 42 *University of Chicago Law Review* 86, 102 note 55, 105 (1974).

5. Transcript of oral argument in *Somersett, supra* note 1, in Granville Sharpe Transcripts, New York Historical Society, quoted in William M. Wiecek, *supra* note 4, at 90.

6. *Somerset, supra* note 1, at 72 (K.B. 1772).

7. The facts of the Scotts' lives were taken from Don E. Fehrenbacher, *Slavery, Law and Politics: The Dred Scott Case in Historical Perspective* 121–9 (Oxford University Press 1981).

8. *Scott v. Emerson*, 15 Mo. 387, 395 (1852).

286   Notes

9. *Dred Scott v. Sandford*, 60 U.S. 393 (1857); Don E. Fehrenbacher, *supra* note 7, at 174–82.

10. *Dred Scott supra* note 9, at 407 (1857).

11. Don E. Fehrenbacher *supra* note 7, at 191.

12. *Supra* note 3, at 49.

13. *Jacobellis v. Ohio*, 378 U.S. 184, 197 (1964)(Stewart, J., concurring).

14. James F. Childress, "The Meaning of the 'Right to Life,'" in *Natural Rights and Natural Law: The Legacy of George Mason* 126 (Univ. Pub. Assoc. 1987)(quoting or citing Ronald Dworkin, *Taking Rights Seriously* xi [Harvard University Press 1977][first quote]; H.L.A. Hart, "Bentham on Legal Rights," *in Oxford Essays in Jurisprudence* 192 [2nd. series, A.W.B. Simpson, ed. 1973][second quote]; Robert Nozick, *Anarchy, State, and Utopia* [Basic Books 1974][third and fourth quotes]; John Finnis, *Natural Law and Natural Rights* [Oxford University Press 1980][fifth quote], and Charles Fried, *Medical Experimentation: Beyond Personal Integrity and Social Policy* [American Elsevier 1974][sixth quote]).

15. *See* W.L. Morison, *John Austin* 164 (Edward Arnold 1982); Joseph William Singer, "The Legal Rights Debate in Analytical Jurisprudence From Bentham to Hohfeld," 1982 *Wis. L. Rev.* 975, 989 note 22; Walter J. Kamba, "Legal Theory and Hohfeld's Analysis of a Legal Right," *Jur. Rev.* 249, 249 (1974).

16. Rex Martin, *A System of Rights* 31 (Oxford University Press 1993).

17. Wesley Newcomb Hohfeld, *Fundamental Legal Conceptions as Applied in Judicial Reasoning* 64 (Walter Wheeler Cook ed. Yale University Press 1919).

18. David Lyons, "Correlativity of Rights and Duties," 4 *Nous* 46 (1970).

19. Richard Tuck, *Natural Rights Theories: Their Origin and Development* 161 (Cambridge University Press 1979).

20. Wesley Newcomb Hohfeld, *supra* note 17, at 38–9, 43; Judith Jarvis Thomson, *The Realm of Rights* 44–56 (Harvard University Press 1990); Jeremy Waldron, in *Theories of Rights* 6 (Jeremy Waldron, ed. Oxford University Press 1984); John Finnis, *supra* note 14.

21. Joel Feinberg, *Social Philosophy* 61, 82 (Prentice Hall 1973); *See Stevenson v. State of Mississippi*, 674 So. 2d. 501, 504 (Miss. 1996).

22. Judith Jarvis Thomson, *supra* note 20, 49, 52.

23. Walter J. Kamba, *supra* note 15, at 253.

24. H.L.A. Hart, *supra* note 14, at 181–2.

25. Isaiah Berlin, "Two Concepts of Liberty," in *Four Essays on Liberty* 121 (Oxford University Press 1969).

26. *Id.* at 124.

27. 7 *The Collected Works of Abraham Lincoln* 301–2 (Roy P. Basler, ed. Princeton University Press 1953–5)(emphases in the original). Isaiah Berlin, *supra* note 25, at 124.

28. James M. McPherson, *Abraham Lincoln and the Second American Revolution* 61–2 (Oxford University Press 1990); Isaiah Berlin, "Introduction," in Isaiah Berlin, *supra* note 25.

29. D.D. Raphael, *Moral Philosophy* 83–7 (2nd. Enlarged ed. Oxford University Press 1994); Isaiah Berlin, *supra* note 25, at 131–54.

30. James M. McPherson, *supra* note 28, at 62; Jeremy Waldron, *The Right to Private Property* 299 (Oxford University Press 1988); Isaiah Berlin, "Two Concepts of Liberty," *supra* note 25, at 131–54, 163.

31. *Id.* at 169.

32. *Id.* at 140–54 (1969); James M. McPherson, *supra* note 28, at 62–3, 137.

33. Wesley Newcomb Hohfeld, *supra* note 17, at 38, 39.

34. *Id.* at 72.

35. *Id.*

36. H.L.A. Hart, *supra* note 14, at 171, 191–2, 196–7.

37. *Union Pac. R. Co. v. Botsford,* 141 U.S. 250, 251 (1891).

38. Walter J. Kamba, *supra* note 15, at 256.

39. *E.g.,* Gary L. Francione, *Rain Without Thunder: The Idealogy of the Animal Rights Movement* 192–3 (Temple University Press 1996).

40. *Union Pac. R. Co. v. Botsford, supra* note 37, at 251.

41. The Supplementary Convention on the Abolition of Slavery, the Slave Trade, and Institutions and Practices Similar to Slavery, *done* Sept. 7, 1956, 18 U.S.T. 3201, T.I.A.S. No. 6418, 266 U.N.T.S. 3 (*entered into force* April 30, 1957) (entered into force for United States, Dec. 6, 1967). *See* Rex Martin, *supra* note 16, at 3 (1993); L.W. Sumner, *The Moral Foundation of Rights* 37–8 (1987).

42. *Harris v. McRae,* 448 U.S. 297, 316–20 (1980).

43. *Compare* 448 U.S., at 313–9 with 448 U.S., at 331 (Brennan, J., dissenting).

44. *City of Beaumont v. Bouillion,* 896 S.W. 2d 143, 148–9 (Tex. 1995); Article 1, secs. 8 and 27 of the Texas Constitution.

45. *Id.* at 149 (referring to Art. I, sec. 29, which provided that "all laws contrary [to the Texas Bill of Rights] . . . shall be void").

46. *Id.* at 148.

47. H.L.A. Hart, "Bentham on Legal Rights," *supra* note 14, at 197, 198 *See* Carl Wellman, *Real Rights* 113 (Oxford University Press 1995); Michael D. Bayles, *Hart's Legal Philosophy: An Examination* 149 (Kluwer Academic Publishers 1992); *Daugherty v. Wallace,* 621 N.E. 2d 1374, 1379 (Ohio Ct. App. 1993), quoting *Sowers v. Ohio Civil Rights Commission,* 252 N.E. 2d 463, 475 (Ohio Ct. Comm. Pleas 1969).

48. William Blackstone, *Commentaries on the Law of England* *129 (1765–9).

49. *Union Pacific R. Co., supra* note 37, quoting *Cooley on Torts,* 29 (emphasis added).

50. Louis Henkin, "International Human Rights as 'Rights,'" 1 *Cardozo L. Rev.* 425, 443 (1979).

51. Hillary Charlesworth, Christine Chinkin, and Shelley Wright, "Feminist Approaches to International Law," 85 *Am. J. Int'l L.* 613, 638 (1991).

52. Rex Martin, *supra* note 16, at 31; L.W. Sumner, *supra* note 41, at 37–8.

53. *Virani v. Jerry M. Lewis Truck Parts & Equipment, Inc.,* 89 F. 3d 574, 577 (9th Cir. 1996), quoting Judith Jarvis Thomson, *supra* note 20, at 59; P.J. Fitzgerald, *Salmond on Jurisprudence* 229 (12th ed., Sweet and Maxwell 1966).

54. *In re Embassy Properties North, Ltd. Partnership,* 196 B.R. 172, 179 note 26 (Bankr. D. Kan. 1996).

55. *Chambers v. Baltimore & Ohio Railroad,* 207 U.S. 142, 148 (1907).

56. *E.g., Grand Int. Brotherhood of L. Engineers v. Mills,* 31 P. 2d 971, 978–9 (Ariz. 1934); *Corum v. University of North Carolina,* 413 S.E. 2d 276, 289, 290 (N.C.), *cert. den. sub nom., Durham v. Corum,* 506 U.S. 985 (1992).

57. *Appeal from a Judgment of the Hungaro-Czechoslovak Mixed Arbitral Tribunal* (*Peter Pazmanv v. Czech.*), 1933 P.C.I.J. (ser. A/B) No. 61, at 231 (Dec. 15).

58. *Golder v. United Kingdom,* Ser. A, vol. 18, paras. 34–5 (Judgment of 21 Feb. 1975).

288   Notes

59. *John F. Kennedy Memorial Hosp. v. Bludworth*, 432 So. 2d 611, 615 (Fla. Dist. Ct. App. 1983), *aff'd* 452 So. 2d 921, 924 (Fla. 1984); Gary L. Francione, *supra* note 39, at 194.

60. Michael Akehurst, *A Modern Introduction to International Law* 72 (Allen and Unwin 1987); M.W. Janis, "Individuals as Subjects of International Law," *Cornell Int'l L. J.* 61, 61–2 (1984).

61. Carlos Manuel Vazquez, "Treat-based Rights and Remedies of Individuals," 92 *Columbia Law Review* 1082, 1088 and note 19 (1992); M.W. Janis, *supra* note 60, at 61, 62–3 and note 14; Jeremy Bentham, *An Introduction to the Principles of Morals and Legislation* 6, 293–6 (J. Burns and H.L.A. Hart eds., London, Athlone P. 1970).

62. Louis B. Sohn, "The New International Law: Protection of the Rights of Individuals Rather than States," 32 *Am. U. L. Rev.* 1, 9 (1982); Richard Deming, *Man and the World: International Law at Work* 157 (Hawthorn Books 1974).

63. Michael Akehurst, *supra* note 60, at 70, 73 (1987); Richard Deming, *supra* note 62, at 56.

64. Naomi Roht-Arriaza, "State Responsibility to Investigate and Prosecute Grave Human Rights Violations in International Law," 78 *Cal. L. Rev.* 449, 474–483 (1990).

65. G.A. Res. 95, U.N. Doc. A/64/Add. 1 (1947); *The Nurnberg Trial*, 6 F.R.D. 69, 110 (Int'l Mil. Tribunal 1946). *See* M.W. Janis, *supra* note 60, at 65–6.

66. *Forti*, 672 F. Supp. 1531,1541–2 (N.D. Ca. 1987);*The Effect of Reservations on the Entry into Force of the American Convention*, Inter-American Court of Human Rights, Advisory Opinion No. OC–2/82 (Sept. 24, 1982).

67. *See* Optional Protocol to the International Covenant on Civil and Political Rights, *entered into force* March 23, 1976, G.A. Res. 2200A, 21 U.N. GAOR Supp. (No.16), at 59, U.N. Doc. A/6316 (1966)(establishing a Human Rights Committee to which individuals may present complaints against states); U.N. Doc. E/CN 4/1982/1, at 31–33 (1982)(two prisoners petitioned the European Commission of Human Rights under Art. 25 of the European Convention of the Protection of Human Rights and Fundamental Freedoms, *opened for signature* Nov. 4, 1950, Europ. T.S. No. 5, 213 U.N.T.S. 222 [*entered into force* Sept. 3, 1953], repr. in Richard B. Lillich, *International Human Rights Instruments* 2nd ed. 500.1 (1990)(allowing "petitions . . . from any person" claiming denial of the right to marry).

68. Universal Declaration of Human Rights, *adopted* Dec. 10, 1948, G.A. Res. 217A (III), 3 U.N. GAOR (Resolutions, part 1) at 71, U.N. Doc. A/810 (1948), repr. in Richard B. Lillich, *International Human Rights Instruments* 2nd ed., at 440.1 (1990)(Article 8); The International Covenant on Civil and Political Rights, adopted December 16, 1966, G.A. Res. 2200, 21 U.N. GAOR Supp. (No. 16) at 52, 6999 U.N.T.S. 171 (*entered into force* March 23, 1976)(*entered into force for U.S.* 1992), repr. in Richard B. Lillich, *International Human Rights Instruments* 2nd ed. 170.2, 170.7 (Articles 2, 3, and 16); American Convention on Human Rights, opened for signature Nov. 22, 1969, O.A.S.T.S. No. 36, (entered into force, July 18, 1978), repr. in Richard B. Lillich, *International Human Rights Instruments* 2nd ed. 190.2 and 190.11(Articles 3, 25[1], and 27); Manuel D. Vargas, "Individual Access to the Inter-American Court of Human Rights," 16 *N.Y.U. J. Int'l L. and Pol.* 601, 604–9 (1984)(Articles 44, 48 [1][f], 61[2] and 63[2]); European Convention for the Protection of Human Rights and Fundamental Freedoms, repr. in Richard B. Lillich, *In-*

ternational Human Rights Instruments 2nd ed. 500.3–4, 500.5 (Articles 6 and 13). *See X and Y v. Netherlands,* 91 Eur. Ct. H. R. (ser. A) (1985)(judgment)(the European Court and Commission of Human Rights held that the Dutch had breached Article 8 by refusing to allow an incompetent woman who had been raped from initiating a criminal prosecution with the aid of her father. The Court said the "fundamental values ... at stake" required it, *id.* at para. 27); *see also,* Case 26/62 *Ven Gend en Loos v. Nederlandse Administratie der Berlastingen,* E.C.R. 1, 23, 2 Comm. Mkt. L. R. 105, 129 (E.C.J. 1963); *Case of Klass and Others (F.R.G.),* 28 Eur Ct. H. R. (ser A.) (1978)(judgment). Article 56 of The African Charter on Human and Peoples' Rights, OAU Doc. CAB/Leg/67/3/Rev.5, repr. in 21 *ILM* 59 (1982), *adopted* June, 1981 (*entered into force* Oct. 21, 1986) permits an aggrieved individual to petition an African Commission on Human Rights, which may refer the complaint to the Assembly constituted by the Charter for an in-depth study and report. The absence of an enforcing court has been said to stem from the conclusion that Western-style courts were not in harmony with the traditional African emphasis upon conciliation, rather that judicial conflict, U. O. Umozurike, "The African Charter on Human and Peoples' Rights," 77 *Amer. J. Inter'l Law* 902, 908–9 (1983).

69. A. Leon Higginbotham, Jr. and F. Michael Higginbotham, "'Yearning to Breath Free': Legal Barriers Against and Options in Favor of Liberty in Antebellum Virginia," 68 *N.Y.U.L.Rev.* 1213, 1234–6 (1993)(discusses the Virginia Freedom Suit Act of 1795); A. Leon Higginbotham, Jr., *In the Matter of Color: Race and the American Legal Process: The Colonial Period* 194–5 (South Carolina), 252–3 (Georgia)(Oxford University Press 1978).

70. H.L.A. Hart, *The Concept of Law* 28 (Oxford University Press 1961).

71. *See* Carl Wellman, *supra* note 47, at 7–8 (1995); Judith Jarvis Thomson, *supra* note 20, at 55 and note 11, 67, 281–285 (1990); L.W. Sumner, *supra* note 41, at 19, 48; Jeremy Waldron, *supra* note 20, at 8; John Finnis, *supra* note 14, at 200–1 (1980); H.L.A. Hart, supra *note* 14, at 180.

CHAPTER SIX

1. Sophocles, "Antigone," in 3 *The Complete Greek Tragedies* 201–202 (David Grene and Richard Lattimore, eds., Elizabeth Wyckoff, trans. Modern Library n.d.).

2. Marshall L. DeRosa, *The Confederate Constitution of 1861: An Inquiry Into American Constitutionalism* 39–40, 63–64 (University of Missouri Press 1991).

3. Robert Jay Lifton, *The Nazi Doctors: Medical Killing and the Psychology of Genocide* 51–75 (BasicBooks, Inc. 1986); A. Mitscherlich and F. Mielke, *The Death Doctors* 233–305 (James Cleugh, trans., Elek Books 1962).

4. Hannah Arendt, *Eichmann in Jerusalem* 24 (Penguin Books 1964).

5. Obituary of Inge Aicher-Scholl in the *New York Times,* p. 40, September 6, 1998, quoting Albert van Schirdung.

6. A. Mitscherlich and F. Mielke, *supra* note 3, at 288–9.

7. Michael Ignatieff, "Human Rights: the Midlife Crisis," *The New York Review of Books* 58, 59 (May 20, 1999).

8. 82 U.N.T.S. 279 (Aug. 8, 1945). *See* Diane F. Orentlicher, "Settling Accounts: The Duty to Prosecute Human Rights Violations of a Prior Regime," 100 *Yale L.J.* 2537, 2556 and note 75, 2560 (1991).

9. Opening Speech of Justice Robert H. Jackson, Nov. 21, 1945, in II *Trial of the Major War Criminals Before the International Military Tribunal* 155 (1947).

10. *United States v. Ohlendorf* (Case No. 9), IV *Trials of War Criminals Before the Nuernberg Military Tribunals Under Control Council Law No. 10* 497 (1950).

11. Hannah Arendt, *supra* note 4, at 135 (emphasis added).

12. "Natural law" refers to universal, or nearly universal, abstract, immutable, and objective principles of justice and ethics that are derivable through reason from an external source such as nature, Randy E. Barnett, "Getting Normative: The Role of Natural Rights in Constitutional Adjudication," 12 *Constitutional Commentary* 93, 107–8 (1995). "Natural rights" refers to those enforceable rights that pre-exist the formation of govenment and are essential to protect the vital interests of members of a society from injury by others or by the state, *id.* at 106–9. "Natural justice" is "that which is founded in equity, in honesty and right," *Kempsey v. Maginnis*, 2 Mich. N.P. 49, 55 (Circ. Ct. 1871).

13. Cicero, *De Re Publica* 3.22.33 (Clinton Walker Keyes, trans., Loeb Classical Library 1928).

14. Cicero, *De Legibus*, 2.5 (Clinton Walker Keyes, trans., Loeb Classical Library 1928).

15. John Mack Farragher, *Daniel Boone: The Life and Legend of an American Pioneer* 332 (Henry Holt 1992), quoting James Fenimore Cooper, *The Deerslayer* (Lightyear Press 1976).

16. Isaiah Berlin, "Two Concepts of Liberty," in *Four Essays on Liberty* 169 (1969).

17. *Planned Parenthood v. Casey*, 505 U.S. 833, 849 (1992)(joint plurality opinion). *See* Peter Linzer, "The *Carolene Products* Footnote and the Preferred Position of Individual Rights: Louis Lusky and John Hart Ely vs. Harlan Fiske Stone," 12 *Const. Comm.* 277, 302–3 (1995).

18. Bryan Wilson, "Introduction," in *Values: A Symposium* 1 (Brenda Almond and Bryan Wilson eds., Humanity Books 1988). Different branches of the social sciences diversely describe similar ideas. Social psychologists often define "beliefs" as *valuations* made about such objects as material objects, living entities, abstract ideas, and views. "Attitudes" are related interconnected networks of beliefs, Janice A. Doyle and Gale M. Sinatra, "Social Psychology Research on Beliefs and Attitudes: Implications for Research on Learning From Text," in *Beliefs About and Instruction With Text* 248–9 (Ruth Garner and Patricia A. Alexander, eds., Lawrence Erlbaum Assoc.1994). Other social scientists may refer to attitudes roughly as theories, Clark A. Chinn and William F. Brewer, "The Role of Anomalous Data in Knowledge Acquisition: A Theoretical Framework and Implications for Science Instruction," 63 *Rev. of Educ. Res.* 1, 39 note 1 (1993). I will use "values" as social psychologists use "beliefs," while I will use beliefs as cognitive psychologists use the term, to mean what social psychologists call attitudes or theories, Janice A. Doyle and Gale M. Sinatra, *supra*, at 249; Clark A. Chinn and William F. Brewer, *supra*, at 40 note 1.

19. Oliver Wendell Holmes, "Ideals and Doubts," 19 *Ill. L. Rev.* 1, 2 (1915).

20. William J. Brennan, Jr., "Reason, Passion, and the 'Progress of the Law,'" 42 *The Rec. of the Ass'n of the Bar of the City of N.Y.* 948, 958 (1987)(forty-second annual Benjamin N. Cardozo Lecture).

21. Oliver Sacks, *An Anthropologist from Mars* 287–8 (Alfred A. Knopf 1995).

22. Antoine Bechara et al., "Deciding Advantageously Before Knowing the Advantageous Strategy," 275 *Science* 1293, 1294 (1997).

23. Clark A. Chinn and William F. Brewer, *supra* note 18, at 15–7. *E.g.*, Janice A. Doyle and Gale M. Sinatra, *supra* note 18, at 261; Lee Ross and Craig Anderson, "Shortcomings in the Attribution Process: On the Origins and Maintenance of Erroneous Social Assessments," in *Judgment Under Uncertainty: Heuristics and Biases* 148–52 (Daniel Kahneman et al., eds. Cambridge University Press 1982).

24. Clark A. Chinn and William F. Brewer, *supra* note 18, at 4–10, 13.

25. Janice M. Dole and Gayle A. Sinatra, *supra* note 18, at 256.

26. Clark A. Chinn and William F. Brewer, *supra* note 18, at 4-at 10–1.

27. *Id.* at 29; Janice A. Doyle and Gale M. Sinatra, *supra* note 18, at 252, 257.

28. These examples of belief preservation are taken from James Reston, Jr., *Galileo* 119, 120, 121, 267 (HarperCollins 1994), Stillman Drake, *Galileo* 38, 44–8 (Hill and Wang 1980), Colin A. Ronan, *Galileo* 24, 112–4, 115 (G.P. Putnam's Sons 1974), Thomas S. Kuhn, *The Copernican Revolution* 225–6 (Harvard University Press 1957), and Giorgio de Santillana, *The Crime of Galileo* 11 (University of Chicago Press 1955).

29. Thomas S. Kuhn, *supra* note 28, at 2.

30. Colin A. Ronan, *supra* note 28, at 135.

31. James Gleick, "The Paradigm Shifts," in *N.Y. Times Magazine*, Dec. 29, 1996, at 25, referring to Thomas S. Kuhn, *The Structure of Scientific Revolutions* (sec. ed. University of Chicago Press 1970).

32. Thomas S. Kuhn, *supra* note 31, at 4, 66–91,110, 148–50,185–6.

33. *Id.* at 150 (emphasis added).*See* Risieri Frondizi, *What is Value?* 159–65 (2nd ed. 1971); Thomas S. Kuhn, "Second Thoughts on Paradigms," in *The Essential Tension* 313–8 (University of Chicago Press 1977).

34. Thomas S. Kuhn, *supra* note 28, at 226.

35. Clark A. Chinn and William F. Brewer, *supra* note 18, at 10; Harold J. Berman, *Law and Revolution: The Formation of the Western Legal Tradition* 22 (Harvard University Press 1983).

36. Thomas S. Kuhn, *supra* note 28, at 150–1, quoting Max Planck, *Scientific Autobiography and Other Papers* 295–6 (F. Gaynor trans., Greenwood Publishing Group 1949).

37. "From Ants to Ethics: A Biologist Dreams About a Unity of Knowledge," *N.Y. Times*, May 12, 1998, at C6 (Professor E.O. Wilson quoting the economist, Paul Samuelson).

38. Fred Kaplan, "Milosevic's yield may disprove dounts on air war," *The Boston Sunday Globe* A 29 (June 6, 1999).

39. Charles Larmore, "Pluralism and Reasonable Disagreement," 11 *Soc. Phil. & Policy* 61, 70 (1994). *See* Cass Sunstein, "Incommensurability and Valuation in Law," 92 *Mich. L. Rev.* 779, 789, 796 and note 58 (1994); Richard Warner, "Incommensurability as a Jurisprudential Puzzle," 68 *Chi-Kent L. Rev.* 147, 157–67 (1992).

40. Cass Sunstein, *supra* note 39, at 789, 839.

41. Albert Einstein, "On Freedom," in *Out of My Later Years* 18 (Littlefield, Adam, and Co. 1967).

42. Oliver Wendell Holmes, Jr., "Natural law," 33 *Harv. L. Rev.* 40, 41 (1918–19). *See* Cass Sunstein, *supra* note 39, at 810–1; Richard Warner, *supra* note 39, at 168.

43. William Harper, "Memoir on Slavery," reprinted in Drew Gilpin Faust, *The Ideology of Slavery* 89 (Louisiana State University Press 1981).

44. *Id.* at 98.

45. Thomas S. Kuhn, *supra* note 28, at 7.

46. Norwood R. Hanson, *Patterns of Discovery : An Inquiry into the Conceptual Foundations of Science* 4–30 (Cambridge University Press 1958).

47. Barbara Tuchman, "Why Policy-Makers Don't Listen," in *Practicing History: Selected Essays* 287 (Ballentine Books 1981); Stephen R. L. Clark, *The Moral Status of Animals* 7 (Oxford University Press 1984).

48. *See generally,* Charles R. Lawrence III, "The Id, the Ego, and Equal Protection: Reckoning with Unconscious Racism," 39 *Stan. L. Rev.* 317 (1987).

49. Program Evaluation—Appellate Judges' Spring Conference 1999 (Washington State); Letter from Justice Faith Ireland to The Honorable Barbara S. Levenson, dated April 8, 1999.

50. Letter from Justice Faith Ireland to Steven M. Wise, dated June 21, 1999; Letter from Steven M. Wise to the Honorable Faith Ireland, May 1, 1999.

51. World Charter for Nature, GA Res. 37/7, Annex, UN GAOR, 37th Sess. Supp. No. 51, UN Doc. A/37/51 (Oct. 28, 1982) (Preamble), repr. in Harold W. Woods, Jr., "The United Nations World Charter for Nature: The Developing Nations' Initiative to Establish Protections for the Environment," 12 *Ecology L. Q.* 977, 992 (1985)(emphasis added).

52. Anthony D'Amato and Sudhir K. Chopra, "Whales: Their Emerging Right to Life," 85 *Amer. J. Int'l L.* 23, 50 (1991).

53. Supplementary Note to the Home Secretary's response to the Animal Procedure's Committee, Interim report on the review of the operation of the Animals (Scientific Procedures) Act 1986 , at 1, para. 11 (November 6, 1997).

54. Part 6, "Use of Animals in Research, Testing, and Teaching," secs 76A and 76B.

55. 16 U.S.C. sec. 1531, *et. seq.* (1988).

56. *Tennessee Valley Authority v. Hill,* 437 U.S. 153, 175, 188 (1978).

57. *Compassion in Dying v. State of Washington,* 79 F. 3d 790, 817 (9th Cir. 1996), *rev. sub nom. Washington v. Glucksburg,* 117 U.S. 2258 (1997).

58. James Rachels, *Created from Animals: The Moral Implications of Darwinism* 89–90 (Oxford University Press 1990), quoting Immanuel Kant, *Lectures on Ethics* 151–2 (Louis Infield, trans., Hackett Publishing Co. 1963).

59. *Quill v. Vacco,* 80 F. 3d 716, 724 (2nd Cir.), *rev. sub nom. Vacco v. Quill,* 117 S. Ct. 2293 (1997); *Compassion in Dying, supra* note 57, 79 F 3d, at 808, *rev'd. sub nom., Washington v. Glucksburg,* 117 S.Ct. 2258 (1997).

60. *Compassion in Dying, supra* note 57, 79 F 3d, at 809.

61. *Quill, supra* note 59, at 724; *Compassion in Dying, supra* note 57, 79 F 3d, at 809–810.

62. *Cruzan v. Director, Missouri Department of Health,* 497 U.S. 261, 280 (1990); *Vacco, supra* note 59, 117 U.S., at 2202; *Glucksburg, supra* note 59 (O'Connor, Stevens, Breyer, Ginsberg, and Souter, J.J., concurring or joining the concurrences of others).

63. Jacob J. Finkelstein, "The Goring Ox: Some Historical Perspectives on Deodands, Forfeitures, Wrongful Death, and Western Notions of Sovereignty," 46

Temple L.Q. 169, 282–283 (1973). *See* Elizabeth Anderson, *Value in Ethics and Economics* 68 (1993).

64. Isaiah Berlin, *supra* note 16, at 171. *E.g.*, Michael Walzer, "Are There Limits to Liberalism?," in *N.Y. Rev. of Books*, October 19, 1995, at 28.

65. Risieri Frondizi, *supra* note 33, at 160.

66. William J. Brennan, Jr., *supra* note 20, at 951.

67. *E.g.*, Steven Weinberg, "The Revolution That Didn't Happen," XLV *New York Review of Books* 48 (October 8, 1998); James W. McAllister, *Beauty and Revolution in Science* 127–39 (Cornell University Press 1996); Lewis Wolpert, *The Unnatural Nature of Science* 93–4, 103, 109, 117 (Harvard University Press 1992).

68. *See generally*, Alan Sokol and Jean Bricmont, *Fashionable Nonsense: Post-Modern Intellectuals' Abuse of Science* (Picador U.S.A. 1998).

69. Steven Weinberg, *supra* note 67, at 51.

70. William J. Brennan, Jr., "A Tribute to Justice Harry A. Blackmun," 108 *Harv. L. Rev.* 1,1 (1994). *See* Martha C. Nussbaum, *Poetic Justice* 67 (Beacon Press 1995).

71. Martha C. Nussbaum, *supra* note 70, at 38, 68.

72. Martha C. Nussbaum, "Emotions as Judgments of Value," 3 *Yale J. Crit.* 201, 203 (1992).

73. "Letter to Albert G. Hodges, dated April 4, 1864," 7 *The Collected Works of Abraham Lincoln* 281 (Rutgers University Press 1990).

74. Susan Sontag, "Why are we in Kosovo," *The New York Times Magazine* 52, 53 (May 2, 1999).

75. Isaiah Berlin, *The Crooked Timber of Humanity* 18 (John Murray, Ltd. 1991); Kenneth C. Randall, *Federal Courts and the International Human Rights Paradigm*, 197–210 (Duke University Press 1990); Isaiah Berlin, *supra* note 16, at 126, 171; Michael Walzer, *supra* note 64, at 30–1; Claude L. Galipeau, *Isaiah Berlin's Liberalism* 65 (Oxford University Press 1994); Colin Wringe, *Children's Rights* 55 (Routledge and Keegan Paul 1981).

76. Isaiah Berlin, *supra* note 16, at 166.

77. Orlando Patterson, *Freedom: Freedom in the Making of Western Culture* ix (Basic Books 1991). *See* Joel Feinberg, *Rights, Justive and the Bounds of Liberty* 268, 285 (1980).

78. Orlando Patterson, *supra* note 77, at ix–63.

79. *Id.* at xii.

80. Liberty rights are different from "liberty-rights," which are a kind of Hohfeldian legal right, discussed in Chap. 5.

81. *Custody of Vaughan,* 664 N.E. 2d 434, 437 (Mass. 1996).

82. The Convention Against Torture and Other Cruel, Inhuman or Degrading Treatment or Punishment, *adopted* December 10, 1984, G.A. Res. 39/46, 39 U.N. GAOR Supp. (No. 51) at 197, U.N. Doc. A/39/51 (1985)(*entered into force* June 26, 1987)(*entered in force* for U.S. 1994), repr. in Richard B. Lillich, *International Human Rights Instruments* 221.1 (2nd ed. 1990).

83. *Reservations to the Convention on the Prevention and Punishment of the Crime of Genocide,* 1951 I.C.J. 15, 23 (May 28)(emphasis added).

84. Report of the United Nations Conference on the Human Environment, U.N. Doc. A/CONF. 48/14/Rev. 1, U.N. Pub. No. E.73.IIA.14, at 4, Principle 1 (1974).

85. Keith Highet, "The Enigma of the Lex Mercatoria," 63 *Tul. L. Rev.* 613, 626 (1989).

86. Article 53 of the Vienna Convention on the Law of Treaties, *adopted* May 22, 1969;*Case Concerning Military and Paramilitary Activities in and Against Nicaragua (Nicar. v. U.S.)*, 1986 I.C.J. 14, 100 (June 27), *judgm't susp. by agreem't*, 1991–2 ICJ Y.B. 149–50 (Sette-Camara, J., concurring); *Siderman de Blake v. Republic of Argentina*, 965 F. 2d 699, 714 (9th Cir. 1992); Ian Brownlie, *Principles of Public International Law* 513 (4th ed., Oxford University Press 1990); Theodor Meron, *On a Hierarchy of International Human Rights* 15–6, 80 *Amer. J. of Intern. Law* 1, 15–6 (1986); Louis B. Sohn, "The New International Law: Protection of the Rights of Individuals Rather than States," 32 *Am. U. L. Rev.* 1, 13 (1982).

87. Constitution of South Africa, in XVII *Constitutions of the Countries of the World* (Albert P. Blaustein and Gisbert H. Flanz, eds. 1971 and 1994 supp.).

88. Constitution of Bolivia of 1967, as amended by the 1994 Law of Reform of the Political Constitution of the State, N. 1585, in II *Constitutions of the Countries of the World* 22 (Albert P. Blaustein and Gisbert H. Flanz, eds. 1971 and January, 1995 supp.).

89. Constitution of Japan, in IX *Constitutions of the Countries of the World* 16 (Albert P. Blaustein and Gisbert H. Flanz, eds. 1971 and April, 1990 supp.).

90. The Basic Law of the Federal Republic of Germany, in VII *Constitutions of the Countries of the World* 113 (Albert P. Blaustein and Gisbert H. Flanz, eds. Oceana Press 1971 and January, 1995 supp.). *See* David P. Currie, *The Constitution of the Federal Republic of Germany* 311 note 266 (University of Chicago Press 1994)(emphasis in the original).

91. Steven M. Wise, "The Eligibility of Nonhuman Animals for Dignty-Rights in a Liberal Democracy," 22 *Vermont Law Review* 864 (1998).

92. *Id.* at 849–64. The quotation is from *Meacham v. Fano*, 4217 U.S. 215, 230 (Stevens, J., dissenting), cited with approval in *Smith v. Organization of Foster Families*, 431 U.S. 816, 846 (1977).

93. Steven M. Wise, *supra* note 91, at 862–4.

94. Randy E. Barnett, *supra* note 12, at 93 (1995).

95. Peter Westen, *Speaking of Equality* 185–229 (Princeton University Press (1990).

96. Kenneth W. Simons, "Equality as a Comparative Right," 65 *Boston University Law Review* 387, 424, 446–447, 479 (1985); Joel Feinberg, *supra* note 77, at 278. *See City of Cleburne v. Ceburne Living Center, Inc.*, 473 U.S. 432, 439 (1985).

97. Henry Fountain, "Proof Positive That People See Colors With the Tongue," *New York Times*, p. F5 (March 30, 1999).

98. Harriet Ritvo, *The Platypus and the Mermaid and Other Figments of the Classifying Imagination* (Harvard University Press 1997).

99. Stephen Jay Gould, *The Mismeasure of Man* 50–69 (W.W. Norton & Co. 1981).

100. Alexis de Tocqueville, 2 *Democracy in America* 97 (1987).

101. *Samaad v. City of Dallas*, 940 F. 2d 925, 941 (5th Cir. 1991). *See* Kenneth W. Simons, "Overinclusion and Underinclusion: A New Model," 36 *UCLA L. Rev.* 447, 465 (1989).

102. Marilyn J. Chambliss, "Why do Readers Fail to Change Their Beliefs After Reading Persuasive Text?" in Ruth Garner and Patricia A. Alexander, eds., *supra* note 18, at 83.

103. Stephen Jay Gould, *supra* note 99, at 42–7.

104. Christopher D. Stone, *Should Trees Have Standing?* 8 (William Kaufman, Inc. 1974).

105. *See* Laurence H. Tribe and Michael C. Dorf, "Levels of Generality in the Definition of Rights," 57 *U. Chi. L. Rev.* 1057, 1106 (1990).

106. *E.g, Logan v. Zimmerman Brush Co.*, 455 U.S. 422, 442 (1982); *Rinaldi v. Yeager*, 384 U.S. 305, 308–9 (1966); *McLaughlin v. Florida*, 379 U.S. 184, 191 (1964). *See* Kenneth W. Simons, *supra* note 101, at 465–7, 501–3.

107. *E.g., New York City Transit Authority v. Beazer*, 440 U.S. 568, 591 (1979).

108. *E.g., Skinner v. Oklahoma*, 316 U.S. 535 (1942).

109. Kenneth W. Simons, *supra* note 101, at 467; Joel Feinberg, *supra* note 77, at 287 note 18.

110. *Romer v. Evans*, 116 S. Ct. 1620, 1627–9 (1996); *Skinner, supra* note 108. *See Thoreson v. Penthouse International, Ltd.*, 563 N.Y.S. 2d 968, 975 (Supr. Ct. 1990).

111. *Romer, supra* note 110, at 1629; Laurence W. Tribe, *American Constitutional Law* 1454, 1627, 1628 (Sec. ed., Foundation Press 1988); Ronald Dworkin, *Law's Empire* 384–5 (Harvard University Press 1986).

112. *Yick Wo v. Hopkins*, 118 U.S. 356, 373, 374 (1886).

113. *McLaughlin v. Florida*, 379 U.S. 184 (1964). *See* Andrew Kull, *The Color-Blind Constitution* 169 (Harvard University Press 1992).

114. *Romer, supra* note 110, at 1628.

115. *Roberts v. City of Boston*, 59 Mass.198, 209–10 (1850); *Plessy v. Ferguson*,163 U.S. 537 (1896). *See* Andrew Kull, *supra* note 113, at 131.

116. *Durkee v. City of Janesville*, 28 Wis. 464, 467 (1871); *Holden v. James*, 11 Mass. 396, 405 (1814); *Roberts, supra* note 115, at 209–10. *See* J. Morgan Kousser, "'The Supremacy of Equal Rights: The Struggle Against Racial Discrimination in Antebellum Massachusetts and the Foundations of the Fourteenth Amendment," 82 *Nw. Univ. L. Rev.* 941, 974, 987 (1988); *Argument of Charles Sumner, Esq. against the Constitutionaliy of Separate Colored Schools, in the Case of Sarah C. Roberts v. The City of Boston* (1849), reprinted in *Abolitionists in Northern Courts* 493 (Paul Finkelman ed., Garland Publishers 1988).

117. *E.g., Chrisafogeorgis v. Brandenberg*, 304 N.E. 2d 88, 92, *reh. den.* (Ill. 1973), quoting *Kwaterski v. State Farm Mutual Automobile Ins. Co.*, 148 N.W. 2d 107, 110 (Wis. 1967).

118. *Puhl v. Milwaukee Automobile Ins. Co.*, 99 N.W. 2d 163, 170, *reh. den.* (Wis. 1960).

119. *E.g., Mone v. Greyhound Lines, Inc.*, 331 N.E. 2d 916, 919–20 (Mass. 1975); *Wiersma v. Maple Leaf Farms*, 543 N.W. 2d 787, 792 (S.D. 1996).

120. Douglas Rae, *Equalities* 59 (Harvard University Press 1981).

121. Alan Gewirth, *Reason and Morality* 121 (University of Chicago Press 1978).

122. *Id.* at 111.

123. Carl Wellman, *Real Rights* 129 (Oxford University Press 1995).

124. *Id.* at 130–1.

125. Deborah Blum, *The Monkey Wars* 150 (Oxford University Press 1994).

CHAPTER SEVEN

1. Theodore F.T. Plucknett, *A Concise History of the Common Law 5th ed.* 13 (Little, Brown & Company 1956).

2. *Id.* at 16–19, 142–149, 232–235; Edward S. Corwin, "The 'Higher Law' Background of American Constitutional Law," 42 *Harv. L. Rev.* 149, 171 (1928).

296   Notes

3. In 1731, Statute 4 Geo. II, c.26 finally made English the exclusive language, of the common law courts, H. Levy-Ullman, *The English Legal Tradition: Its Sources and History*, 123, 124 (M. Mitchell, trans. MacMillan Brooks 1935); Roger Cotterell, *The Politics of Jurisprudence* 34–35 (University of Pennsylvania Press 1992).
4. Theodore F.T. Plucknett, *supra*, note 1, at 18–19, 256–257.
5. *Id.* at 257, 342.
6. Arthur R. Hogue, *Origins of the Common Law* 5 (Liberty Press 1966).
7. *Norway Plains Company v. Boston and Maine Railroad*, 67 Mass. (1 Gray) 263, 267 (1854).
8. 27 *American L. Rev.* 622 (1892).
9. 1 Coke Rep. 88b (1581).
10. Gerald J. Postema, *Bentham and the Common Law Tradition* 11 (Oxford University Press 1986).
11. Edward Levi, *An Introduction to Legal Reasoning* 1, 2 (University of Chicago Press 1949).
12. Peter Westen, *Speaking of Equality* 219–23 (Princeton University Press 1990).
13. Joel Levin, *How Judges Reason* 159 (Peter Lang Publishing 1992). *See* Moliere, "Le Bourgeois Gentilhomme," Act II, Scene iv (1670).
14. *Cf.* P.S. Atiyah and R.S. Summers, *Form and Substance in Anglo-American Law* 411 (Oxford University Press 1987). On the influence of religious training on one influential Justice of the United States Supreme Court, *see* George Kannar, "The Constitutional Catechism of Antonin Scalia," 99 *Yale L. J.* 1297, 1309–20 (1990).
15. Scott Brewer, "Exemplary Reasoning: Semantics, Pragmatics, and the Rational Force of Legal Argument by Analogy," 109 *Harv. L. Rev.* 923, 932 (1996).
16. Melvin Aron Eisenberg, *The Nature of the Common Law* 14–26 (Harvard University Press 1988); Robert S. Summers, "Two Types of Substantive Reasons: the Core of a Theory of Common Law Justification," 63 *Cornell L. Rev.* 707, 718 (1978); Harry H. Wellington, "Common Law Rules and Constitutional Double Standards: Some Notes on Adjudication," 83 *Yale L. J.* 221, 223–5 (1973).
17. *Planned Parenthood v. Casey*, 505 U.S. 833, 849 (1992)(joint plurality opinion). *See* Peter Linzer, "The *Carolene Products* Footnote and the Preferred Position of Individual Rights: Louis Lusky and John Hart Ely vs. Harlan Fiske Stone," 12 *Const. Comm.* 277, 302–3 (1995).
18. *Donoghue v. Stevenson*, L.R. App. Cas. 562 (H.L. 1932).
19. Scott Brewer, *supra* note 15, at 952–3; Julius Stone, *Legal Systems and Lawyer's Reasonings* 273 (Stanford University Press 1964).
20. Richard A. Posner, *The Problems of Jurisprudence* 91 (Harvard University Press 1990).
21. Sterling Harwood, *Judicial Activism – A Restrained Defense* 80 (Austin & Winfield 1994), quoting Richard A. Wasserstron, *The Judicial Decision – Toward a Theory of Legal Justification* 39n (Stanford University Press 1961). *See generally*, Henry M. Hart and Albert M. Sacks, *The Legal Process – Basic Problems in the Making and Application of Law* 545–98 (William N. Eskridge, Jr. and Philip P. Frickey, eds., Foundation Press 1994).
22. Martin P. Golding, "Substantive Interpretation in Common law Elaboration," at 29 (article on file with the author).
23. *Burnet v. Coronado Oil and Gas Co.*, 285 U.S. 393, 406 (1932).
24. 1 Laurence H. Tribe, *American Constitutional Law* 82–3 (3rd ed. Foundation Press 1999).

25. *Planned Parenthood v. Casey, supra* note 17, at 849.

26. 410 U.S. 113 (1973).

27. P.S. Atiyah and R.S. Summers, *supra* note 14, at 115, quoting *Jones v. DPP,* (1962) AC 635, 711 (Lord Devlin); Joseph Raz, *The Authority of Law* 51 (1979).

28. John Henry Merryman, *The Civil Law Tradition* 48–9 (sec. ed., Stanford University Press 1985). *See London Tramways Co. v. London County Council,* A.C. 375 (1898).

29. John Henry Merryman, *supra* note 28, at 48.

30. Kathleen M. Sullivan, "The Supreme Court 1991 Term – Foreward: The Justices of Rules and Standards," 106 *Harv. L. Rev.* 22, 58 (1992).

31. Martin P. Golding, Book Review of *The Nature of the Common Law* by Melvin Aron Eisenberg, 43 *Rutgers L. Rev.* 1261, 1269 (1991); Arthur Goodhart, "Determining the Ratio Decidendi of a Case," 40 *Yale L. J.* 161 (1930).

32. *Comeau v. Harrington,*130 N.E. 2d 554 (Mass. 1955).

33. Henry M. Hart and Albert M. Sacks, *supra* note 21, at 587.

34. *Bayliss v. Bishop of London,* 1 Ch. 127, 137 (1913) (Farwell, LJ)(emphasis added).

35. Melvin Aron Eisenberg, *supra,* note 16, at 77–8.

36. *MacPherson v. Buick Motor Co.,* 111 N.E. 1050, 1053 (N.Y. 1916).

37. Robert S. Summers, "Form and Substance in Legal Reasoning," in *Legal Reasoning and Statutory Interpretation* 11 (J. van Dunne, ed. 1989).

38. P.S. Atiyah and R.S. Summers, *supra* note 14, at 5.

39. Oliver Wendell Homes. "The Path of the Law," 10 *Harvard L. Rev.* 457, 469 (1897).

40. Edmund Ursin, "Judicial Creativity and Tort Law," 49 *Geo. Wash. L. Rev.* 229, 252 (1980).

41. Michael C. Dorf, "The Supreme Court 1997 Term – Forward: The Limits of Socratic Deliberation," 112 *Harv. L. Rev.* 4, 16 n.60 (1998).

42. Leonard W. Levy, *The Law of the Commonwealth and Chief Justice Shaw* 24 (Harper Torchbooks 1957).

43. *People v. Hall,* 4 Cal. 399, 405 (1854); *Motion to Admit Miss Lavinia Goodell to the Bar of this Court,* 39 Wis. 232, 245–6 (1875).

44. 1 Laurence H. Tribe, *supra* note 24, at 80–1; Kathleen M. Sullivan, *supra* note 30, at 74.

45. Melvin Aron Eisenberg, *supra* note 16, at 26–37; Robert S. Summers, *supra* note 16, at 707, 717–8, 722–4; Harry H. Wellington, *supra* note 16, at 223–5.

46. Richard A. Posner, "Legal Reasoning From the Top Down and From the Bottom Up: the Question of Unenumerated Fundamental Rights," 59 *Univ. of Chi. L. Rev.* 433, 434 (1992).

47. P.S. Atiyah and R.S. Summers, *supra* note 14, at 287.

48. Warren Hoge, "Lord Denning, 100, a Populist Who Enlivened British Courts," *New York Times,* page A13 (March 6, 1999).

49. *Cartright's Case,* 2 Rushworth 468 (1569), reprinted in I *Judicial Cases concerning American Slavery and the Negro* 9 (Helen Tunnicliff Catterall, ed. Negro University Press 1926)("*Judicial Cases*").

50. *Butts v. Penny,* 83 Eng. Rep. 518, 518 (K.B. 1677).

51. *Gelly v. Cleve,* reprinted in *Judicial Cases, supra* note 49, at 10.

52. *Chamberline v. Harvey,* 87 Eng. Rep. 598 (K.B. 1697).

53. *Id.* at 599 (K.B. 1697)(arguments of counsel).

54. *Id.* at 600.

55. *Id.* at 601.

56. *Chamberline v. Harvey*, 90 Eng. Rep. 830 (K.B. 1697); 91 Eng. Rep. 994 (K.B. 1697).

57. *Smith v. Brown and Cooper*, 91 Eng. Rep. 566, 566 (K.B. 1701).

58. *Smith v. Brown and Cooper, supra* note 57, at 566.

59. *Id.* at 567.

60. *Smith v. Gould*, 91 Eng. Rep. 567, 567 (K.B. 1705); 92 Eng. Rep. 338, 338 (K.B. 1706).

61. *Id.*, 92 Eng. Rep. at 338.

62. Hugh Thomas, *The Slave Trade* 474 (Simon & Schuster 1997).

63. *Id.* at 474.

64. 27 Eng. Rep. 47 (Ch. 1749).

65. *Id.*.

66. William M. Wiecek, "Somerset: Lord Mansfield and the Legitimacy of Slavery in the Anglo-American World," 42 *Univ. of Chi.L.Rev.* 86, 101 (1974).

67. *Omichund v. Barker*, 1 Atk. 21, 33 (K.B. 1744); Melvin Aron Eisenberg, *supra*, note 16, at 81.

68. *Jones v. Randall*, 98 Eng. Rep. 954, 955 (K.B. 1774); *Fisher v. Prince*, 97 Eng. Rep. 876, 876 (K.B. 1762).

69. Bernard L. Shientag, "Lord Mansfield—A Modern Assessment," 10 *Fordham L. Rev.* 345, 351 (1941).

70. *Somerset v. Stewart*, 98 Eng. Rep. 499 (K.B. 1772).

71. A. Leon Higginbotham, Jr., *In the Matter of Color* 337 (Oxford University Press 1978).

72. *Id.* at 501.

73. *Id.* at 502, 503.

74. *Id.* at 500.

75. *Id.*

76. *Id.*

77. *Id.*

78. *Id.* at 507.

79. *Id.*

80. *Id.*

81. *Id.*

82. *Id.* at 505, 506.

83. *Id.*

84. *Id.* at 509.

85. William M. Wiecek, *supra* note 66, at 101; Theodor F.T. Plucknett, *supra* note 1, at 350; Bernard L. Shientag, *supra* note 69, at 350–4.

86. *Somerset, supra* note 70, at 510.

87. U.S. Constitution, Article IV, Section 2(3).

88. William E. Nelson, "The Impact of the Antislavery Movement Upon Styles of Judicial Reasoning in Nineteenth century America," 87 *Harv. L. Rev.* 513, 525, 538–9 (1974).

89. *Id.* at 525–38.

90. Henry Mayer, *All on Fire – William Lloyd Garrison and the Abolition of Slavery* 317 (St. Martin's Press 1998); Leonard W. Levy, *supra* note 42, at 118–82.

91. *Prigg v. Pennsylvania*, 41 U.S. 539, 611 (1842).

92. *Sims' Case*, 61 Mass. 285, 313 (1851).

93. *Id*. at 318.

94. William E. Nelson, *supra* note 88, at 543.

95. *Id*. at 544.

96. *Id*. at 538–9, 544.

97. *Id*. at 538–9.

98. *Id*. at 547—62.

99. Oliver Wendell Holmes, *The Common Law* 5 (Mark deWolfe Howe, ed. Little, Brown and Co. 1963).

100. *Dietrich v. Inhabitants of Northampton*, 138 Mass. 14, 15–6 (1884).

101. *Walker v. Railway Co.*, 28 L.R. Ir. 69 (1891).

102. *Id*. (O'Brien, J.).

103. *Id*.

104. *Allaire v. St. Luke's Hospital*, 56 N.E. 638 (Ill. 1900).

105. *Id*. at 640 (Boggs, J.)(dissenting).

106. *Id*. at 641.

107. *Id*. at 641.

108. *Id*. at 642.

109. *Id* at 641, 642.

110. *Drobner v. Peters*, 133 N.E. 567, 568 (N.Y. 1921).

111. *Lipps v. Milwaukee Electric R. & L. Co.*, 159 N.W. 916, 917 (Wis. 1916).

112. *Magnolia Coca Cola Bottling Co. v. Jordan*, 78 S.W. 945 (Ct. of App. of Tex. 1935).

113. *Id*. at 950.

114. *Montreal Tramways v. Leveille*, 4 D.L.R. 337 (Can. 1933).

115. *Id*. at 345.

116. *Bonbrest v. Kotz*, 65 F. Supp. 138, 140 (D.D.C. 1942).

117. *Id*. at 143.

118. *Id*. at 142–143.

119. *Id*. at 142.

120. *Prosser and Keeton on Torts* 368 (West Publishing Co. 5th ed. 1984).

121. Article 1, sec. 16 of the Ohio Constitution.

122. *Williams v. Marion Rapid Transport*, 87 N.E. 2d 334, 336 (Ohio 1949).

123. *Id*. at 338, 339.

124. *Id*. at 340.

125. *E.g.*, *Farley v. Sartin*, 466 S.E. 2d 522, 534–5 (W. Va. 1995); *Summerfield v. Superior Court*, 698 P. 2d. 712, 722–3 (Ariz. 1985).

126. *Verkennes v. Corniea*, 38 N.W. 2d 838 (Minn. 1949).

127. *Woods v. Lancet*, 102 N.E. 2d 691, 692 (N.Y. 1951).

128. *Id*. at 694, quoting *United Australia, Ltd. v. Barclay Bank, Ltd.* A.C. 1, 29 (1941).

129. *Id*. at 694, quoting *Funk v. United States*, 290 U.S. 371, 382 (1933).

130. *Id*. at 695, quoting *Woods v. Lancet*, 105 N.Y. S. 2d 417, 418 (App. Div. 1951)(Heffernan, J., dissenting).

131. Henry M. Hart and Albert M. Sacks, *supra* note 21, at 476.

132. *Amann v. Faidy*, 114 N.E. 2d 412 (Ill. 1953).

133. *Id*. at 418.

134. *Id*. at 416, 417, 418.

135. *Kwaterski v. State Farm Mutual Automobile Insurance Co.*, 148 N.W. 2d 107, 111 (Wis. 1969).

136. *Libbee v. Permamente Clinic*, 518 P. 2d 636, 638 (Ore. 1974).

137. *Smith v. Brennan*, 157 A. 2d 497, 502 (N.J. 1960); *Presley v. Newport Hospital*, 365 A. 2d 748, 753 (R.I. 1976).

138. *Scott v. McPheeters*, 92 P. 2d 678, 681, 683 (Cal. Ct. of App. 1939).

139. *Puhl v. Milwaukee Automobile Insurance Co.*, 99 N.W. 2d 163, 171 (Wis. 1960).

140. *Farley v. Sartin*, 466 S.E. 2d 522, 529 (W.Va. 1995).

141. *Scott, supra* note 138, at 683.

142. *Tucker v. Howard L. Carmichael & Sons, Inc.*, 63 S.E. 2d 909, 910 (Ga. 1951).

143. *Smith, supra* note 137, at 501, quoting *State v. Culver*, 129 A. 2d 715, 721 (N.J. 1957).

144. *See* cases collected in *Farley, supra* note 140, at 528 note 13. *See also* Richard E. Wood., "Wrongful Death and the Stillborn Fetus: A Common Law Solution to a Statutory Dilemma," 43 *Univ. of Pittsburgh L. Rev.* 819, 821 (1982).

145. *Farley, supra* note 140, *at* 533–4. In *Porter v. Lassiter*, 87 S.E. 2d 100 (Ga. App. 1955), the Georgia Court of Appeals said that a fetus must only be "quick," that is be capable of movement within the womb. Even judges who refuse damages to the estate of a nonviable fetus killed in the womb does not mean that the judges did not believe she was not a legal person. Some courts have said that the purpose of a wrongful death statute is to compensate those who were capable of having an independent life, which only a viable fetus could have, *Thibert v. Milka*, 646 N.E. 2d 1025, 1026 (Mass. 1995).

146. *Bliss v. Passeni*, 95 N.E. 2d 206, 207 (Mass. 1950).

147. *Id.* at 207.

148. *Cavanaugh v. First National Stores, Inc.*, 107 N.E. 2d 307, 308 (Mass. 1952).

149. *Id.*

150. *Keyes v. Construction Service, Inc.*, 165 N.E. 2d 912, 913 (Mass. 1960).

151. *Id.* at 915.

152. Melvin Aron Eisenberg, *supra* note 16, at 80–1.

153. Robert E. Keeton. "Judicial Law Reform – A Perspective on the Performance of Appellate Courts," 44 *Texas L. Rev.* 1254, 1260 (1966).

154. *See* Kathleen M. Sullivan, *supra* note 30, at 75–112 for examples just within American constitutional law.

155. *Byrn v. New York City Health and Hospitals Corp.*, 335 N.Y.S. 390, 397 (N.Y. 1972)(Burke, J., dissenting).

Chapter Eight

1. *E.g.*, Stephen Budiansky, *If Lions Could Talk: Animal Intelligence and the Evolution of Consciousness* (The Free Press 1998); Daniel C. Dennett, "Animal Consciousness: What Matters and Why," in 62 *Social Research* 691–710 (Fall 1995): Peter Carruthers, *The Animals Issue* (Cambridge University Press 1992).

2. Ted Honderich, "Mind, Brain, and Self-Conscious Mind," in Colin Blakemore and Susan Greenfield, eds., *Mindwaves* 445 (Basil Blackwell 1987).

3. Bernard J. Baars, *In the Theatre of Consciousness: The Workspace of the Mind* 16 (Oxford University Press 1997).

4. Colin McGinn, "Can we ever understand consciousness?," XLVI *The New York Review of Books* 44, 44 (June 10, 1999).

5. "Consciousness," definitions 4a, 5a, and 6, 3 *Oxford English Dictionary* 757 (sec. ed. 1988); "mind," definition 17a, 9 *Oxford English Dictionary* 799 (sec. ed. 1988); "thinking," definition 1a, 17 *Oxford English Dictionary* 949 (sec. ed. 1988); "thought," definitions 1a, b, and c, *id.* at 983.

6. John R. Searle, *The Mystery of Consciousness* 121–2 (NYREV, Inc. 1997)(emphasis in the original).

7. Charles Darwin, *The Descent of Man* 494 (1871)(repr. Modern Library 1982).

8. Letter from Descartes to Henry More (February 5, 1649)(Anthony Kenny, trans), reprinted in *Animal Rights and Human Obligations* 66 (Tom Regan and Peter Singer, eds., Prentice-Hall, Inc. 1976).

9. Daniel C. Dennett, *Kinds of Minds: Toward an Understanding of Consciousness* 164 (BasicBooks 1996).

10. Daniel C. Dennett, *Consciousness Explained* 405–6 (Little, Brown & Co. 1991).

11. Simon Baron-Cohen, *Mindblindness: An Essay on Autism and Theory of Mind* 121 (MIT Press 1995).

12. Elliot Sober, "Morgan's Canon," in *The Evolution of Mind* 229 (Denise Dellarosa Cummins and Colin Allen, eds. Oxford University Press 1998); C. Lloyd-Morgan, *An Introduction to Comparative Psychology* 53 (Walter Scott 1894).

13. Bernard E. Rollin, *The Unheeded Cry: Animal Consciousness, Animal Pain and Science* 75–8 (Oxford University Press 1989).

14. C. Lloyd-Morgan, *The Animal Mind* 89 (Edward Arnold 1930).

15. Frans B.M. de Waal, *Good-Natured: The Origins of Right and Wrong in Humans and Other Animals* 64 (Harvard University Press 1996).

16. Elliot Sober, *supra* note 12, at 225.

17. *Id.* at 231–4.

18. *Id.* at 65.

19. Donald R. Griffin, "From cognition to consciousness," 1 *Animal Cognition* 3, 11 (1998).

20. Frans B.M. de Waal, "The Chimpanzee's Sense of Social Regularity and Its Relation to the Human Sense of Justice," 34 *American Behavioral Scientist* 335, 341 (Jan./Feb. 1991). *See* Frans de Waal, *supra* note 15, at 64–5.

21. Daniel J. Povinelli and Steve Giambrone, "Inferring other minds: Failure of the argument by analogy," in *Philosphical Topics* 1 (G. Massey and B.D. Massey, eds. 1999) (in press).

22. Daniel J. Povinelli et al., Book Review of *Reaching Into Thought: The Minds of the Great Apes* (Anne E. Russon et al., eds., Cambridge University Press 1996), 2 *Trends in Cognitive Sciences* 158 (emphasis added).

23. Daniel J. Povinelli and Steve Giambrone, *supra* note 21, at 23.

24. Lawrence Weiskrantz, *Consciousness Lost and Found* 82 (Oxford University Press 1997).

25. Donald R. Griffin, *Animal Minds* 4 (University of Chicago Press 1992); Mark Bekoff, "Playing with play: What can we learn about cognition, negotiation, and evolution?," in Denise Dellarosa Cummins and Colin Allen, eds., *supra* note 12, at 178.

26. Mark Bekoff, *supra* note 25, at 229–31, 240; Bernard E. Rollin, *supra* note 13, at 79.

27. Marian Stamp Dawkins, *Through Our Eyes Only? The Search for Animal Consciousness* 176 (W.H. Freeman 1993)(emphasis added).

28. Sue Savage-Rumbaugh et al., *Apes, Language, and the Human Mind* 107 (Oxford University Press 1998)(emphasis in the original).

29. Daniel C. Dennett, "The intentional stance in theory and practice," in *Machiavellian Intelligence: Social Expertise and the Evolution of Intellect in Monkeys, Apes, and Humans* 181–2 (Richard Byrne and Andrew Whiten, eds. Oxford University Press 1988).

30. Francis Crick and Christof Koch, "The Problem of Consciousness," in 267 *Scientific American* 152 (September, 1992); Gerald M. Edelman, *The Remembered Present: A Biological Theory of Consciousness* 4–5 (Basic Books 1989); William James, *Psychology: The Briefer Course* 152 (Henry Holt 1893); David Ballin Klein, *The Concept of Consciousness: A Survey* 34–5 (University of Nebraska Press 1984).

31. Donald R. Griffin, *supra* note 25, at 4–5; John R. Searle, *supra* note 6, at 89–90; Owen Flanagan, *Consciousness Reconsidered* 1–3 (MIT Press 1992); Mario Bunge, *The Mind-Body Problem: A Psychological Approach* 6 (Pergamon 1980).

32. Gerald M. Edelman, *Bright Air, Brilliant Fire: On the Matter of Mind* 112 (Basic Books 1992); Owen J. Flanagan, Jr., *The Science of the Mind* 28–30, 33–35 (MIT Press 1984).

33. William James, *supra* note 30, at 26; Owen Flanagan, *supra* note 31, at 153–175; Gerald M. Edelman, *supra* note 32, at 111.

34. Compare Owen Flanagan, *supra* note 31, at 41, 169–175 *with* Daniel C. Dennett, *supra* note 10, at 253–254, 356.

35. Giulio Tononi and Gerald M. Edelman, "Consciousness and Complexity," 283 *Science* 1846, 1846 (1998).

36. For an argument over the nature and properties of qualia, *compare* Owen Flanagan, *supra* note 31, at 62–85 *with* Daniel C. Dennett, *supra* note 10, at 369–411 and Gerald M. Edelman, *supra* note 32, at 113–116.

37. Peter Carruthers, *supra* note 1, at 170; *id.* at 186.

38. Alison Jolly, "Conscious Chimpanzees? A Review of Recent Literature," in Carolyn A. Ristau, *Cognitive Ethology: The Minds of Other Animals* 232 (Lawrence Erlbaum Associates 1991).

39. Terrence W. Deacon, *The Symbolic Species: The co-evolution of language and the brain* 442 (W.W. Norton 1997).

40. Marian Stamp Dawkins, *Animal Suffering: The Science of Animal Welfare* 25 (Chapman and Hall 1980).

41. Susan A. Greenfield, *Journey to the Centers of the Mind* 78, 82, 132 (W.H. Freeman and Co. 1995); John R. Searle, *supra* note 6, at 83.

42. Bernard A. Baars, *supra* note 3, at 33.

43. Donald R. Griffin, *supra* note 19, at 5.

44. *Id.* at 13.

45. Richard Sorabji, *Animal Minds and Human Morals: The Origin of the Western Debate* 18, 30, 50, 64, 87 (Cornell University Press 1993).

46. Conversation with Donald R. Griffin, April 11, 1999.

47. *The International Dictionary of Psychology* (N.S. Sutherland, ed., Continuum 1989).

48. Gerald M. Edelman, *supra* note 32 at 122.

49. Donald R. Griffin, *supra* note 25, at 10. Both definitions are taken from Thomas Natsoulas' "Consciousness," in 33 *American Psychologist* 906–913 (1978).

50. Donald R. Griffin, *supra* note 25, at 11–12, 248–249.

51. M. Meijsing, "Awareness, self-awareness and perception: An essay on animal consciousness," in M. Dol et al., *Animal Consciousness and Animal Ethics* 48, 57 (Van Gorcum 1997).

52. Owen Flanagan, *supra* note 31, at 194–195. *E.g.* John R. Searle, *supra* note 6, at 142–143.

53. Antonio R. Damasio, *Descartes' Error: Emotion, Reason, and the Human Brain* 165 (Bard 1998)(1994).

54. Gerald M. Edelman, *supra* note 32 at 124.

55. N.S. Sutherland, ed., *supra* note 47.

56. Paul E. Griffiths, "Thinking about consciousness," 397 *Nature* 117, 118 (1999)(reviewing *Consciousness and Human Identity* [John Cornwall, ed. Oxford University Press 1998]).

57. Susan Greenfield, *supra* note 41, at 196.

58. Steven Pinker, *How the Mind Works* 146 (W.W. Norton 1997).

59. James Gorman, "Consciousness Studies: From Stream to Flood," in *The New York Times*, p.C5 (April 29, 1997).

60. Stephen Priest, *Theories of Mind* 102–14 (Houghton Mifflin 1991); U.T. Place, "Is Consciousness a Brain Process?," in 47 *British Journal of Psychology* 44–50 (1956).

61. Patricia Churchland, *Neurophilosophy: Toward a Unified Science of the Mind: Brain* 299–312 (MIT Press 1986); Owen J. Flanagan, Jr., *supra* note 32, at 219–221; Paul Churchland, "Eliminative Materialism and the Propositional Attitudes," 78 *Journal of Philosophy* 67–90 (1981).

62. Stephen Priest, *supra* note 60, at 35–64; Gilbert Ryle, *The Concept of Mind* (Barnes and Noble 1949).

63. John R. Searle, *supra* note 6, at 111–112, 116; Donald R. Griffin, *Animal Thinking* 8 (Harvard University Press 1984).

64. Stephen Priest, *supra* note 60, at 133–149; Hilary Putnam, "Psychological Predicates," in W.H. Capitan and D.D. Merrill, eds., *Art, Mind, and Religion* 37–48 (University of Pittsburgh Press 1967).

65. Thomas Nagel, "The Mind Wins!," in *New York Review of Books* XL(5):37–41 (March 4, 1993); John R. Searle, *supra* note 6, at 11, 14, 28, 89–98.

66. Owen Flanagan, *supra* note 31, at 2–3, 220–221.

67. Stephen Priest, *supra* note 60, at 1–34; Karl Popper and John Eccles, *The Self and Its Brain* (Springer Internat 1977).

68. Colin McGinn, *The Problem of Consciousness* 20–21 (Basil Blackwell 1991).

69. Thomas Nagle, *The View from Nowhere* (Oxford University Press 1986).

70. Elizabeth Pennisi, "Worming secrets from the *C. elegans* Genome," 282 *Science* 1972, 1974 (1998).

71. Nicholas Wade, "Human or chimp? 50 Genes are the key," *New York Times* p. D1 (October 20, 1998).

72. Morris Goodman et al., "Primate Phylogeny and Classification Elucidated at the Molecular Level," in *Evolutionary Theory and Practice: Modern Perspectives* 193, 207 (S. P. Wasser, ed., Kluwer Academic Publishers 1999).

73. Maryellen Ruvolo, "Molecular Phylogeny of the Hominids," 14 *Molecular and Biological Evolution* 248 (1996); David Pilbeam, "Genetic and Morphological Records of the Hominoidea and Hominid Origins: A Synthesis," 5 *Molecular Phylogentics and Evolution* 155, 164 (1996).

74. Marian Stamp Dawkins, *supra* note 27, at 7–9, 169–173; Owen Flanagan, *supra* note 31, at 40–46; Gerald M. Edelman, *supra* note 32, at 20; Bernard J. Baars, *A Cognitive Theory of Consciousness* 347–356 (Cambridge University Press 1988); David Ballin Klein, *supra* note 30, at 102–107; Stephen Walker, *Animal Thought* 39–46, 387–388 (Routledge & Kegan Paul 1983).

75. Bernard A. Baars, *supra* note 3, at 27–33.

76. Gerald M. Edelman, *supra* note 32, at 122–3, 151.

77. Susan A. Greenfield, *supra* note 41, at 126.

78. Donald R. Griffin, "Progress Toward a Cognitive Ethology," in Carolyn A. Ristau, ed., *supra* note 38, at 15.

79. Gerald M. Edelman, *supra* note 32, at 17 (10^{10}); Jean-Pierre Changeux, *Neuronal Man: The Biology of Mind* 51 (Oxford University Press 1985) (3×10^{10}); Owen Flanagan, *supra*, note 31, at 35–36 (10^{11}); Patricia Churchland, *supra*, note 61, at 36 (10^{12}–10^{14}).

80. Giulio Tononi and Gerald M. Edelman, *supra* note 35, at 1847, 1849; Susan A. Greenfield, *supra* note 41, at 128; *Lives in the Balance: The Ethics of Using Animals in Biomedical Research. The Report of a Working Party of the Institute of Medical Ethics* 48 (Jane A. Smith and Kenneth M. Boyd, eds. Oxford University Press 1991).

81. Irene M. Pepperberg, "A Communicative Approach to Animal Cognition: A Study of Conceptual Abilities of an African Grey Parrot," in Carolyn A. Ristau, ed., *supra* note 38, at 159–160; Theodore Xerophon Barber, *The Human Nature of Birds* 109 (St. Martin's Press 1993).

82. Richard Passingham, "Brain," in David McFarland, *ed.*, *The Oxford Companion to Animal Behavior* 45 (Oxford University Press 1987).

83. Jean-Pierre Changeux, *supra* note 79, at 159, quoting A.R. Luria, *Higher Cortical Functions in Man* (Basic Books 1980).

84. Jean-Pierre Changeux, *supra* note 79, at 159.

85. Katerina Semendeferi et al., "The evolution of the frontal lobes: a volumetric analysis based on three-dimensional reconstructions of magnetic resonance scans of human and ape brains," 32 *Journal of Human Evolution* 375, 375 (1997).

86. *Id.* at 380–7.

87. Gerald M. Edelman, *supra* note 32, at 17; Owen Flanagan, *supra* note 31, at 37; Jean-Pierre Changeux, *supra* note 79, at 52; Patricia Churchland, *supra* note 61, at 40.

88. Owen Flanagan, *supra* note 31, at 37; Paul M. Churchland, *A Neurocomputational Perspective: The Nature of Mind and the Structure of Science* 132 (MIT Press 1989).

89. Stephen Walker, *supra* note 74, at 339.

90. Stephen Jay Gould, *The Mismeasure of Man* 35 (W.W. Norton 1981); Winthrop D. Jordan, *White Over Black: American Attitudes Toward the Negro 1550–1812* 220–1(University of North Carolina Press 1968).

91. Stephen Jay Gould, *Teeth and Horses's Toes* 241 (W.W. Norton 1983).

92. Ernst Mayr, *Principles of Systematic Zoology* 56 (McGraw-Hill 1969); G.E.R. Lloyd, "The Development of Aristotle's Theory of the Classification of Animals, 6 *Phronesis* 59, 73 (Cambridge University Press 1961).

93. Stephen Jay Gould, *Dinosaur in a Haystack* 424 (Crown Publishers 1995).

94. Ernst Mayr, *Systematics and the Origin of Species* 108 (Harvard University Press 1942).

95. Stephen Jay Gould, *The Flamingo's Smile* 263 (W.W. Norton 1985); Arthur O. Lovejoy, *The Great Chain of Being* 234 (Harper Torchbook 1960).

96. Above and below the family, for example, are the superfamily and subfamily. The three full ranks are the legion and cohort, inserted after class, and tribe, slipped in between family and genus. *See* Malcolm C. McKenna and Susan K. Bell, *Classification of Mammals Above the Species Level* 20 (Columbia University Press 1998); Daniel Boorstin, *The Discoverers* 420–46 (Vantage Books 1985); Peter J. Bowler, *Evolution: The History of an Idea* 60–1 (University of California Press 1984).

97. Charles Darwin, *On the Origin of Species* 20 (1859).

98. Stephen Jay Gould, *The Panda's Thumb* 206–7 (W.W. Norton 1980).

99. Elizabeth E. Watson et al., "Homo genus: A Review of the Classification of Humans and the Great Apes," *The Proceedings of the Dual Congress 1998*: 4 (Angelo Pontecorboli 1999)(in press) quoting Linneaus, *Systema Naturae* (10th ed. 1778).

100. Frank Spencer, "Pithekos to Pithecanthropus: An Abbreviated Review of Changing Scientific Views on the Relationship of the Anthropoid Apes to Homo," in *Ape, Man, Apeman: Changing Views since 1600* 16 (Raymond Corbey and Bert Theunissen, eds. Dept. of Prehistory, Leiden University 1995).

101. Ernst Mayr, *supra* note 94, at 96.

102. Roger Lewin, "Family Relationships are a Biological Conundrum," 242 *Science* 671 (November 4, 1988).

103. Ernst Mayr, *supra* note 94, at 284.

104. *Id.* at 270.

105. *Id.* at 238–42.

106. Morris Goodman et al., *supra* note 72, at 194.

107. Ernst Mayr, *supra* note 94, at 279; George Gaylord Simpson, *Principles of Animal Taxonomy* 214–5 (Columbia University Press 1961).

108. Roger Lewin, *supra* note 102, at 671; Julian Huxley, "The Three Types of Evolutionary Process," 180 *Nature* 454, 455 (1957).

109. Michael P. Ghiglieri, *The Dark Side of Man: Tracing the Origins of Male Violence* 70 (Perseus Books 1999).

110. Stephen Jay Gould, *supra* note 93, at 254.

111. *Id.* at 398–9.

112. Roger Lewin, *supra* note 102, at 671.

113. Morris Goodman, "Epilogue: A Personal Account of the Origins of a New Paradigm," 5 *Molecular Phylogenetics and Evolution* 269, 272 (1996).

114. Elizabeth. E. Watson et al., *supra* note 99, at 6-7.

115. Morris Goodman, "The Genomic Record of Humankind's Evolutionary Roots," 64 *American Journal of Human Genetics* 31, 33, 38 (1999); Morris Goodman et al., *supra* note 72, at 200, 209; Elizabeth Watson et al., *supra* note 99; Morris Goodman et al., "Toward a Phylogenetic Classification of Primates Based on DNA Evidence Complemented by Fossil Evidence," 9 *Molecular Phylogenetics and Evolution* 585, 594, 596 (1998); Jared Diamond, *The Rise and Fall of the Third Chimpanzee* 27 (HarperCollins 1991).

116. David Pilbeam, *supra* note 73, at 155; Stephen Jay Gould, *supra* note 93, at 248–59.

117. *E.g.*, Sherrie Lyons, "Taxonomy Recapitulates Society," 279 *Science* 38 (1998); George Lakoff, *Women, Fire, and Dangerous Things: What Categories Reveal About the Mind* (University of Chicago Press 1985).

118. Londa Schiebinger, "Why mammals are called mammals: gender politics in eighteenth century natural history," 90 *The American History Review* 382 (1993).

119. Michael D. Rugg et al., "Dissociation of the neural correlates of implicit and explicit memory, 392 *Nature* 595–8 (1998).

120. Robert E. Clark and Larry R. Squire, "Classical Conditioning and Brain Systems: The Role of Awareness, 280 *Science* 77, 77 (1998).

121. *Id.* at 80 note 3.

122. *Id.* at 77.

123. *Id.* at 78–9; Daniel L. Schacter, "Memory and Awareness," 280 *Science* 59, 60 (1998).

124. Daniel L. Schacter, *supra* note 123, at 59.

125. Robert E. Clark and Larry R. Squire, *supra* note 120, at 81 note 28.

126. *Id.* at 80 note 10; *id.* at 81 note 24.

127. Daniel L. Schacter, *supra* note 123, at 60.

128. Robert E. Clark and Larry R. Squire, *supra* note 120, at 80.

129. *Id.* at 77 (emphasis added).

130. Peter Carruthers, *supra* note 1, at 172–3, 182–3.

131. Lawrence Weiskrantz, *supra* note 24, at 23.

132. *Id.* at 17.

133. *Id.* at 87–8.

134. Stephen Jay Gould, *Ontogeny and Phylogeny* 2, 267 (Harvard University Press 1977). See Sue Taylor Parker and Michael L. McKenney, *Origins of Intelligence: The Evolution of Cognitive Development in Monkeys, Apes, and Humans* 342–4 (Johns Hopkins University Press 1999).

135. Janet Wilde Astington, *The Child's Discovery of the Mind* (Harvard University Press 1993), Henry M .Wellman, *The Child's Theory of Mind* (MIT Press 1990); Stanley I. Greenspan, *The Growth of the Mind* (Addison-Wesley 1997).

136. Peter Mitchell, *Introduction to Theory of Mind: Children, Autism and Apes* 130–49 (Arnold 1997).

137. Daniel C. Dennett, "The intentional stance in theory and in practice," Richard Byrne and Andrew Whiten, eds., *supra* note 29, at 185–6.

138. Janet Wilde Astington, *supra* note 135, at 55–60; Josef Perner et al., "Prelief: The conceptual origins of belief and pretence," in *Children's Early Understanding of Mind: Origins and Development* 261 (C. Lewis and P. Mitchell, eds., Lawrence Erlbaum 1994); A.S. Lillard, "Young children's conception of pretense: Action or mental representational state?," 64 *Child Development* 372–86 (1993); Alan M. Leslie, "Pretence and representation: The origins of theory of mind," 94 *Psychological Review* 412–26 (1987).

139. David G. Premack and Gary Woodruff, "Does the chimpanzee have a theory of mind?" 1 *Behavioral and Brain Sciences* 515, 515 (1978).

140. Juan Carlos Gómez, "Are apes persons? The case for primate intersubjectivity," in *Etica and Animali* (Special issue on primate personhood)(in press); C. Trevarthen, "The foundations of intersubjectivity: Development of interpersonal

and cooperative understanding in infants," in *The Social Foundations of Language and Thought* 316–42 (W.W. Norton 1980).

141. Juan Carlos Gómez, "Primate theories of primate minds: conceptual and methodological issues,' in *Enfance* (special issue)(in press); Juan Carlos Gómez, "Nonhuman primate theories of (non-human primate) minds; some issues concerning the origins of mindreading," in *Theories of Theories of Mind* 330, 341–2 (Cambridge University Press 1996).

142. David J. Povinelli et al., "Can Chimpanzees Guess What Others Know?," in 18 *American Journal of Primatology* 161 (1989).

143. Angeline Lillard, "Ethnopsychologies: Cultural variations in theories of mind," 123 *Psychological Bulletin* 3, 3 (1998).

144. Janet Wilde Astington, *supra* note 135, at 18–9 (emphasis in the original).

145. Robert Seyfarth, "Sir Oran Haut-ton finds his voice," 395 *Science* 29, 30 (1998), reviewing Sue Savage-Rumbaugh et al., *supra* note 28.

146. Angeline Lillard, *supra* note 143, at 13.

147. *Id.* at 26–7.

148. Henry W. Wellman, *supra* note 135, at 276–7.

149. Angeline Lillard, *supra* note 143, at 4–5; Vittorio Gallese and Alvin Goldman, "Mirror neurons and the simulation theory of mind-reading," 2 *Trends in Cognitive Science* 493, 496–7 (December, 1998); Janet Astington, "What is theoretical about the child's theory of mind?: a Vygotskian view of its development," in *Theories of Theories of Mind* 184–8 (Peter Carruthers and Peter K. Smith, eds., Cambridge University Press 1996); Peter Carruthers and Peter K. Smith, "Introduction," in *id.* at 4–5.

150. Vittorio Gallese and Alvin Goldman, *supra* note 149, at 496.

151. Michael Tomasello and Josef Call, *Primate Cognition* 59, 108, 142, 404 (Oxford University Press 1997); Alison Gopnik and Andrew N. Meltzoff, "Minds, bodies, and persons: Young children's understanding of the self and others as reflected in imitation and theory of mind research," in *Self-Awareness in Animals and Humans: Developmental Perspectives* 170–1 (Sue Taylor Parker et al., eds. Cambridge University Press 1994); Daniel Hart and Suzanne Fegley, "Social imitation and the emergence of a mental model of self," in *id.* at 151.

152. C. Trevarthan, "Communication and cooperation in early infancy," in *Before Speech: The Beginnings of Human Communication* 321–47 (M. Bullowa, ed., Cambridge University Press 1979).

153. Karen Wynn, "An Evolved Capacity for Number, " in Denise Dellarosa Cummins and Colin Allen, eds., *supra* note 12, at 109–16; Stanislas Daehaene, *The Number Sense* 52–5 (Oxford University Press 1997); Karen Wynn, "Addition and subtraction by human infants," 358 *Nature* 749 (1992).

154. Stanislas Daehaena, *supra* note 153, at 56–7.

155. Michael Tomasello and Josef Call, *supra* note 151, at 404.

156. S.L. Williams et al., "Comprehension Skills of Language-Competent and Nonlanguage-Competent Apes," 17 *Language & Communication* 301 302 (1997).

157. James E. King et al., "Evolution of Intelligence, Language, and Other Emergent Processes for Consciousness: A Comparative Perspective," in *Toward a Science of Consciousness II* 383, 387 (Stuart R. Hameroff et al. eds., M.I.T. Press

1998); David Premack and Ann James Premack, "Infants attribute value (plus and minus) to the goal-directed actions of self-propelled objects," 9 *Journal of Cognitive Neuroscience* 848–56 (1997); Simon Baron-Cohen, *supra* note 11, at 48; Michael Tomasello and Josef Call, *supra* note 151, at 102, 323, 362, 405, 408; Janet Wilde Astington, *supra* note 135, at 21–4, 41–2.

158. Michael Tomasello, "Joint attention as social cognition,: in *Joint Attention: Its Origins and Role in Development* 126 (Chris Moore and Philip J. Dunham, eds., Lawrence Erlbaum 1995); Jerome Bruner, "From joint attention to the meeting of minds: An Introduction," in *id.* at 4.

159. Daniel J. Povinelli and Todd M. Preuss, "Theory of Mind: evolutionary history of a cognitive specialization," in 18 *Trends in Cognitive Science* 418, 422 (1995); Janet Wilde Astington, *supra* note 135, at 41, 99.

160. M. Carpenter, K. Nagell, and M. Tomasello, "Social cognition, joint attention, and communicative competence from 9 to 15 months of age," 63 *Monographs of the Society for Research in Child Development* (1998); G. Butterworth, "Origins of mind in perception and action," in Chris Moore and Philip J. Durham, eds., *supra* note 158 at 29.

161. Simon Baron-Cohen, *supra* note 11, at 48; Daniel J. Povinelli and Todd M. Preuss, *supra* note 159, at 422.

162. Michael Tomasello and Josef Call, *supra* note 151, at 39.

163. *Id.* at 58, 408, 409.

164. *Id.* at 38–9.

165. *Id.*at 337; Daniel J. Povinelli and Todd M. Preuss, *supra* note 159, at 422, 423; Daniel Hart and Suzanne Fegley, "Social imitation and the emergence of a mental model of self," in Sue Taylor Parker et al., eds., *supra* note 151, at 152, 159; Gordon G. Gallup, Jr., "Self-recognition; Research strategies and experimental design," in Sue Taylor Parker et al., eds., *supra* note 151, at 39.

166. Daniel J. Povinelli and Todd M. Preuss, *supra* note 159, at 422.

167. Janet Wilde Astington, *supra* note 135, at 52.

168. Michael Tomasello and Josef Call, *supra* note 151, at 409; Daniel J. Povinelli and Todd M. Preuss, *supra* note 159, at 422; Janet Wilde Astington, *supra* note 135, at 52–3.

169. Michael Tomasello and Josef Call, *supra* note 151, at 362, 409.

170. Janet Wilde Astington, *supra* note 135, at 102; Michael Tomasello and Josef Call, *supra* note 151, at 93.

171. *Id.* at 42, 59, 95.

172. Marc Hauser & Susan Carey, "Building a Cognitive Creature from a Set of Primitives: Evolutionary and Developmental Insights," in Denise Dellarosa Cummins and Colin Allen, eds., *supra* note 11, at 8–61; Stanislas Daehaene, *supra* note 153, at 53.

173. Malinda Carpenter et al., "Fourteen-through–18-month-old infants differentially imitate intentional and accidental actions," 21 *Infant Behavior & Development* 315 (1998).

174. Michael Tomasello and Josef Call, *supra* note 151, at 409.

175. Sue Taylor Parker and Anne E. Russon, "On the wild side of culture and cognition in great apes," in Anne E. Russon et al., eds., *supra* note 22, at 431.

176. James E. Reaux et al., "A longitudinal investigation of chimpanzees' understanding of visual perception," 70 *Child Development* 275 (March/April 1999);

John H. Flavell et al., "Young children's knowledge about visual perception: Further evidence for the level 1-level 2 distinction," 17 *Developmental Psychology* 99–103 (1981).

177. David Premack and Ann James Premack, *supra* note 157, at 854.

178. Henry M. Wellman, "From Desires to Beliefs: Acquisitions of a Theory of Mind," in *Natural Theories of Mind: Evolution, Development and Simulation of Everyday Mindreading* 28–32 (Andrew Whiten, ed. Basil Blackwell 1991).

179. Michael Tomasello and Josef Call, *supra* note 151, at 337, 338; Daniel J. Povinelli and Todd M. Preuss, *supra* note 159, at 420; Janet Wilde Astington, *supra* note 135, at 82, 84–5.

180. Michael Tomasello and Josef Call, *supra* note 151, at 411.

181. David Premack, "'Does the chimpanzee have a theory of mind?' revisited," in Richard Byrne and Andrew Whiten, eds., *supra* note 29, at 164.

182. James E. Reaux et al., *supra* note 176; John H. Flavell et al., *supra* note 176.

183. Josep Call and Michael Tomasello, "A nonverbal false belief task: The performance of children and great apes," 70 *Child Development* 381, 387 (1999).

184. Henry M. Wellman, *supra* note 178, at 24; G.J. Hogrefe, H. Wimmer, and J. Perner, "Ignorance versus false belief," 57 *Child Development* 567–82 (1986).

185. Alison Gopnik and Janet W. Astington, "Children's understanding of representational change, and its relation to the understanding of false belief and the appearance-reality distinction," 59 *Child Development* 26 (1988).

186. Simon Baron-Cohen, *supra* note 11, at 60; Simon Cohen-Baron, "Precursors to a Theory of Mind: Understanding Attention in Others," in Andrew Whiten, ed., *supra* note 178, at 244–50.

187. Simon Baron-Cohen, "Are autistic children 'behaviourists'? An examination of their mental-physical and appearance-reality distinctions," 19 *Journal of Autism and Developmental Disorders* 570–600 (1989).

188. Francesca Happe, *Autism: An introduction to psychological theory* 35–7 (Harvard University Press 1995); Simon Baron-Cohen, *supra* note 11, at 72–83; Alan Leslie, "The Theory of Mind Impairment in Autism: Evidence for a Modular Mechanism of Development?, in Andrew Whiten, ed., *supra* note 178, at 65–74; A.H. Atwood et al., "The understanding and use of interpretational gestures by autistic and Down's syndrome children," 18 *Journal of Autistic and Developmental Disorders* 241 (1988); G. Dawson and F.C. McKissick, "Self-recognition in autistic children," 14 *Journal of Autism and Developmental Disorders* 38 (1984).

189. M. Chandler and S. Hala, "The role of personal involvement in the assessment of early false belief skills," in *Children's Early Understanding of Mind: Origins and Development* 413–25 (C. Lewis and P. Mitchell, eds. Lawrence Erlbaum 1994): B. Sodian, "Early deception and the conceptual continuity claim," in *id.* at 385–401.

190. Sharon Griffin, "Young children's awareness of their inner world: A neostructural analysis of the development of intrapersonal intelligence," in *The Mind's Staircase: Exploring the Conceptual Underpinnings of Children's Thought and Knowledge*, 189–196 (Robbie Cass, ed., Lawrence Erlbaum Associates 1992).

191. Alison Gopnik and Andrew N. Meltzoff, *supra* note 151, at 175.

192. Simon Baron-Cohen, *supra* note 11, at 70–1.

193. Alison Gopnik and Andrew N. Meltzoff, *supra* note 151, at 178.

194. Daniel J. Povinelli and Todd M. Preuss, *supra* note 159, at 420; Alison Gopnik and Andrew N. Meltzoff, *supra* note 151, at 175; Henry M. Wellman, *supra* note 178, at 20.

195. J.S. DeLoache, "Rapid change in the symbolic functioning of very young children," 238 *Science* 1556 (1987).

196. Michael Tomasello and Josef Call, *supra* note 151, at 228, 242; Richard W. Byrne and Andrew Whiten, "Computation and Mindreading in Primate Tactical Deception," in Andrew Whiten, ed., *supra* note 178, at 127.

197. E. Sue Savage-Rumbaugh and Duane M. Rumbaugh, "Perspective on Consciousness, Language, and Other Emergent Processes in Apes and Humans," in Stuart R. Hameroff et al. eds., *supra* note 157, at 538.

198. Simon Baron-Cohen, "The development of a theory of mind in autism: deviance and delay?, 14 *Psychiatric Clinics of North America* 33 (1991); Alison Gopnik and Janet W. Astington, *supra* note 185.

199. Alison Gopnik and Andrew N. Meltzoff, *supra* note 151, at 180; Gyorgy Gergely, "From self-recognition to theory of mind," in Sue Taylor Parker et al., eds., *supra* note 151, at 53.

200. Janet W. Astington and Alison Gopnik, "Developing Understanding of Desire and Intention," in Andrew Whiten, ed., *supra* note 178, at 39.

201. Michael Tomasello and Josef Call, *supra* note 151, at 308.

202. *Id.* at 155; Karen Wynn, "Children's understanding of counting," 36 *Cognition* 155 (1990).

203. *Id.* at 155–8.

204. Susan R. Leekam, "Jokes and Lies: Children's Understanding of Intentional Falsehood," in Andrew W. Whiten, ed., *supra* note 178, at 161; Josef Perner, "Higher-order beliefs and intentions in children's understanding of social interaction," in Janet W. Astington et al., eds., *Developing Theories of Mind* 271–90 (Cambridge University Press 1988).

205. Carl Nils Johnson, "Theory of mind and the structure of conscious experience," in Janet W. Astington et al., eds., *supra* note 204, at 54–55.

206. Edward O. Wilson, *Biophilia* 129 (Harvard University Press 1984).

207. Gerald M. Edelman, *supra* note 32, at 218.

208. John Horgan, "The Mastermind of Artificial Intelligence," in 269 *Scientific American* 35, 38 (1993).

209. Tim Beardsley, "Here's Looking at You," in 280 *Scientific American* 39 (1999); Daniel C. Dennett, *supra* note 1, at 694.

210. Stephen Budiansky, *supra* note 1, at 64.

211. Tim Beardsley, "Humans Unite!," 280 *Scientific American* 35 (1999).

212. Daniel C. Dennett, *supra* note 1, at 694 (emphasis in the original).

213. Oliver Wendell Holmes, *The Common Law* 7 (Mark deWolfe Howe, ed. Little, Brown and Co. 1963)(1881).

214. Daniel C. Dennett, *supra* note 1, at 695, 703 (emphasis in the original).

215. Thomas Nagel, "What we have in mind when we say we're thinking," *Wall Street Journal* p. 14 (November 7, 1991)(reviewing Daniel Dennett's *Consciousness Explained*); John R. Searle, *supra* note 6, at 121–2.

216. Donald R. Griffin, *supra* note 25, at 239–40; Ludwig Wittgenstein, *Philosphical Investigations* 223 (Basil Blackwell 1958).

217. Stephen Budiansky, *supra* note 1, at 193.

218. Paul Bloom, "Some issues in the evolution of language," in Denise Dellarosa Cummins and Colin Allen, eds., *supra* note 12, at 215–6.

219. Steven Pinker, *The Language Instinct: How the Mind Creates Language* 70–3 (William Morrow and Co. 1994); Antonio Damasio, *supra* note 53, at 107.

220. Oliver Sacks, *An Anthropologist on Mars* 266–7 (Alfred A. Knopf 1995); Russell T. Hulbert et al., "Sampling the inner experience of autism: A preliminary report," 24 *Psychological Medicine* 385–95 (1994); Russell T. Hulbert, "Sampling normal and schizophrenic inner experience," 54–5, 60, 101–2 (Plenum Press).

221. Susan Schaller, *A Man Without Words* 138–9, 144, 180, 189 (Summit Books 1991).

222. Harlan Lane, *The Wild Boy of Aveyron* 243 (Harvard University Press 1976), quoting Ferdinand Berthier, *L'Abbé de L'Epée* 38 (Michel Levy Frères 1852).

223. Oliver Sacks, *Seeing Voices* 38, 39 (University of California Press 1989).

224. *Id.* at 40 (emphasis in the original).

225. *Id.* at 41.

226. *Id.* at 46.

227. *Id.* at 40–1.

228. Oliver Sacks, "Foreword" in Susan Schaller, *supra* note 221, at 13.

229. Susan Schaller, *supra* note 221, at 174.

230. *Id.* at 142. *See id.* at 171–2.

231. *Id.* at 132.

232. *Id.* at 134.

233. *Id.* at 146.

234. *Id.*

235. *Id.* at 181–8.

236. Stephen Jay Gould, *supra* note 93, at 249.

237. Gerald M. Edelman, *supra* note 32, at 125; *see* Roger K.R. Thompson et al., "Language-Naive Chimpanzees (*Pan troglodytes*) Judge Relations Between Relations in a Conceptual Matching-to-Sample Task," 23 *Journal of Experimental Psychology* 31 (1997).

238. Lawrence Weiskrantz, "Afterthoughts," in *Thought Without Language* 507 (Lawrence Wesikrantz, ed., Oxford University Press 1988).

239. Lawrence Weiskrantz, *supra* note 24, at 167.

240. *Id.* at 230, 251. *See id.* at 96.

241. Sue Savage-Rumbaugh and Roger Lewin, *Kanzi: The Ape at the Brink of the Human Mind* 259 (John Wiley and Sons 1994), quoting Matt Cartmill, "Human Uniqueness and Theoretical Content in Paleoanthropology," 11 *International Journal of Primatology* 173, 187 (1990).

CHAPTER NINE

1. Nicholas K. Humphrey, "The social function of intellect," in *Growing Pains in Ethology,* 303–17 (P.P.G. Bateson and R. A. Hinde, eds. 1976); Alison Jolly, "Lemur social behavior and primate intelligence," 153 *Science* 501–6 (1966).

2. Frans de Waal, *Chimpanzee Politics: Power and Sex Among the Apes* 212 (Johns Hopkins University Press, 1982).

3. Frans de Waal, *Chimpanzee Politics: Power and Sex Among the Apes* xv (rev. ed. Johns Hopkins University Press 1998).

4. *Machiavellian Intelligence II: Extensions and Evaluations* (Andrew Whiten and Richard W. Byrne, eds., Cambridge University Press 1997); *Machiavellian Intelligence: Social Expertise and the Evolution of Intelligence in Monkeys, Apes, and Humans* (Richard Byrne and Andrew Whiten, eds., Oxford University Press 1988).

5. Frans de Waal, *supra* note 2, at 218 note 5.

6. Frans de Waal, *Peacemaking Among Primates* 270 (Harvard University Press 1989).

7. See *id.* at 273–8; Jeffrey Moussaieff Masson, *The Wild Child* (Free Press Paperbacks 1997); Harlan Lane, *The Wild Boy of Aveyron* (Harvard University Press 1979).

8. Michael Tomasello, *The Cultural Origins of Human Cognition*, Chap. 2, at 6 (Harvard University Press)(in press)(manuscript). See *id.* at 3–6.

9. *Id.* at Chap. 2, at 2. See *id.* , Chap. 2, at 6–12; Chap. 3, at 32.

10. *Id.*, Chap. 2, at 36; Chap. 6, at 12; Chap. 7, at 1–2.

11. *Id.*, Chap. 7, at 14. *E.g.*, Angeline Lillard, "Developing a cultural theory of mind," 8 *Current Directions in Psychological Science* 57 (1999); Angeline Lillard, "Ethnopsychologies: Cultural variations in theories of mind," 123 *Psychological Bulletin* 3 (1998).

12. Michael Tomasello, *supra* note 8, Chap. 6, at 25 (emphasis added). See *id.*, Chap. 6, at 33. In light of the lack of uniformity of theory of mind across cultures and because "nativist" explanations do not fit the data even within a single culture, I will ignore such explanations for theory of mind.

13. Winthrop N. Kellogg and Luella A. Kellogg, *The Ape and the Child* 12-13 (Hafner Publishing Co., 1933) (emphasis in the original).

14. Roger S. Fouts and Stephen Tukel Mills, *Next of Kin: What Chimpanzees Have Taught Me About Who We Are* 150 (William Morrow 1997).

15. Primo Levi, *The Reawakening* 25–6 (Touchstone 1995).

16. Michael Tomasello, *supra* note 8, Chap. 1, at 6; Chap. 4, at 36; Chap. 6, at 26; Melinda A. Carpenter et al., "Social Cognition, Joint Attention, and Communicative Competence from 9 to 15 Months of Age," 63 (4) *Monographs of the Society for Research in Child Development* 1 (1998); Janet Astington, "What Is Theoretical About the Child's Theory of Mind? A Vygotskyan View of the Development," in *Theories of Theories of Mind* 188–94 (Peter Carruthers and Peter K. Smith, eds., Cambridge University Press 1996).

17. Sue Savage-Rumbaugh et al., *Apes, Language, and the Mind* 198 (Oxford University Press 1998); S. L. Williams et al., "Comprehension Skills of Language-Competent and Nonlanguage-Competent Apes," 17 *Language & Communication* 301, 313–4 (1997).

18. Richard Davenport, "Some behavioral disturbances of great apes in captivity," in *The Great Apes* 341, 351 (D. Hamburg and E.R. McCown, eds., Benjamin/Cummings 1979).

19. Gordon G. Gallup et al., "Capacity for self-recognition in differentially reared chimpanzees," 21 *Psychological Record* 68–74 (1971); S.D. Hill, "Responsiveness of young nursey-reared chimpanzees to mirrors," 33 *Proceedings of the Louisiana Academy of Sciences* 77–82 (1970).

20. Andrew Whiten et al., "Imitative learning of Artificial Fruit Processing in Children (*Homo sapiens*) and Chimpanzees (*Pan troglodytes*), 110 *Journal of Experimental Psychology* 3, 13 (1996).

21. S.L. Williams et al., *supra* note 17.

22. *Id.* at 313–4; Candida C. Peterson and Michael Siegal, "Deafness, Conversation, and Theory of Mind,"36 *Journal of Child Psychiatry and Psychology* 459, 471–2 (1995).

23. Candida C. Peterson and Michael Siegal, *supra* note 22, at 469–72; R.P. Hobson, "Blindness and psychological development 0–10 years" (Paper presented at the Mary Kitzinger Trust Symposium, September 1995, University of Warwick), discussed in Peter Mitchell, *Introduction to Theory of Mind: Children, Autism and Apes* 93–4 (Arnold 1997); R. Mayberry, "The cognitive develpment of deaf children: Recent insights," in 7 *Handbook of Neuropsychology* 51–68 (S. Segalowitz and I. Rapin, eds., Elsevier 1995).

24. Sue Taylor Parker and Michael L. McKinney, *Origins of Intelligence: The Evolution of Cognitive Development in Monkeys, Apes, and Humans* 159 (Johns Hopkins University Press 1999).

25. H. Lyn Miles, "Symbolic Communication with and by Great Apes," in *The Mentalities of Gorillas and Orangutans* 197, 200 (Sue Taylor Parker, et al, eds., Cambridge University Press 1999).

26. Josep Call and Michael Tomasello, "The effect on humans on the cognitive development of apes, "in *Reaching Into Thought: The Minds of the Great Apes* 390–4 (Anne E. Russon et al., Cambridge University Press 1996).

27. Personal communication from Sue Savage-Rumbaugh, dictated October 12, 1999.

28. *Id.* at 394.

29. *Id.* at 391.

30. Personal communication from Michael Tomasello to Jerome Bruner dated January 26, 1994, quoted in Jerome Bruner, "From joint attention to the meeting of minds: An introduction," in *Joint Attention: Its Origins and Role in Development* 8 (Chris Moore and Philip J. Durham, eds., Lawrence Erlbaum 1995)(emphasis added).

31. Michael Tomasello, *supra* note 8, Chap. 2, at 21. *See* Josep Call and Michael Tomasello, *supra* note 26, at 392–3. In earlier work, Tomasello mistakenly attributed the term "socialization of attention" to Lev Vygotsky, when it was actually an apt phrase of his own turning, Personal communication from Michael Tomasello, dated May 15, 1999.

32. *Id.* at 379–95.

33. Katherine Nagell et al., "Processes of Social Learning in the Tool Use of Chimpanzees (*Pan troglodytes*) and Human Children (*Homo sapiens*)," 107 *Journal of Comparative Psychology* 174 (1993).

34. Michael Tomasello, E. Sue Savage-Rumbaugh, and A.C. Kruger, "Imitative learning of actions on objects by children, chimpanzees, and enculturated chimpanzees," 64 *Child Development* 1688–1705 (1993). Details about Chantek's enculturation can be found in H. Lyn Miles, "The cognitive foundation for reference in a signing orangutan," in *"Language" and Intelligence in Monkeys and Apes: Comparative Developmental Perspectives* 511–39 (Sue T. Parker and Karen R. Gibson, eds. Cambridge University Press 1990).

35. Josep Call and Michael Tomasello, "Distinguishing Intentional from Accidental Actions in Orangutans (*Pongo pygmaeus*), Chimpanzees (*Pan troglodytes*),

and Human Children (*Homo sapiens*)," 112 *Journal of Comparative Psychology* 192, 197–9, 201, 204 (1998).

36. *Compare* Sue Savage-Rumbaugh, *et. al, supra* note 17, at 57, 59, 64–5 (Oxford University Press 1998); Sue Savage-Rumbaugh, "Why are we afraid of apes with language?," in *The Origin and Evolution of Intelligence* 43, 57–62 (A.B. Scheibel and J.W. Schopf, eds., Jones and Barllett 1997); E. Sue Savage-Rumbaugh and Duane M. Rumbaugh, "Perspectives on consciousness, language, and other emergent processes in apes and humans," *Towards a Science of Consciousness* 539, 547 (Stuart R. Hameroff et al. eds., MIT Press 1998) *with* Josep Call and Michael Tomasello, "A nonverbal false belief task: The performance of chimpanzees and human children," 70 *Child Development* 381 (1999).

37. Sue Savage-Rumbaugh et al., *supra* note 17, at 198.

38. Sue Savage-Rumbaugh, "Scientific Schizophrenia With Regard to the Language Act," in *Piaget, Evolution, and Development* 145, 149–50 (J. Langer and M. Killen, eds., Lawrence Erlbaum 1998). *See* Michael Tomasello, "Joint attention as social cognition," in Chris Moore and Philip J. Durham, eds., *supra* note 30, at 123; Jerome Bruner, *supra* note 28, at 9–10; Janet Astington, *supra* note 16, at 187–8; Judy Dunn, "Understanding Others: Evidence From Naturalistic Studies of Children," In *Natural Theories of Mind* 51–9 (Andrew Whiten, ed., Basil Blackwell 1991).

39. Sue Savage-Rumbaugh, "Ape Language: Between a Rock and a Hard Place," in *Origins of Language: What Nonhuman Primates Can Tell Us* (Barbara King, ed. Sar Press 1999).

40. Personal communication from Sue Savage-Rumbaugh, dated May 15, 1999.

41. Personal communication from Sue Savage-Rumbaugh, dated May 24, 1999; Personal communication from Sue Savage-Rumbaugh, dated April 24, 1999.

42. Janet Astington, *supra* note 16, at 194–8; M. Appleton and V. Reddy, "Teaching three-year-olds to pass false belief tests," 5 *Social Development* 275–91 (1996).

43. Eugene Linden, *Silent Partners: The Legacy of the Ape Language Experiments* 226 (Times Books 1986).

44. Herbert S. Terrace et al., "Can an Ape Create a Sentence?" 206 *Science* 892 (1979); Herbert S. Terrace, *Nim: A Chimpanzee Who Learned Sign Language* 370–87 (Washington Square Press 1979).

45. Herbert S. Terrace, *supra* note 44, at 30. *See id.* at 6.

46. *Id.* at 45.

47. *Id.* at 107, 118.

48. *Id.* at 162.

49. *Id.* at 205, 207.

50. *Id.* at 57, 165, 185, 205, 391.

51. *Id.* at 124.

52. *Id.* at 81.

53. *Id.* at 70–1, 76, 87, 90.

54. *E.g., id.* at 171–7.

55. *Id.* at 89, 147.

56. *Id.* at 73, 75, 83, 121.

57. *Id.* at 289–90.

58. *Id.* at 312.

59. Herbert S. Terrace et al., *supra* note 44, at 901 note 7.

60. *Compare id.* generally, *with* William C. Stokoe, "Comparative and Developmental Sign Language Studies: A Review of Recent Advances," in R. Allen Gard-

ner and Beatrix Gardner, *Teaching Sign Language to Chimpanzees* 308–15 (R. Allen Gardner, *et. al.*, eds., State University of New York Press 1989); R. Allen Gardner, *id.*, generally.

61. Herbert S. Terrace, *supra* note 44, at 145.

62. *Id.* at 337–8. *See id.* at 339.

63. *Id.* at 342.

64. Thomas E. Van Cantford and James B. Rimpau, "Sign language studies with children and chimpanzee," 34 *Sign Language Studies* 14, 20 (1982); Roger Fouts et al., "Studies of Linguistic behavior in apes and children," in *Understanding Language Through Sign Language Research* 173 (Academic Press 1978).

65. David Premack, *Gavagai! Or the Future History of the Animal Language Controversy* 29 (MIT Press 1986).

66. Thomas E. Van Cantford and James B. Rimpau, *supra* note 64, at 38.

67. Herbert S. Terrace, *supra* note 44, at 332.

68. *Id.* at 123.

69. Herbert S. Terrace, *supra* note 44, at 208. *See id.* at 335.

70. Philip Lieberman, *The Biology and Evolution of Language* 244–5 (Harvard University Press 1984). *See* Roger Fouts and Stephen Tukel Mills, *supra* note 14, at 273–8; Jeffrey Moussaieff Masson, *The Wild Child* (Free Press Paperbacks 1997); Harlan Lane, *The Wild Boy of Aveyron* (Harvard University Press 1979).

71. Chris O'Sullivan and Carey Page Yeager, "Communicative Contect and Linguistic Competence: The Effects of Social Setting on a Chimpanzee's Coversational Skill," in R. Allen Gardner et al., eds., *supra* note 60, at 268–79.

72. Peter Mitchell, *supra* note 23, at 53.

73. Michael Tomasello, "Chimpanzee Social Cognition," in Daniel J. Povinelli and Timothy J. Eddy, "What young chimpanzees know about seeing," 61 *Monographs of the Society for Research in Child Development* 161 (2, Serial No. 247); Daniel J. Povinelli and Steve Giambrone, "Inferring Other Minds: Failure of the Argument by Analogy," *Philosophical Topics* (G. Massey and B.D. Massey, eds. 1999)(in press)(special issue on zoological philosophy).

74. Daniel J. Povinelli and Timothy J. Eddy, *supra* note 73, at 26–7.

75. *Id.* at 27.

76. *Id.* at 27.

77. *Id.*

78. *Compare* Sue Savage-Rumbaugh et al., *supra* note 17, at 210–1 and Michael Tomasello and Josep Call, *Primate Cognition* 390–4 (Oxford University Press 1997) *with* Daniel J. Povinelli and Steve Giambrone, *supra* note 73 (in press); Daniel J. Povinelli and Timothy J. Eddy, *supra* note 73, at 179–81.

79. Katherine Nagel et al., *supra* note 33, at 184. *E.g.*, Josep Call and Michael Tomasello, *supra* note 35, at 391 ("[n]egative results are always difficult to explain"); Daniel J. Povinelli et al., "Do Rhesus monkeys (*Macaca mulatta*) attribute knowledge and ignorance to others?," 105 *Journal of Comparative Psychology* 318, 324 (1991)("it is always difficult to interpret negative results"); David Premack, "'Does the chimpanzee have a theory of mind?' revisited," in Richard Byrne and Andrew Whiten, *supra* note 4, at 178 ("[n]egative outcomes seldom lend themselves to diagnosis)."

80. Daniel J. Povinelli and Steve Giambrone, supra note 73 (in press).

81. Katherine Nagel et al., *supra* note 33, at 184.

82. Daniel J. Povinelli and Timothy J. Eddy, *supra* note 73, at 183.

83. Elizabeth Pennisi, "Are our primate cousins 'conscious'?" 284 *Science* 2073, 2076 (1999).

CHAPTER TEN

1. Frans de Waal, *Good-Natured: The Origins of Right and Wrong in Humans and Other Animals* 209 (Harvard University Press 1996).

2. Tetsuro Matsuzawa, "Chimpanzee intelligence in nature and in captivity: isomorphism of symbol use," in William C. McGrew et al., *Great Ape Societies* 196–8 (Cambridge University Press 1996); Tetsuro Matsuzawa, "Form perception and visual acuity in a chimpanzee," 55 *Folia Primatolgia* 24–32 (1990).

3. Michael Tomasello and Josep Call, *Primate Cognition* v (Oxford University Press 1997).

4. *Id.* at 311, 405.

5. *E.g.*, Christophe Boesch and Michael Tomasello, "Chimpanzee and Human Cultures," in 39 *Current Anthropology* 592–611 (1998); *Chimpanzee Cultures* (Richard W. Wrangham et al. eds., Harvard University Press 1994); W.C. McGrew, *Chimpanzee Material Culture* (Cambridge University Press 1992).

6. Christophe Boesch and Michael Tomasello, *supra* note 5, at 592, 604.

7. A. Whiten et al., "Cultures in chimpanzees," 399 *Nature* 682 (1999). *See* "Chimps, Yes. But they've got culture," *New York Times*, at D5 (June 22, 1999).

8. Frans B.M. de Waal, "Cultural primatology comes of age," 399 *Nature* 636 (1999).

9. David J. Wolfson, "Beyond the Law: Agribusiness and the Systemic Abuse of Animals Raised for Food or Food Production," 2 *Animal Law* 123 (1996). *See generally, Animals and Their Legal Rights* (Animal Welfare Institute 4th ed. 1990).

10. R. Adams and J. Martin, "Pain," in *Harrison's Principles of Internal Medicine* 7–15 (R. Reterdorf, et al., eds. 10th ed. 1983).

11. Margaret Rose and David Adams, "Evidence for Pain and Suffering on Other Animals," in *Animal Experimentation–The Consensus Changes* 42, 47–9 (Gill Langley ed., Chapman and Hall 1989); K.J.S. Anand and P.R. Hickey, "Pain and its Effects on the Human Neonate and Fetus," 316 *New England Journal of Medicine* 1321, 1322 (1987); Lawrence Kruger and Barbara E. Rodin, "Peripheral mechanisns involved in pain," in *Animal Pain: Perception and Alleviation* 1–26 (Ralph L. Kitchell and Howard H. Erickson, eds., American Physiological Society 1983).

12. Margaret Rose and David Adams, *supra* note 11, at 48, 49.

13. R. Adams and J. Martin, *supra* note 10, at 7–15.

14. Margaret Rose and David Adams, *supra* note 11, at 51, 56; K.J.S. Anand and P.R. Hickey, *supra* note 11, at 1323.

15. Margaret Rose and David Adams, *supra* note 11, at 51–2; Jeanne D. Talbot, "Multiple Representations of Pain in Human Cerebral Cortex," 251 *Science* 1335 (1991).

16. Margaret Rose and David Adams, *supra* note 11, at 55–6, 59–60.

17. Eric J. Cassell, *The Nature of Suffering and the Goals of Medicine* 32, 35, 36 (Oxford University Press 1991).

18. Roselyn Rey, *The History of Pain* 293 (Harvard University Press 1998)(1993); K.J.S. Anand and P.R. Hickey, *supra* note 11, at 1321.

19. J. Englund and J. Alexander, "The Experience of Pain in Children," in *Pain: A Sourcebook for Nurses and Other Professionals* 453–76 (Ada Janox, ed., Little, Brown and Co. 1977).

20. K.J.S. Anand and P.R. Hickey, *supra* note 11, at 1326.

21. Jane Goodall, *The Chimpanzees of Gombe–Patterns of Behavior* 587 (Harvard University Press 1986).

22. Emil W. Menzel, Jr., "Chimpanzee Spatial Memory Organization," 182 *Science* 943 (1973); Emil W. Menzel, Jr., "Group behavior in young chimpanzees: Responsiveness to cumulative novel exchanges in a large outdoor enclosure," 74 *Journal of Comparative and Physiological Psychology* 46 (1971).

23. Sue Savage-Rumbaugh, et al., *Apes, Language, and the Human Mind* 36–44 (Oxford University Press 1998).

24. S. Gagnon and F.Y. Dore, "Search behavior in various breeds of adult dogs *(Canis familiaris)*; object permanence and olfactory clues," 106 *Journal of Comparative Psychology* 58 (1992).

25. Claude Dumas, "Object permanence in cats *(Felis catus)*: An ecological approach to the study of invisible displacements," 106 *Journal of Comparative Psychology* 404 (1992); G. Schino, "Object concept and mental representation in *Cebus apella* and *Macaca fascicularis*," 31 *Primates* 537 (1990); S. Gagnon and F.Y. Dore, *supra* note 25; Irene M. Pepperberg et al., "Development of Piagetian Object Permanence in a Grey Parrot *(Psittacus erithacus)*," 111 *Journal of Comparative Psychology* 63, 63, 70 (1997).

26. Sue Taylor Parker and Michael L. McKinney, *Origins of Intelligence: The Evolution of Cognitive Development in Monkeys, Apes, and Humans* 42 (Johns Hopkins University Press 1999).

27. *Id.* at 49–9, 63.

28. David Premack, *Gavagai!—Or the Future History of the Animal Language Controversy* 11 (MIT Press 1986).

29. Emil W. Menzel et al., "Chimpanzee *(Pan troglodytes)* spatial problem solving with the use of mirrors and televised equivalents of mirrors," 99 *Journal of Comparative Psychology* 211 (1985).

30. J. Veauclair et al., "Rotation of mental images in baboons when the visual input is directed to the left cerebral hemisphere," 4 *Psychological Science* 99 (1993).

31. Tetsuro Matsuzawa, "Nesting cups and metatools in chimpanzees," 4 *Behavioral and Brain Sciences* 570 (1991).

32. Michael Tomasello and Josep Call, *supra* note 3, at 65.

33. Valerie Kuhlmeier et al., "Scale model comprehension by chimpanzees," *Journal of Comparative Psychology* (1999)(in press).

34. Andrew Whiten and Richard Byrne, "The emergence of metarepresentation in human ontogeny and primate phylogeny," in *Natural Theories of Mind* 269–72 (Andrew Whiten, ed., Basil Blackwell 1991).

35. Richard Wrangham and Dale Peterson, *Demonic Males: Apes and the Origins of Human Violence* 253–5 (Mariner Books 1996).

36. *Id.* at 268–70; M.L.A. Jensvold and R.S. Fouts, "Imaginary Play in Chimpanzees *(Pan troglodytes)*," 8 *Human Evolution* 217–27 (1993); Sue Savage-

Rumbaugh and Kelly McDonald, "Deception and social manipulation in symbol-using apes," in *Machiavellian Intelligence–Social Expertise and the Evolution of Intellect in Monkeys, Apes and Humans* 232 (Richard Byrne and Andrew Whiten, eds. Oxford University Press 1988).

37. Sue Savage-Rumbaugh et al., *supra* note 23, at 59–61.

38. Michael Tomasello and Josep Call, *supra* note 3, at 118–22.

39. *Id.* at 127.

40. Tetsuro Matsuzawa, "Spontaneous sorting in human and chimpanzee," in Sue T. Parker and Karen R. Gibson,*"Language" and Intelligence in Monkeys and Apes* 451, 453–68 (Cambridge University Press 1990).

41. Keith J. Hayes and Catherine J. Nissen, "Higher mental functions of a home-raised chimpanzee," in *Behavior of Nonhuman Primates* 59, 88 (A.M. Schrier and and F. Stollnitz, eds., Academic Press 1971).

42. Masayuki Tanaka, "Object Sorting in Chimpanzees *(Pan troglodytes):* Classification Based on Physical Identity, Complementarity, and Familiarity," 109 *Journal of Comparative Psychology* 151 (1995).

43. Michael Tomasello and Josep Call, *supra* note 3, at 122–4.

44. David Premack, *supra* note 28, at 22–3.

45. Roger K.R. Thompson et al., "Language-naive chimpanzees *(Pan troglodytes)* judge relations between relations in a conceptual matching-to-sample task," 23 *Journal of Experimental Psychology (Animal Behavior Processes)* 31, 41–2 (1997).

46. *Id.*

47. Patrizia Potí et al., "Sensorimotor precursors of logicomathematical operations in human enculturated and language reared chimpanzees," *Contemporary Psychology* (in preparation); Jonas Langer, "Phylogenetic and ontogenetic origins of cognition: Classification," in J. Langer and M. Killen, *Piaget, Evolution and Development* 35–54 (Lawrence Erlbaum 1998). Patrizia Potí, "Logical structures of young chimpanzees' object grouping," 18 *International Journal of Primatology* 33–59. See Sue Taylor Parker and Michael McKinney, *supra* note 26 at 82–3.

48. Eric Korn, "The Meaning of Mngwotngwotiki," *London Review of Books* 16 (January 10, 1991), reviewing Thomas Crump, *The Anthropology of Numbers* (Cambridge University Press 1990); Thomas Crump, *supra* at 5, 9, 25, 72.

49. Marc Hauser & Susan Carey, "Building a Cognitive Creature from a Set of Primitives–Evolutionary and Developmental Insights," in *The Evolution of Mind* 68–73 (Denise Dellarosa Cummins and Colin Allen, eds., Oxford University Press 1998).

50. S. Dehaene et al., "Sources of mathematical thinking: Behavioral and brain-imaging evidence," 284 *Science* 970 (1999); Marc Hauser & Susan Carey, *supra* note 49, at 82; Karen Wynn, "An Evolved Capacity for Number," in Denise Dellarosa Cummins and Colin Allen, eds., *supra* note 49, at 107–113.

51. Michael J. Beran et al., "Chimpanzee *(Pan troglodytes)* Counting in a Computerized Testing Paradigm," 48 *The Psychological Record* 3, 4–5 (1998); Sarah T. Boysen, "Representation of Quantities by Apes," in 26 *Advances in the Study of Behavior* 435, 436–7 (Academic Press 1997).

52. Daniel F. Rice et al., "Ordinality and Counting by a Chimpanzee *(Pan troglodytes)"* (in progress); Michael J. Beran et al., *supra* note 51; Sarah T. Boysen,

"Counting as the Chimpanzee Views It," in *Cognitive Aspects of Stimulus Control* 367 (W.K. Honig and J.G. Fetterman, eds. Lawrence Erlbaum 1991).

53. Tetsuro Matsuzawa, "Chimpanzee intelligence in nature and in captivity: isomorphism of symbol use," *supra* note 2, at 199.

54. Michael J. Beran et al., *supra* note 51; Michael Tomasello and Josep Call, *supra* note 3, at 161; Duane M. Rumbaugh et al., "Chimpanzee Counting and Rhesus Monkey Ordinality Judgments," in *Primatology today* 701 (Akiyosi Ehara et al., eds., Elsevier Science Publishers, B.V. [Biomedical Division] 1991); Duane M. Rumbaugh et al., "Lana Chimpanzee Learnes to Count by 'NUMATH': A Summary of a Videotaped Experiment report," 39 *The Psychological Record* 459 (1989); Tetsuro Matsuzawa, "Use of numbers by a chimpanzee," 315 *Nature* 57 (1985).

55. Sarah T. Boysen et al., "Indicating Acts During Counting by a Chimpanzee *(Pan troglodytes),*" 109 *Journal of Comparative Psychology* 47 (1995).

56. S.T. Boysen, "Counting in Chimpanzees: Nonhuman principles and emergent properties of number," in *The Development of Numerical Competence: Animal and Human Models* 39 (S.T. Boysen and E.J. Capaldi, eds., Lawrence Erlbaum 1993); Sarah T. Boysen and Gary G. Berntson, "Numerical Competence in a Chimpanzee *(Pan troglodytes),* 103 *Journal of Comparative Psychology* 23 (1989).

57. Sarah T. Boysen, *supra* note 51, at 449–450.

58. *Id.* at 449–58; S.T. Boysen et al., "Quantity-Based Interference and Symbolic Representations in Chimpanzees *(Pan troglodytes),* 22 *Journal of Experimental Psychology: Animal Behavior Processes* 76 (1996); Sarah T. Boysen and Gary G. Berntson, "Responses to Quantity: Perceptual Versus Cognitive Mechanisms *(Pan troglodytes),*" 21 *Journal of Experimental Psychology: Animal Behavior Processes* 82 (1995).

59. Sarah T. Boysen, et al, "Overcoming response-bias using symbolic representation of number by chimpanzees *(Pan troglodytes)* 27 *Animal Learning and Behavior* 229 (1999).

60. Stanislas Dehaene, *The Number Sense–How the Mind Creates Mathematics* 25–6 (Oxford University Press 1997); Duane M. Rumbaugh et al., "Summation in the Chimpanzee *(Pan troglodytes);* 13 *Journal of Experimental Psychology: Animal Behavior Processes* 107 (1987). Rumbaugh et al., did not claim that Sherman and Austin were adding.

61. Gary Woodruff and David Premack, "Primitive Mathematical Concepts in the Chimpanzee: Proportionality and Numerosity," 293 *Nature* 568 (1981).

62. Stanislas Dehaene, *supra* note 60, at 25.

63. Karen Wynn, *supra* note 50, at 121–2.

64. David Premack and Ann James Premack, *The Mind of an Ape* 77–82, 143 (W.W. Norton 1983); Gary Woodrull et al., "Conservation of Liquid and Solid Quantity by the Chimpanzee, 202 *Science* 991 (1978).

65. S.J. Muncer, "'Conservations' with a Chimpanzee," 16 *Developmental Psychology* 1 (1983).

66. Jane Goodall, *Through a Window–My Thirty Years with the Chimpanzees of Gombe* 19 (Houghton-Mifflin 1990).

67. Sue T. Parker and Robert W. Mitchell, "The mentalities of gorillas and orangutans in phylogenetic perspective," in *The Mentalities of Gorillas and Orangutans: Comparative Perspectives* 403 (Sue Taylor Parker et al., eds., Cambridge University Press 1999).

68. Sayoshi Hirata et al., "Use of Leaves as Cushions to Sit on Wet Ground by Wild Chimpanzees," 44 *American Journal of Primatology* 215–9 (1998); Rosalind Alp. "Stepping-Sticks' and 'Seat-Sticks': New Types of Tools used by Wild Chimpanzees *(Pan troglodytes)* in Sierra Leone," 41 *American Journal of Primatology* 45–51 (1997).

69. Hedwige Boesch-Achermann and Christophe Boesch, "Tool Use in Wild Chimpanzees: new light from dark forests," 2 *Current Directions in Psychological Science* 18 (1993).

70. Christophe Boesch and Hedwige Boesch, "Tool use and tool making in wild chimpanzees," 54 *Folia Primatologica* 86 (1990).

71. Michael Tomasello and Josep Call, *supra* note 3, at 78.

72. Sue Taylor Parker and Michael L. McKinney, *supra* note 26, at 56. Tetsuro Matsuzawa, "Field Experiments on Use of Stone Tools by Chimpanzees in the Wild," in Richard W. Wrangham et al. eds., *supra* note 5, at 351, 362, 366; Tetsuro Matsuzawa, "Chimpanzee intelligence in nature and in captivity: isomorphism of symbol use," *supra* note 2, at 201–6; Tetsuro Matsuzawa and Gen Yamakoshi, "Comparison of chimpanzee material culture between Bossou and Nimba, West Africa," in *Reaching Into Thought–The Minds of the Great Apes* 211, 214–7 (Anne E. Russon et al., eds., Cambridge University Press 1996).

73. W.C. McGrew, *supra* note 5, at 163–5.

74. Michael Tomasello and Josep Call, *supra* note 3, at 71–7; Jane Goodall, *supra* note 21, at 535–64.

75. Deborah Custance and Kim A. Bard, "The comparative and developmental study of self-recognition and imitation: The importance of social factors," in *Self-Awareness in Animals and Humans–Development Perspectives* 214 (Sue Taylor Parker et al., eds., Cambridge University Press 1990).

76. Richard Byrne, *The Thinking Ape–Evolutionary Origins of Intelligence* 154 (Oxford University Press 1995); Jane Goodall, *supra* note 21, at 588.

77. M. Mathieu et al., "Piagetian causality in two house-reared chimpanzees," 34 *Canadian Journal of Psychology* 179 (1980).

78. *Id.*

79. Luca Lemongelli et al., "Comprehension of Cause-Effect Relations in a Tool-using task by Chimpanzees *(Pan troglodytes)*," 109 *Journal of Comparative Psychology* 18, 25 (1995).

80. Sue Savage-Rumbaugh and Roger Lewin, *Kanzi–The Ape on the Brink of the Human Mind* 214 (John Wiley & Sons, Inc. 1994).

81. Richard Byrne, *supra* note 76, at 104; Kathy D. Schick and Nicholas Toth, *Making Silent Stones Speak–Human Evolution and the Dawn of Technology* 136–9 (Simon & Schuster 1993).

82. Nicholas Toth, et al., "Pan the Tool-maker: Investigations into the Stone-Tool-Making and Tool-Using Capabilities of a Bonobo *(Pan paniscus)*," 20 *Journal of Archeological Science* 81, 87–8 (1993). Tomasello and Call err when they state that Kanzi was *taught* how to produce stone flakes by throwing a stone against a hard floor, Michael Tomasello and Josep Call, *supra* note 3, at 81.

83. Kathy D. Schick et al., "Continuing investigations into the stone tool-making and tool-using capacities of a bonobo *(Pan paniscus)*," 26 *Journal of Archaeological Science* 821, 831 (1999).

84. Roger S. Fouts and Stephen Tukel Mills, *Next of Kin–What Chimpanzees Have Taught Me About Who We Are* 52–3 (William Morrow 1997); Maurice K. Temerlin, *Lucy—Growing up Human* 104–7 (Science and Behavior Books, Inc. 1975).

85. *E.g.*, Juan Carlos Gómez, "Ostensive behavior in great apes: The role of eye contact," in Anne E. Russon et al., eds., *supra* note 72, at 131.

86. Alison Jolly, "Conscious Chimpanzees? A Review of Recent Literature," in *Cognitive Ethology–The Minds of Other Animals* 231–3 (Carolyn A. Ristau, ed., Lawrence Erlbaum Associates 1991), Sue Savage-Rumbaugh and Kelly McDonald, in Richard Byrne and Andrew Whiten, eds., *supra* note 36, at 223–227.

87. Vittorio Gallese and Alan Goldman, "Mirror neurons and the simulation theory of mind-reading," 2 *Trends in Cognitive Science* 493, 495 (December, 1998).

88. *Id.* at 497–8.

89. *Id.* at 495, 498, 500.

90. Steven Pinker, *How the Mind Works* 331 (W.W. Norton 1997).

91. Andrew Whiten, "The Machiavellian Mindreader," in *Machiavellian Intelligence II–Extensions and Evaluations* 154–5 (Andrew Whiten and Richard W. Byrne, eds., Oxford University Press 1997).

92. Juan Carlos Gómez, "Primate theories of primate minds: conceptual and methodological issues," *Enfance* (Special issue)(in press); Juan Carlos Gómez, "Are apes persons? The case for primate intersubjectivity," *Etica and Animali* 9 (1998), pp. 51–63 (Special issue on primate personhood); Juan Carlos Gómez, in Anne E. Russon et al., eds., *supra* note 72, at 146; Juan Carlos Gómez, "Non-human primate theories of (non-human primate) minds; some issues concerning the origins of mind-reading," in *Theories of Theories of Mind* 330, 341–2 (Peter Carruthers and Peter K. Smith, eds., Cambridge University Press 1996); Juan Carlos Gómez, "Mutual awareness in primate communication: A Gricean approach," in Sue Taylor Parker et al., *supra* note 75, at 72–7.

93. Juan Carlos Gómez, "Mutual awareness in primate communication: A Gricean approach," in Sue Taylor Parker et al., *supra* note 75, at 65–8, 74.

94. Juan Carlos Gómez, "Some thoughts about the evolution of LADS, with special reference to TOM and SAM," in *Language and Thought: Interdisciplinary Themes*, 76, 80 (P. Carruthers and J. Boucher, eds., Cambridge University Press 1998).

95. Alison Jolly, *supra* note 86, at 235, discussing Frans B.M. de Waal, "Games Pygmy Chimpanzees Play," in *Zoonooz* (1987); Frans B. M. de Waal, "Imaginative Bonobo Games," in 59 *Zoonooz* 6–10 (1986).

96. Gordon G. Gallup, "Toward a comparative psychology of mind," in *Animal Cognition and Behavior* 473 (R.L. Melgren, ed., North Holland Press 1983).

97. Personal communication from Juan Carlos Gómez, dated June 3, 1999; Juan Carlos Gómez, "Mutual awareness in primate communication: A Gricean approach," in Sue Taylor Parker et al., *supra* note 75, at 76–7.

98. *Compare* Juan Carlos Gómez, "Mutual awareness in primate communication: A Gricean approach," in Sue Taylor Parker et al., *supra* note 75, *with* Alison Gopnik and Andrew N. Meltzoff, "Minds, bodies, and persons: Young children's understanding of the self and others as reflected in imitation and theory of mind research," in *id.* at 166; Louis J. Moses, "Foreword," in "Chimpanzee intelligence in nature and in captivity: isomorphism of symbol use," in *id.* at xiii–xiv; Andrew N. Meltzoff and Alison Gopnik, "The role of imitation in understanding persons

and developing a theory of mind," in *Understanding Other Minds–Perspectives from Autism* 335–66 (S. Baron-Cohen, H. Tager-Flusberg, and D.J. Cohen, eds., Oxford University Press 1993).

99. Alison Jolly, *The Evolution of Primate Behavior, Second Edition* 397 (MacMillan Publishing Co. 1985).

100. R. Allen Gardner and Beatrix T. Gardner, "Teaching Sign Language to a Chimpanzee," 165 *Science* 644 (1969).

101. *See generally*, Edward Klima and Ursula Bellugi, *The Signs of Language* (Harvard University Press 1979).

102. R. Allen Gardner and Beatrix T. Gardner, "Comparative psychology and language acquisition," *Annals of the New York Academy of Sciences* 37, 54 (1978).

103. R. Allen Gardner and Beatrix T. Gardner, "A Cross-Fostering Laboratory," in *Teaching Sign Language to Chimpanzees* 7 note* (R. Allen Gardner et al., eds., State University of New York Press 1989); Beatrix T. Gardner, et al., "The Shapes and Uses of Signs in a Cross-Fostering Laboratory," in *id.* at 86. *See* Herbert S. Terrace, *Nim: A Chimpanzee Who Learned Sign Language* 370–87 (Washington Square Press 1979), for a good discussion and examples.

104. H. Lyn Miles, "ME CHANTEK: The development of self-awareness in a signing orangutan, " in Sue Taylor Parker et al., *supra* note 75, at 264.

105. *Id.* at 219.

106. Tetsuro Matsuzawa, "Chimpanzee intelligence in nature and in captivity: isomorphism of symbol use," *supra* note 2, at 201; Shoji Itakura, "Symbolic representation of possession in a chimpanzee," in Sue Taylor Parker et al., eds., *supra* note 75, at 240–6; Shoji Itakura and Tetsuro Matsuzawa, "Acquisition of personal pronouns by a chimpanzee," in *Language and Communication: Comparative Perspectives* 347–63 (H. Roitblat, eds., Lawrence Erlbaum 1993).

107. Personal Communication from Tetsuro Matsuzawa, dated April 12, 1999.

108. Gordon Gallup, Jr., "Chimpanzees: self-recognition," 167 *Science* 86 (1970). Michael Tomasello and Josep Call, *supra* note 3, at 332–3, compile these experiments.

109. Charles W. Hyatt and William D. Hopkins, "Self-awareness in bonobos and chimpanzees: a comparative perspective," in Sue Taylor Parker et al., eds., *supra* note 75, at 248–52.

110. Dorothy L. Cheney and Robert M. Seyfarth, *How Monkeys See the World* 242 (University of Chicago Press 1990); John H. Flavell, *Cognitive Development, 2nd ed.* 155–156 (Prentice Hall, Inc. 1985); M. Lewis and J. Brooks-Gunn, *Social Cognition and the Acquisition of Self* (Plenum Press 1979).

111. A. Linn et al., "Development of self-recognition in chimpanzees," 106 *Journal of Comparative Psychology* 120 (1992).

112. Sarah T. Boysen et al.,"Shadows and mirrors: Alternative avenues to the development of self-recognition in chimpanzees," in Sue Taylor Parker et al., eds., *supra* note 75, at 234–7.

113. Sue Savage-Rumbaugh et al., *supra* note 23, at 35.

114. *Id.* at 267.

115. P.A. Cameron and G.G. Gallup. Jr., "Shadow self-recognition in human infants," 11 *Infant Behavior and Development* 465 (1988).

116. Jerome Kagan, *The Second Year: The Emergence of Self-Consciousness* (Harvard University Press 1981).

117. Sarah T. Boysen et al., *supra* note 59, at 228–34.

118. Karyl B. Swartz and Sîan Evans, "Not all chimpanzees *(Pan troglodytes)* show self-recognition," 32 *Primates* 483 (1991). See Karyl B. Swartz et al., "Comparative Aspects of Mirror Self-Recognition in Great Apes," in Sue Taylor Parker and Robert W. Mitchell, *supra* note 67, at 288, for a summary of all investigations into mirror self-recognition in great apes.

119. Daniel J. Povinelli et al., "Absence of knowledge attribution and self-recognition in young chimpanzees *(Pan troglodytes)*, 108 *Journal of Comparative Psychology* 74 (1994).

120. Sue Taylor Parker and Michael L. McKinney, *supra* note 26, at 155.

121. *Id.*

122. Michael Cole and Barbara Means, *Comparative Studies of How People Think* (Harvard University Press 1981).

123. Richard Byrne, *supra* note 76, at 86.

124. Elliott Sober, "Morgan's Canon," in Denise Dellarosa Cummins and Colin Allen, eds., *supra* note 49, at 237.

125. Andrew Whiten, *supra* note 91, at 162–3, 164–5.

126. D.J. Povinelli and D.J. Eddy, "Chimpanzees: Joint visual attention," 7 *Psychological Science* 129–35 (1996).

127. Michael Tomasello et al., "Five primate species follow the gaze of conspecifics," 55 *Animal Behavior* 1063 (1998).

128. Shoji Itakura and Masayuki Tanaka, "Use of Experimenter-Given Cues During Object-Choice tasks by Chimpanzees *(Pan troglodytes)*, an Orangutan *(Pongo pygmaeus)*, and Human Infants *(Homo sapiens)*, 112 *Journal of Comparative Psychology* 119, 120–1 (1998).

129. Josep Call and Michael Tomasello, "Distinguishing Intentional from Accidental Actions in Orangutans *(Pongo pygmaeus)*, Chimpanzees *(Pan troglodytes)*, and Human Children *(Homo sapiens)*," 112 *Journal of Comparative Psychology* 200–04 (1998).

130. *Id.* at 195.

131. *Id.* at 201.

132. *Id.* at 197–8.

133. *Id.* at 202.

134. *Id.*

135. Connie Russell et al., "Social Referencing by Young Chimpanzees *(Pan troglodytes)*, 111 *Journal of Comparative Psychology* 185–92 (1997).

136. Michael Tomasello and Josep Call, "Social cognition of monkeys and apes," 37 *Yearbook of Physical Anthropology* 273, 287 (1994).

137. Mark A. Krause and Roger S. Fouts, "Chimpanzee *(Pam troglodytes)* Pointing: Hand Shapes, Accuracy, and the Role of Eye Gaze," 111 *Journal of Comparative Psychology* 330 (1997); David A. Leavens et al., "Indexical and Referential Pointing in Chimpanzees *(Pan troglodytes)*," 110 *Journal of Comparative Psychology* 346 (1996).

138. Mark A. Krause and Roger S. Fouts, *supra* note 137, at 331–5.

139. Sue Savage-Rumbaugh and Roger Lewin, *supra* note 80, at 91

140. *Id.* at 134.

141. Sue Savage-Rumbaugh, "Why are we afraid of apes with language?" in *The Origin and Evolution of Intelligence* 56, 57 (A.B. Scheibel and J.W. Schopf, eds. Jones and Bartlett 1997).

142. Roger Fouts and Stephen Tukel Mills, *supra* note 84, at 31.

143. Sue Savage-Rumbaugh, *supra* note 141, at 50–1; Sue Savage-Rumbaugh and Roger Lewin, *supra* note 80, at 79.

144. M. Miyowa, "Imitation of facial gestures by an infant chimpanzee," 37 *Primates* 207–13 (1996).

145. Robert Boyd and Peter Richerson, "Why culture is common but cultural evolution is rare," 88 *Proceedings of the British Academy* 77–93 (1996).

146. Elisabetta Visaberghi and Luca Limongelli, "Acting and understanding: Tool use revisited through the minds of capuchin monkeys," in Anne E. Russon et al., *supra* note 72, at 72–3; James R. Anderson, "Chimpanzees and capuchin monkeys: Comparative cognition," in *id.* at 23, 43–4; Deborah Custance and Kim A. Bard, "The comparative and developmental study of self-recognition and imitation: The importance of social factors," in Sue Taylor Parker et al., eds., *supra* note 75, at 207–23; Michael Tomasello et al., "Imitative learning of actions on objects by children, chimpanzees, and enculturated chimpanzees," 64 *Child Development* 1688–1705 (1993); Andrew Whiten and R. Ham, "On the nature and evolution of imitation in the animal kingdom: Reappraisal of a century of research," 21 *Advances in the Study of Behavior* 239, 259, 262–7 (1992).

147. Frans de Waal, *Chimpanzee Politics–Power and Sex among Apes* 130 (rev. ed., Johns Hopkins Univ. Press 1998).

148. Keith J. Hayes and Cathy Hayes, "Imitation in a home-raised chimpanzee," 45 *Journal of Comparative Physiological Psychology* 450–9 (1952).

149. Roger Fouts and Stephen Tukel Mills, *supra* note 84, at 13, 35.

150. Christophe Boesch and Michael Tomasello, *supra* note 5, at 601.

151. *Id.*

152. Herbert S. Terrace, *supra* note 103, at 208.

153. *Id.* at 212, 214.

154. Herbert S. Terrace et al., "Can an Ape Create a Sentence?" 206 *Science* 892, 896, 900 (1979); Herbert S. Terrace, *supra* note 103, at 208, 209, 273;

155. Andrew Whiten, "Imitation of the sequential structure of actions by chimpanzees (*Pan troglodytes*)," 112 *Journal of Comparative Psychology* 279 (1998).

156. A. Whiten et al., *supra* note 7, at 686.

157. *E.g.*, Michael Tomasello and Josep Call, *supra* note 3, at 304, 307–7; W.C. McGrew, *supra* note 5.

158. B. Rogoff et al., "Questioning assumptions about culture and individuals," 16 *Behavioral and Brain Sciences* 533 (1993); D. Olson and J. Astington, "Cultural learning and educational process," 16 *Behavioral and Brain Sciences* 531 (1993).

159. Christophe Boesch and Michael Tomasello, *supra* note 5, at 602.

160. Christophe Boesch, "Three approaches for assessing chimpanzee culture," in Anne E. Russon et al., eds., *supra* note 72, at 430; Christophe Boesch, "Aspects of transmission of tool use in wild chimpanzees," in *Tools, language, and cognition in human evolution* 171–83 (K.R. Gibson and T. Ingold, eds., Cambridge University Press 1992); Christophe Boesch, "Teaching among wild chimpanzees," 41 *Animal Behavior* 530–2 (1991).

161. David Premack, "On the control of human differences" in *Assessing Individual Differences in Human Behavior* 329, 333 (Lubenski and Davis eds., Davis-Black 1995). Premack defines "theory of mind" here as "attribution of mental states to others."

162. *Id.*

163. Personal communication from Gabriel Waters of the Chimpanzee and Human Communications Institute, dated June 14, 1999.

164. Roger S. Fouts, "Transmission of a Human Gestural Language in a Chimpanzee Mother-Infant Relationship," in *The Ethological Roots of Culture* 257–69 (R.A. Gardner et al., eds., Kluwer Academic Publishers 1994).

165. Roger Fouts and Stephen Tukel Mills, *supra* note 84, at 154.

166. Jonathan Wake, "Scientists teach chimpanzees to speak English," *London Sunday Times*, p. 1 (July 25, 1999).

167. Robert W. Mitchell, "Deception and concealment as strategic script violation in great apes and humans," in Sue Taylor Parker and Robert W. Mitchell, eds., *supra* note 67, at 299–307.

168. Richard W. Byrne and Andrew Whiten, "Primate tactical deception," in Andrew Whiten, ed., *supra* note 34, at 128; Marc D. Hauser, "Minding the behavior of deception," in Andrew Whiten and Richard W. Byrne, eds., *supra* note 91, at 122.

169. Peter Mitchell, *Introduction to Theory of Mind—Children, Autism and Apes* 39 (Arnold 1997).

170. Andrew Whiten, *supra* note 91, at 157.

171. Richard W. Byrne and Andrew Whiten, "Primate tactical deception," in Andrew Whiten, ed., *supra* note 34, at 131.

172. Andrew Whiten and Richard W. Byrne, in Richard Byrne and Andrew Whiten, eds., *supra*, note 34, at 220.

173. Richard W. Byrne and Andrew Whiten, "Primate tactical deception," in Andrew Whiten, ed., *supra* note 34, at 131.

174. Jane Goodall, *supra* note 21, at 577.

175. *Id.* at 578.

176. *Id.* at 581–582.

177. Frans de Waal, supra note 147, at 36–7.

178. *Id.* at 62.

179. Dorothy L. Cheney and Robert M. Seyfarth, *supra*, note 110, at 195–196.

180. Frans de Waal, *supra* note 147, at 34–5.

181. Emil Menzel, Jr., "A group of young chimpanzees in a one-acre field," in 5 *Behavior of Nonhuman Primates* 83, 134-5 (M. Schrier and F. Stolnitz, eds., Academic Press 1974).

182. Richard W. Byrne and Andrew Whiten, in Andrew Whiten, ed., *supra* note 34, at 131 (emphases in the original).

183. Jane Goodall, *supra* note 21, at 581.

184. Sue Savage-Rumbaugh and Kelly McDonald, *supra* note 36, at 228.

185. *Id.* at 233.

186. *Id.* at 229–230.

187. *Id.* at 230.

188. Roger Fouts and Stephen Tukel Mills, *supra* note 84, at 151.

189. Maurice K. Temerlin, *supra* note 84, at 122–123 .

190. Personal communication from Tetsuro Matsuzawa, dated July 2, 1999, quoting from the abstract by S. Hirata and T. Matsuzawa, "Chimpanzees' understanding of other chimpanzees' knowledge," Napoli Social Learning Conference (June 1996).

191. Frans de Waal, *supra* note 1, at 48.

192. *Id.*

193. Daniel J. Povinelli et al., "Comprehension of role reversal in chimpanzees: evidence of empathy?," 43 *Animal Behavior* 633–40 (1992).

194. Letter from Charles Sedgwick, D.V.M. to the author, dated April 22, 1998.

195. Roger Fouts and Stephen Mills Tukel, *supra* note 84, at 179–80.

196. *Id.* at 291.

197. Frans de Waal, *supra* note 1, at 83.

198. *Id.* at 83.

199. Frans de Wall and Frans Lanting, *Bonobo—The Forgotten Ape* 156 (University of California Press 1997).

200. *Id.*

201. *Scientific American Frontiers* (broadcast, January 20, 1999).

202. Personal communication from Michael A. Huffman, dated February 16, 1999. *See* Personal communication from Michael A. Huffman, dated February 11, 1999; Personal communication from Michael A. Huffman, dated January 14, 1999; Michael A. Huffman et al., "African great ape self-medication: A new paradigm for treating parasite diseases with natural medicines?," *Towards Natural Medicine Research in the Twenty-First Century* 115–23 (H. Ageta et al., eds Elsevier Science B.V. 1998): Michael A. Huffman, "Current evidence for self-medication in primates: A multi-disciplinary perspective," 40 *Yearbook of Physical Anthropology* 171–200 (1997); Michael A. Huffman, "Seasonal trends in intestinal nematode infection and medicinal plant use among chimpanzees in the Mahale Mountains, Tanzania," 38 *Primates* 111–25 (1997); Michael A. Huffman et al., "Ethnobotany and zoopharmacognosy of *Vernonia amygdalina,* a medicinal plant used by humans and chimpanzees," in 2 *Compositae: Biology and Utilization* 351–60 (P.D.S. Caligari and D.J.N. Hinde, eds., The Royal Botanic Gardens 1996); Michael A. Huffman, "Leaf-swallowing by chimpanzees: A behavioral adaptation for the control of stringyle nematode infections," 17 *International Journal of Primatology* 475–503 (1996).

203. Sue Savage-Rumbaugh and Roger Lewin, *supra* note 80, at 142 J-K.

204. *Id.* at 87–8.

205. Richard Byrne, *supra* note 76, at 166.

206. Terrence W. Deacon, *The Symbolic Species–The Co-Evolution of language and the Brain* 84 (W.W. Norton 1997). *See* Duane Rumbaugh et al., "Toward a new outlook on primate learning and behavior: Complex learning and emergent processes in comparative perspective," 38 *Japanese Psychological Research* 113, 118 (1996).

207. E. Sue Savage-Rumbaugh et al., "Language perceived; *Paniscus* branches out," in William C. McGrew et al., *supra* note 2, at 173–84.

208. C. Boesch, "Symbolic Communication in wild chimpanzees?" 6 *Human Evolution* 81, 83 (1991).

209. *Id.* at 84.

210. *Id.* at 84.

211. *Id.* at 86.

212. Personal Communication from Tetsuro Matsuzawa, dated April 12, 1999.

213. See the Language Research Center's video, "Bonobo People" (1993), with footage shot by NHK of Japan.

214. Sue Savage-Rumbaugh and William H. Fields, "Linguistic, Cultural, and Cognitive Capacities of Bonobos (*pan paniscus*) in Changing Times: Reflections on the Development of Self and Culture," in *Culture and Psychology* (Maria Lyria and Cynthia Lightfoot, eds.) (in press) (special issue).

215. Steven Pinker, *The Language Instinct—How the Mind Creates Language* 335–41, 468 (William Morrow and Co. 1994).

216. Sue Savage-Rumbaugh et al., *supra* note 23, at 51–2.

217. Beatrix T. Gardner and R. Allen Gardner, "Comparing the early utterances of child and chimpanzee, in 8 *Minnesota Symposium on Child Psychology* 24 (Pick, ed. University of Minnesota Press 1974).

218. Roger Fouts and Stephen Tukel Mills, *supra* note 84, at 123.

219. *Id.* at 267.

220. Personal Communication from Roger S. Fouts, dated February 25, 1999; Mark Bodamer et al., "Functional Analysis of Chimpanzee (Pan troglodytes) Private Signing," 9 *Human Evolution* 281–96 (1994).

221. Herbert S. Terrace, *supra* note 103, at 104–5.

222. *Id.* at 215.

223. *Id.* at 246.

224. *Id.* at 247.

225. Sue Savage-Rumbaugh, *supra* note 141.

226. Tetsuro Matsuzawa, "Chimpanzee intelligence in nature and in captivity: isomorphism of symbol use," *supra* note 2, at 198.

227. Shoji Itakura, *supra* note 106, at 243–4; Tetsuro Matsuzawa, *supra* note 40, at 451–2; Tetsuro Matsuzawza, "Spontaneous pattern construction in a chimpanzee," in Paul G. Heltne et al., *Understanding Chimpanzees* 252–65 (Harvard University Press 1989).

228. Herbert S. Terrace, *supra* note 103, at 225–7.

229. Terrence W. Deacon, *supra* note 206, at 27.

230. Steven Pinker, *supra* note 215, at 341. *E.g.*, Steven Mithen, *The Prehistory of the Mind* 86–7 (Thames and Hudson 1996); Joel Wallman, *Aping Language* (Cambridge University Press 1992).

231. Philip Lieberman, *The Biology and Evolution of Language* 256–329 (Harvard University Press 1984).

232. Cathy Hayes, *The Ape in Our House* (Harper & Brothers 1951).

233. Jonathan Wake, *supra* note 166, at 1.

234. Sue Savage-Rumbaugh and Roger Lewin, *supra* note 80, at 67, 174.

235. Roger Fouts and Stephen Tukel Mills, *supra* note 84, at 101.

236. Roger Fouts et al., "Transfer of signed responses in American Sign Language from vocal English stimuli to physical object stimuli by a chimpanzee (pan)," 7 *Learning and Motivation* 458–75 (1976).

237. E. Sue Savage-Rumbaugh and Duane Rumbaugh, "Perspectives on Consciousness, Language, and Other Emergent Processes in Apes and Humans," in *Towards a Science of Consciousness* 539, 540–1 (Stuart R. Hameroff et al. eds., MIT Press 1998).

238. E. Sue Savage-Rumbaugh et al., "Language Comprehension in Ape and Child," 58 *Monograph of the Society for Research in Child Development* 17–23 (1993).

239. Elizabeth Bates, "Commentary" in *id.* at 222.

240. *Id.* at 229–38; E. Sue Savage-Rumbaugh et al., "How does evolution design a brain capable of learning language?," in *id.* at 243–51.

241. Roger Fouts and Stephen Tukel Mills, *supra* note 84; R. Allen Gardner et al., eds., *supra* note 100; Herbert S. Terrace, *supra* note 103.

242. William C. Stokoe, "Comparative and Developmental Sign Language Studies: A Review of Recent Advances," in R. Allen Gardner et al., *supra* note 103, at 308–15; R. Allen Gardner and Beatrix T. Gardner, "A Cross-Fostering Laboratory," in R. Allen Gardner et al., *supra* note 103, at 15–25; R. Allen Gardner and Beatrix T. Gardner, "Feedforward versus feedbackward: An ethological alternative to the law of effect," 11 *Behavioral and Brain Sciences* 429, 443–7 (1988).

243. S.N. Cianelli and R.S. Fouts, "Chimpanzee to Chimpanzee American Sign Language. Communication During High Arousal Interactions," 13 *Human Evolution* 1067 (1998); Roger Fouts and Stephen Tukel Mills, *supra* note 84, at 134, 135, 175, 176, 211, 240, 333; Deborah H. Fouts, "The Use of Remote Video Recordings to Study the Use of American Sign Language by Chimpanzees When No Humans are Present," in *The Ethological Roots of Culture* 271–81 (R. A. Gardner et al., eds., Kluwer Academic Publishers 1994 ; Roger S. Fouts, "Chimpanzee language and elephant tails: A theoretical synthesis," in *Language in Primates–Perspectives and Implications* 63–75 (Springer-Verlag 1983).

244. Sue Savage-Rumbaugh and Roger Lewin, *supra* note 80, at 67.

245. E. Sue Savage-Rumbaugh et al., *supra* note 238, at 15.

246. *Id.* at 10.

247. *Id.* at 127.

248. *Id.* at 146–7.

249. *Id.* at 45.

250. Sue Savage-Rumbaugh and Roger Lewin, *supra* note 80, at 135.

251. Personal communication from Roger Fouts, dated February 2, 1999; Roger Fouts and Stephen Tulek Mills, *supra* note 84, at 21, 29, 78.

252. Sue Savage-Rumbaugh, "Behavior and Mental Abilities of Primates," videotaped presentation at the Tufts University Center for Animals and Public Policy, November 14, 1998.

253. Sue Savage-Rumbaugh and William H. Fields, *supra* note 214, at 10.

254. Sue Savage-Rumbaugh and Roger Lewin, *supra* note 80, at 138.

255. E. Sue Savage-Rumbaugh et al., *supra* note 238, at 24.

256. *Id.* at 45.

257. Sue Savage-Rumbaugh et al., *supra* note 23, at 209; Sue Savage-Rumbaugh et al., "Spontaneous Symbol Acquisition and Communicative Use By Pygmy Chimpanzees," in 115 *Journal of Experimental Psychology: General* 211, 214 (1986).

258. Roger Fouts and Stephen Tukel Mills, *supra* note 84, at 162–5.

259. Personal communication from Tetsuro Matsuzawa, dated April 11, 1999 (discussing unpublished findings).

260. E. Sue Savage-Rumbaugh et al., *supra* note 238, at 125, 140, 142, 152.

261. Sue Savage-Rumbaugh, *supra* note 252.

262. Sue Savage-Rumbaugh, "Scientific Schizophrenia With Regard to the Language Act," in *Piaget, Evolution, and Development* 158 (J. Langer and M. Killen, eds., Lawrence Erlbaum 1998).

263. Sue Savage-Rumbaugh et al., *supra* note 23, at 68–73; E. Sue Savage-Rumbaugh et al., *supra* note 238, at 44–97, 100, 111–210.

264. Sue Savage-Rumbaugh and Roger Lewin, *supra* note 80, at 171–2.

265. Peter W. Mitchell, *supra* note 169, at 24.

266. *Id.* at 157–62; E. Sue Savage-Rumbaugh et al., *supra* note 238, at 42.

267. Sue Savage-Rumbaugh and Roger Lewin, *supra* note 80, at 144.

268. Personal communication from Tetsuro Matsuzawa, dated April 11, 1999; Tetsuro Matsuzawa, "Chimpanzee intelligence in nature and in captivity: isomorphism of symbol use," *supra* note 2, at 198; Tetsuro Matsuzawa, *supra* note 227, at 254–9.

269. Tetsuro Matsuzawa, "Chimpanzee intelligence in nature and in captivity: isomorphism of symbol use," *supra* note 2, at 205.

270. Sue Savage-Rumbaugh and William H. Fields, "Language and Culture: A Trans-generational weaving," *LOS Forum* No. 27 at 16, 25 (Fall 1998). Sue Savage-Rumbaugh and William H. Fields, *supra* note 214.

271. Sue Savage-Rumbaugh and Roger Lewin, *supra* note 80, at 177.

272. E. Sue Savage-Rumbaugh et al., *supra* note 238, at 45.

273. Karen E. Brakke and E. Sue Savage-Rumbaugh, "The development of language skills in bonobo and chimpanzee: I. Comprehension," 2 *Language and Communication* 121, 131 (1995).

274. Sue Savage-Rumbaugh, *supra* note 262, at 146–50.

275. *Id.* at 148.

276. *Id.* at 155–60.

277. *Id.* at 157.

278. Deborah Blum, "All in the Family," *New York Times Book Review* 11 (October 12, 1997)(review of Roger Fouts and Stephen Tukel Mills, *supra* note 84).

279. Sue Savage-Rumbaugh and William H. Fields, *supra* note 214 (in press).

280. Sue Savage-Rumbaugh et al., *supra* note 23, at 39; E. Sue Savage-Rumbaugh and Duane Rumbaugh, *supra* note 237, at 542.

281. E. Sue Savage-Rumbaugh and Duane Rumbaugh, *supra* note 237, at 543.

282. Sue Savage-Rumbaugh, *supra* note 262, at 146–50. See H. Lyn Miles, "Symbolic communication with and by great apes," in Sue Taylor Parker et al., *supra* note 67, at 205–6 for a list of the different categories of words that chimpanzees and bonobos use.

283. Sue Savage-Rumbaugh and William H. Fields, *supra* note 214 (in press).

284. Personal communication from Sue Savage-Rumbaugh, dated May 24, 1999; Sue Savage-Rumbaugh, *supra* note 262, at 160.

285. E. Sue Savage-Rumbaugh and Duane Rumbaugh, *supra* note 237, at 543.

286. Sue Savage-Rumbaugh, et al., *supra* note 23, at 57, 59, 64–5; E. Sue Savage-Rumbaugh and Duane Rumbaugh, *supra* note 237, at 543; Sue Savage-Rumbaugh, *supra* note 141, at 57–62.

287. J. Chamberlin, "Two psychologists win $1 million grants," 30 *Monitor* 47 (April, 1999).

288. Daniel J. Povinelli and Steve Giambrone, "Inferring Other Minds: Failure of the Argument by Analogy," *Philosophical Topics* (G. Massey and B.D. Massey, eds. 1999)(in press)(special issue on zoological philosophy); Daniel J. Povinelli and Timothy J. Eddy, "What young chimpanzees know about seeing," 61 *Monographs of the Society for Research in Child Development* (2, Serial No. 247).

289. Juan Carlos Gómez, *supra* note 94, at 83; Juan Carlos Gómez, "Primate theories of primate minds: conceptual and methodological issues," *supra* note 92 (in press) Juan Carlos Gómez, *supra* note 85, at 142–3; Juan-Carlos Gómez, "Nonhuman primate theories of (nonhuman primate) minds: Some issues concerning the origins of mindreading," in Peter Carruthers and Peter K. Smith, eds., *supra* note 92, at 330–43.

290. Juan Carlos Gómez, "Primate theories of primate minds: conceptual and methodological issues," *supra* note 92 (in press).

291. Juan Carlos Gómez, "Assessing theory of mind with nonverbal procedures: Problems with training methods an an alternative "key" procedure," 21 *Behavioral and Brain Sciences* 119–20 (1998).

292. Juan Carlos Gómez, "Primate theories of primate minds: conceptual and methodological issues," *supra* note 92 (in press).

293. Personal communication from Sue Savage-Rumbaugh, dated May 15, 1999.

294. Personal communication from Sue Savage-Rumbaugh, dated May 24, 1999.

295. Telephone conversation with Sue Savage-Rumbaugh, May 17, 1999; Personal communication from Sue Savage-Rumbaugh, dated May 15, 1999.

296. Dan Shillito, Rob Shumaker, Gordon Gallup, and Ben Beck (Ph.D dissertation of Dan Shallito) (in progress). The facts I relate are based upon a telephone conversation I had with Dan Shillito on May 23, 1999, a videotape that shows a sampling of Indah's trials, and personal communications from Dan Shillito, dated June 9 and July 5, 1999.

297. Personal communication from Marc Hauser, dated May 24, 1999.

298. Personal communication from Marc Hauser, February 5, 1999; *Scientific American Frontiers* (first broadcast January 20, 1999).

299. Daniel J. Povinelli et al., 2 *Trends in Cognitive Science* 158, 159 (April, 1998)(reviewing Anne E. Russon, ed., *supra* note 72); Daniel Povinelli, "Chimpanzee theory of mind? the long road to strong inference," in Peter Carruthers and Peter K. Smith eds., *supra* note 92, at 326.

300. Daniel J. Povinelli and Timothy J. Eddy, "Factors Influencing Young Chimpanzees' (*Pan troglodytes*) Recognition of Attention," 110 *Journal of Comparative Psychology* 336, 344 (1996).

301. Janet Astington, "What is theoretical about the child's theory of mind?: a Vygotskian view of its development," in Peter Carruthers and Peter K. Smith eds., *supra* note 92, at 198.

302. Angeline Lillard, "Developing a cultural theory of mind," 8 *Current Directions in Psychological Science* 57, 58 (1999).

303. Janet Astington, *supra* note 301, at 189.

304. Angeline Lillard, Ethnopsychologies: Cultural variations in theories of mind," 123 *Psychological Bulletin* 3, 13 (1998).

305. Janet Astington, *supra* note 301, at 189.

306. Angeline Lillard, *supra* note 302, at 58–9; Angeline Lillard, *supra* note 304, at 3.

307. Judy Dunn, "Understanding Others: Evidence From Naturalistic Studies of Children," In Andrew Whiten, ed., *supra* note 34, at 51–9.

308. Juan-Carlos Gómez, in Peter Carruthers and Peter K. Smith, eds., *supra* note 92, at 330–1.

309. *Id.* at 331; Andrew Whiten, "Grades of mindreading," in *Children's Early Understanding of Mind: Origins and Development* 66 (Charlie Lewis and Peter Mitchell, eds., Lawrence Erlbaum Associates 1994).

310. Stephen Budiansky, *If a Lion Could Talk–Animal Intelligence and the Evolution of Consciousness* 194 (The Free Press 1998).

311. Steven Pinker, *supra* note 215, at 369..

312. Daniel J. Povinelli et al., *supra* note 299, at 159; J. Chamberlin, *supra* note 287.

313. Daniel J. Povinelli et al., *supra* note 299, at 159.

314. Virginia Morell, "The Sixth Extinction," 195 *National Geographic* 42, 47 (February 1999).

CHAPTER ELEVEN

1. Henry Mayer, *All on Fire: William Lloyd Garrison and the Abolition of Slavery* (St. Martin's Press 1998).

2. Roger Fouts and Steven Tukel Mills, *Next of Kin: What Chimpanzees Have Taught Me About Who We Are* 285 (William Morrow 1997).

3. *Id.* at 213–4, 330.

4. *Id.* at 286; Eugene Linden, *Silent Partners: The Legacy of the Ape Language Experiments* 159–61 (Times Books 1986).

5. Roger Fouts and Steven Tukel Mills, *supra* note 2, at 354–8.

6. 1 *Adams Family Correspondence* 359 (L.H., ed., Harvard University Press 1963).

7. Letter from Lord Acton to Bishop Mandell Creighton, dated April 3, 1887.

8. William Lee Miller, *Arguing About Slavery: The Great Battle in The United States Congress* 230, 257, 267–8 (Alfred A. Knopf 1996).

9. Lloyd L. Weinreb, *Oedipus at Fenway Park* 110 (Harvard University Press 1994).

10. *Dred Scott v. Sandford*, 60 U.S. 393 (1856).

11. L.W. Sumner, *The Moral Foundation of Rights* 206 (Oxford University Press 1987).

12. *See* the definition of "inconceivable," VII *Oxford English Dictionary* 812 (sec. ed. Oxford University Press 1989).

13. Edward O. Wilson, *The Diversity of Life* 38 (Harvard University Press 1992); Ernst Mayr, *The Growth of Biological Thought* 273 (Harvard University Press 1982).

14. Gorillas, orangutans, and gibbons, too.

15. M.J. Morwood et al., "Fission track ages of stone tools and fossils on the East Indonesian island of Flores," 392 *Nature* 173 (1998); Ann Gibbons, "Ancient island tools suggest *Homo erectus* was a seafarer," 279 *Science* 1635 (1998); Terrence W. Deacon, *The Symbolic Species: The co-evolution of language and the brain* 368–70 (W.W. Norton 1997); C.C. Swisher III et al., "Latest *Homo erectus* of Java: Potential Contemporaneity with *Homo sapiens* in Southeast Asia," 274 *Science* 187 (Thames and Hudson, Ltd. 1996); Steven Mithen, *The Prehistory of the Mind* 20, 25, 26, 29, 31, 116–20, 135–6, 142, 144, 206 (1996); Richard Leakey, *The Origin of Mankind* xiv, 73–7, 132, 135 (Basic Books 1994); Roger Lewin, *The Origin of Modern Humans* 30–3, 164–9 (W. H. Freeman & Co. 1993); Kathy D. Schick and Nicholas Toth, *Making Silent Stones Speak: Human Evolution and the Dawn of Technology* 219, 229, 280–3

(Simon & Schuster 1993); Niles Eldridge and Ian Tattersall, *The Myths of Human Evolution* 145 (Columbia University Press 1982).

16. Constance Holden, "No last word on language origins," 282 *Science* 1455 (1998); Constance Holden, "How much like us were the Neandertals?," 282 *Science* 1456 (1998); Paul Mellars, "The fate of the Neanderthals," 395 *Nature* 539, 540 (1998); Donald Johanson, "Reading the minds of fossils," *Scientific American* (March 1998)(reviewing Ian Tattersall, *Becoming Human: Evolution and Human Uniqueness* [1998]); Terrence W. Deacon, *supra* note 15, at 372; Steven Mithen, *supra* note 15, at 25, 26, 119, 125–30, 129, 132, 147–150; Richard Leakey, *supra* note 15, at 125, 132–3, 155; Roger Lewin, *supra* note 15, at 84–5, 122, 67, 169, 179; Philip Lieberman, *Uniquely Human: The Evolution of Speech, Thought, and Selfless Behavior* 110 (Harvard University Press 1991).

17. John Noble Wilford, "Discovery Suggests Man is a Bit Neanderthal," *New York Times*, at 1, 21 (April 25, 1999).

18. Ian Tattersall, *Becoming Human: Evolution and Human Uniqueness* 150 (Harcourt Brace 1998); Daniel Lieberman, "Sphenoid shortening and the evolution of modern cranial shape," 393 *Nature* 158 (1998); Mathias Kring et al., "Neanderthal DNA sequences and the origin of modern humans," 90 *Cell* 19–30 (1997); Ian Tattersall, *The Last Neanderthal* 10–5, 17 (Macmillan 1995).

19. Wole Soyinka, "Every Dictator's Nightmare," *New York Times Magazine* 90 (April 18, 1999).

20. *Id.* at 90, 91.

21. *E.g., Planned Parenthood v. Casey*, 505 U.S. 833, 851 (1992)(joint plurality opinion); *Cruzan v. Director, Missouri Dept. of Health*, 497 U.S. 261, 289 (1990)(O'-Connor, J., concurring); Ronald Dworkin, *Taking Rights Seriously* 166 (Harvard University Press 1977).

22. *People v. Kevorkian,* 527 N.W. 2d 714, 727 note 14 (Mich. 1994); *Natanson v. Kline,* 350 P. 2d 1093, 1104 (Kan. 1960).

23. *Matter of Guardianship of L.W.,* 482 N.W. 2d 60, 69 (Wis. 1992), quoting *Eichner v. Dillon,* 426 N.Y.S. 2d 517, 542 (App. Div.), *mod.* 52 N.Y. 2d 363 (1980). *E.g., Gray v. Romeo,* 697 F. Supp. 580, 587 (D.R.I. 1988); *Matter of Moe,* 432 N.E. 2d 712, 719 (Mass. 1982); *Delio v. Westchester Cty. Med. Ctr.,* 516 N.Y.S. 2d 677, 686 (N.Y. App. Div. 1987).

24. Jordan J. Paust, "Human Dignity as a Constitutional Right: A Jurisprudentially Based Inquiry into Criteria and Content," 27 *Howard L. J.* 50–162 (1984).

25. *Youngberg v. Romeo,* 457 U.S. 307, 315–6, 325 (1982).

26. *See, e.g., Matter of Tavel,* 661 A. 2d 1061, 1069 (Del. 1995); *In re Guardianship of Barry,* 445 So. 2d 365, 370 (Fla. Dist. Ct. App. 1984); *DeGrella by and through Parrent v. Elston,* 858 S.W. 2d, 698, 709 (Ky. 1993); *Care and Protection of Beth,* 587 N.E. 2d 1377, 1382 (Mass. 1982), and cases cited.

27. *People v. Kevorkian, supra* note 22, at 727 note 41.

28. *In the Matter of Guardianship of Eberhardy,* 307 N.W. 2d 881, 898 (Wis. 1981)(Gallow, J., dissenting).

29. *Care and Protection of Beth, supra* note 26, at 1382 (emphases added).

30. *Id.* (emphases added).

31. *Matter of Conroy,* 486 A. 2d 1209,1229 (N.J. 1985).

32. *Superintendent of Belchertown State School v. Saikewicz,* 370 N.E. 2d 417, 431 (Mass. 1977).

33. Louise Harmon, "Falling Off the Vine: Legal Fictions and the Doctrine of Substituted Judgment, 100 *Yale L.J.* 1, 58–9, 68 (1990); Lawrence H. Tribe, *American Constitutional Law* 1369 (sec. ed. Foundation Press 1988).

34. *In re Grady*, 426 A. 2d 467, 480–1 (N.J. 1981)(emphases added). *See, e.g., In re L.H.R.*, 321 S.E. 2d 716 (Ga. 1984)(four month old in chronic vegetative state); *Strunk v. Strunk*, 445 S.W. 2d 145 (Ky, 1969)(twenty-seven year old with an I.Q. of thirty-five and a mental age of six); *In re Barry, supra* note 26 (anencephalic ten month old with no cognitive brain function).

35. *Conservatorship of Drabick*, 245 Cal. Rpt. 840, 855, *cert. den. sub nom., Drabick v. Drabick*, 488 U.S. 958 (1988)(emphasis added).

36. Barbara Herman, *The Practice of Moral Judgment* 62 and note 25 (Harvard University Press 1993).

37. Ellen Langer, *The Power of Mindful Learning* 4 (Addison Wesley 1997); Barbara Herman, *supra* note 36, at 229.

38. Tom Regan, *The Case for Animal Rights* 84–5 (University of California Press 1983); James Rachels, *Created From Animals* 140, 147 (Oxford University Press 1990); William A. Wright, "Treating animals as ends," *J. Value Inquiry* 353, 357, 362 (1993); Christopher Cherniak, *Minimal Rationality* 3–17 (M.I.T. Press 1985).

39. *People v. Nauton*, 34 Cal. Rptr. 2d 861, 864 (Ct. App. 1994).

40. *Commonwealth v. Martin*, 683 N.E. 2d 280, 283 (Mass. 1997).

41. *Guardianship of Doe*, 583 N.E. 2d 1263, 1268 (Mass. 1992); *id.* at 1272–3 (Nolan, J., dissenting); *id.* at 1275 (O'Connor, J., dissenting).

42. *International Shoe Co. v. Washington*, 326 U.S. 310, 316 (1945); *Tauza v. Susquehanna Coal Co.*, 115 N.E. 915, 917 (N.Y. 1917); *Pramatha Nath Nullick v. Pradyumna Kumar Mullick*, 52 Indian L. R. 245, 250 (India 1925).

43. Lawrence M. Friedman, *A History of American Law* 19 (Simon & Schuster 1973).

44. Jeremy Bentham, "Elements of Packing as Applied to Juries," in 5 *The Works of Jeremy Bentham* 92 (J. Bowring, ed., 1843)(emphasis in the original).

45. Norma Basch, *In The Eyes of the Law* 17 (Cornell University Press 1982). *E.g., Motion to Admit Miss Lavinia Goodall to the Bar of this Court*, 39 Wis. 232, 244–6 (1875); *Bradwell v. Illinois*, 83 U.S. 130, 141 (Bradley, J., concurring).

46. Michael J. Meyer, "Kant's Concept of Dignity and Modern Political Thought," in 8 *History of Eur. Ideas* 320–1 (1987).

47. A. John Simmons, *The Lockean Theory of Rights* 193 note 79 (Princeton University Press 1992); James Rachels, *supra* note 38, at 173–223.

48. H. Tristram Englehardt, Jr. , *The Foundation of Biotethics* (sec. ed. Oxford University Press 1996). *See* Carl Wellman, *Real Rights* 140 (Oxford University Press 1995); Jed Rubenfeld, "On the Legal Status of the Proposition that 'Life Begins at Conception,'" 43 *Stan. L. Rev.* 599, 612 (1991). *See* S.I. Benn, "Abortion, Infanticide, and Respect for Persons," in *The Problem of Abortion* 92 (Joel Feinberg, ed., Wadsworth Publishers 1973).

49. John Rawls, *A Theory of Justice* 509–10 (Harvard University Press 1971). *See* John Noonan, Jr., "An almost absolute value in history," in *The Morality of Abortion* 51 (Harvard University Press, 1974).

50. Diana T. Meyers, *Inalienable Rights: A Defense* 129 (Columbia University Press 1985); A.I. Melden, *Rights and Persons* 222–3 (University of California Press 1980).

51. Frederick A. King et al., "Primates," 240 *Science* 1475, 1475, 1476, 1479, 1481 (1988).

52. State of the Union Address, December 1, 1862, in 5 *Collected Works of Abraham Lincoln* 537 (Roy P. Basler, ed. Princeton University Press 1953).

53. Carl Cohen, "The Case for the Use of Animals in Biomedical Research," 317 *New. Eng. J. Med.* 865, 866 (1986).

54. James Rachels, *supra* note 38, at 187.

55. Carl Cohen, "Race, Lies, and 'Hopwood,'" 101 *Commentary* 39 (June, 1996).

56. Laurence H. Tribe, *supra* note 33, at 1589. *See id.* at 1527–8.

57. *Wittmer v. Peters*, 87 F. 3d 916, 918 (7th Cir.), *reh. and sugg. for reh. En Banc den.* (1996); Laurence H. Tribe, *supra* note 33, at 1523.

58. "In Poll, Americans Reject Means But Not Ends of Racial Diversity," *New York Times*, p. A1 (December 16, 1997).

59. *Hopwood v. State of Texas*, 78 F. 3d 932, 945–6 (5th Cir.), *cert den.*, 116 S.Ct. 2581 (1996), quoting Richard A. Posner, "The DeFunis Case and the Constitutionality of Preferential Treatment of Racial Minorities," 1974 *Sup. Ct. Rev.* 12 (1974).

60. *Lassiter v. Northampton Election Board*, 360 U.S. 45 (1959). *See In re Gault*, 387 U.S. 1, 13 (1967).

61. *In re Borgogna*, 175 Cal. Rptr. 588, 595 (Ct. App. 1981). *See Youngberg, supra* note 25.

62. *Youngberg, supra* note 25, at 321; *Parham v. J.R.*, 442 U.S. 584 (1979).

63. *Cruzan v. Director, Missouri Dept. of Health*, 497 U.S. 261, 290–1 (1990); *Commonwealth v. A Juvenile*, 449 N.E. 2d 654, 657 (Mass. 1983).

64. *Fulton v. Shaw*, 25 Va. 597, 599 (1827).

65. *Korematsu v. United States*, 323 U.S. 214 (1944). *See* Michu Weglyn, *Years of Infamy: The Untold Story of America's Concentration Camps* (University of Washington Press 1976).

66. "Redress for War Internees Ended," *New York Times*, p. A15 (February 15, 1999).

67. Robert Jay Lifton, *The Nazi Doctors: Medical Killing and the Psychology of Genocide* (Basic Books 1986); A. Mitscherlich and F. Mielke, *The Death Doctors* (James Cleugh, trans. Elek Books Limited 1962)(1949).

68. Ralph Blumenthal and Judith Miller, "Japan Rebuffs Requests for Information About its Germ-Warfare Atrocities," *New York Times*, p. A10 (March 4, 1999); Sheldon H. Harris, *Factories of Death* (Routledge 1994); Robert Whymant, "The Butchers of Harbin," *The Boston Globe*, p. 49 (September 5, 1992); "Japanese Germ Use in War is Studied," *The Boston Globe*, p. 88, (April 16, 1992); Christopher S. Wren, "2 Wooden Plaques Evoke Manchuria Deathcamp," *New York Times*, p. 2 (March 22, 1983).

69. *United States v. Holmes*, 26 Fed. Cas. 360, No. 15, 383 (3rd. Cir. 1842).

70. *Regina v. Dudley & Stephens*, 14 Q.B 273 (1884).

71. 163 U.S. 537 (1986).

72. 347 U.S. 497 (1954).

73. 347 U.S. 483 (1954).

74. Ronald Dworkin, *supra* note 21, at 221; *Bolling, supra* note 72, at 499.

75. *Id.* at 189–90 (Oxford University Press 1994).

76. Richard Kluger, *Simple Justice* 605, 606 (Alfred A. Knopf 1975).

77. *Brown, supra* note 73, at 494, 495.

78. *See Casey, supra* note 21, at 866 (plurality opinion); *Bolling, supra* note 73, at 499; *Truax v. Corrigan*, 257 U.S. 312, 332 (1921); *Simonds v. Simonds*, 380 N.E. 2d. 189, 192 (NY 1978). *See also* Richard Rorty, "Pragmatism, Relativism, and Irrationalism," in *Consequences of Pragmatism* 172 (University of Minnesota Press 1982); Ronald Dworkin, "Sex, Death, and the Courts," *N.Y. Rev. of Books* (August 8, 1996); Ronald Dworkin, *supra* note 21, at 178–85, 219–27.

79. Karl Jacoby, "Slaves by Nature? Domestic Animals and Human Slaves," *Slavery and Abolition: A Journal of Slave and Post-Slave Studies* 89 (April 1994).

80. Richard A. Posner, *The Problems of Jurisprudence* 347–8 (Harvard Universoty Press 1990).

81. Personal communication from Richard A. Posner, dated March 15, 1999.

82. Personal communication from Richard A. Posner, dated March 26, 1999, quoting Bruce Ackerman, *Social Justice in the Liberal State* 80 (Yale University Press 1980). *See* Richard A. Posner, *The Problems of Jurisprudence* 336 (Harvard University Press 1990).

83. Bruce Ackerman, *supra* note 82, at 73–5.

84. *See generally*, Richard Sorabji, *Animal Minds & Human Morals: The Origins of the Western Debate* (Cornell University Press 1993).

85. 7 U.S.C. sec. 2143(a).

86. D. Keleman, "Beliefs about purpose: On the origins of teleological thought," *The Evolution of the Hominid Mind* (Michael Corballus and S. Lea, eds., Oxford University Press 1998).

87. *Id. at 289.*

88. *Id.* at 282–3.

89. *Id.* at 280–1, 287.

90. *Id.* at 288.

91. Claudia Dreifus, "Ex-Priest Takes the Blasphemy Out of Evolution," *New York Times*, at D5 (April 27, 1999).

92. Steven Weinberg, "A Designer Universe?," XLVI *The New York Review of Books* 46 (October 21, 1999).

93. 1 Corinthians 13:11.

94. VI *Oxford English Dictionary* 445 (Second Ed. 1989).

95. *Merriam-Webster's Collegiate Dictionary* (10th ed. 1997).

96. *Convention on the Prevention and Punishment of the Crime of Genocide*, December 9, 1948, 78 U.N.T.S. 277 (entered into force Jan 12, 1951)(republished in Richard B. Lillich, *International Human Rights Instruments* 130.1 (2nd. ed 1990)(Article 2).

97. *Reservations to the Convention on the Prevention and Punishment of the Crime of Genocide*, 1951 I.C.J. 15, 23.

98. Andrew Whiten et al., "Culture in chimpanzees." 399 *Nature* 682 (1999); Christophe Boesch and Michael Tomasello, "Chimpanzee and Human Cultures," 39 *Current Anthropology* 592 (1998); *Chimpanzee Cultures* (Richard W. Wrangham et al., eds., Harvard University Press 1994); W.C. McGrew, *Chimpanzee Material Culture* (Cambridge University Press 1992).

99. Frans de Waal, *Chimpanzee Politics: Power and Sex Among the Apes* (rev. ed. Johns Hopkins University Press 1998).

Chapter Twelve

1. I speak here of an *entitlement* to legal rights. As I noted in Chapter 11, judges may decide to bestow legal rights upon those who are not entitled to them.

2. S. Itakura, "An exploratory study of gaze-monitoring in nonhuman primates," 38 *Japanese Psychological Record* 174–179 (1996).

3. Juan-Carlos Gómez, "Non-human primate theories of (nonhuman primate) minds: Some issues concerning the origins of mindreading," in *Theories of Theories of Mind* 338–41 (Peter Carruthers and Peter K. Smith, eds., Cambridge University Press 1996).

4. Francine G.P. Patterson and Ronald H. Cohn, "Self-recognition and self-awareness in lowland gorillas," in *Self-Awareness in Animals and Humans: Developmental Perspectives* 273–90 (Sue Taylor Parker et al., eds., Cambridge University Press 1994)(gorillas); H. Lyn White Miles, "ME CHANTEK: The development of self-awareness in a signing orangutan," in Sue Taylor Parker et al., *supra* at 254–72.

5. Juan Carlos Gómez, "Visual behavior as a Window for Reading the Minds of Others in Primates," in *Natural Theories of Mind* 195–207 (Andrew Whiten, ed. Blackwell 1991).

6. Francine Patterson and Wendy Gordon, "The case for the personhood of gorillas," in *The Great Ape Project* 59, 60, 61 (Paola Cavalieri and Peter Singer, eds. 4th Estate 1993).

7. Nathan J. Emery, "Gaze following and Joint Attention in Rhesus Monkeys (*Macaca mulatta*), 111 *Journal of Comparative Psychology* 286–92 (1997); Robert L. Thompsom and Susan L. Boatright-Horowitz, "The question of mirror-mediated self-recognition in apes and monkeys: Some new results and reservations," in Sue Taylor Parker et al., eds. *supra* note 4 at 330–49; S. Itakura, "Use of a mirror to direct their responses in Japanese monkeys (*Macaca fuscata fuscata*), 28 *Primates* 343 (1987).

8. M. Mathieu et al., "Piagetian object permanence in *Cebus capucinus, Lagothrica flavicauda,* and *Pan troglodytes,*" 24 *Animal Behavior* 585 (1976).

9. Robert W. Mitchell and James R. Anderson, "Pointing, Withholding Information, and Deception in Capuchin Monkeys (*Cebus apella*), 11 *Journal of Comparative Psychology* 351 (1997).

10. Gregory Charles Westergaard, "What capuchin monkeys can tell us about the origins of hominid material culture," 3 *Journal of Material Culture* 5 (1998).

11. Dorothy L. Cheney and Robert L. Seyfarth, "The representation of social relations by monkeys," 37 *Cognition* 167 (1992).

12. Marc D. Hauser, *Wild Minds: How Animals Think* (Henry Holt, forthcoming 2000); Personal communication from Professor Marc D. Hauser, February 5, 1999; *Scientific American Frontiers* (first broadcast January 20, 1999). Hauser sets the stage in Marc Hauser & Susan Carey, "Building a cognitive creature from a set of primitives: Evolution and developmental insights," in *The Evolution of Mind* 51–101 (Denise Dellarosa Cummins and Colin Allen, eds., Oxford University Press 1998).

13. Elizabeth M. Brannon and Herbert S. Terrace, "Response," 283 *Science* 1852 (1999); Elizabeth M. Brannon and Herbert S. Terrace, "Ordering of the Numerosities 1 to 9 by Monkeys, 282 *Science* 746 (1998); Marc D. Hauser et al., "Numerical

representations in primates," 93 *Proceedings of the National Academy of Sciences* 1514–17 (1996); David A. Washburn and Duane M. Rumbaugh, "Ordinal Judgments of numerical symbols by macaques (*Macaca mulatta*)," 2 *Psychological Science* 190 (1991).

14. Kenneth Marten and Suchi Psarakos, "Evidence of self-awareness in the bottlenose dolphin," in Sue Taylor Parker et al., *supra* note 4, at 361–79.

15. Susan Milius, "The search for animal inventors: How innovative are other species?," 155 *Science News* 364, 365 (June 5, 1999); "Whale chat," *Discover* 32 (Jun, 1999).

16. Irene M. Pepperberg et al., "Mirror use by African Grey Parrots (Psittacus erithacus), 109 *Journal of Comparative Psychology* 182 (1995); Daniel J. Povinelli, "Failure to find mirror self-recognition in elephants (*Elephas maximum*) in contrast to their use of mirror cues to discover hidden food," 103 *Journal of Comparative Psychology* 103 (1987).

17. Irene M. Pepperberg et al., "Development of Piagetian Object Permanence in a Grey Parrot (*Psittacus erithacus*), 111 *Journal of Comparative Psychology* 63–75 (1997); S. Gagnon and F.Y. Dore, "Cross-sectional study of object permanence in domestic puppies (*Canis familiaris*), 108 *Journal of Comparative Psychology* 220–32 91994).

18. Daniel C. Dennett, *Kinds of Minds: Towards an Understanding of Consciousness* 164–66 (BasicBooks 1996).

19. Gavin R. Hunt, "Manufacture and use of hook-tools by New Caledonian crows," 379 *Nature* 249 (1996).

20. Bernd Heinrich, "An experimental investigation of insight in common ravens (*Corvus corvax*), 112 *The Auk* 994–1003 (1995).

21. Nicola S. Clayton & Anthony Dickinson, "Episodic-like memory during cache recovery by scrub jays," 395 *Nature* 272–4 (1998); Kathryn Jeffery and John O'Keefe, "Worm holes and avian space-time," 395 *Nature* 215–5 (1998).

22. Irene M. Pepperberg, ""A communicative approach to animal cognition: A study of conceptual abilities of an African grey parrot," in *Cognitive Ethology* (Carolyn A. Ristau, ed., Lawrence Erlbaum 1991); David Stipp, "Einstein Bird Has Scientists Atwitter Over Mental Feats," *Wall Street Journal*, pp. A1, A7 (May 9, 1990).

ABOUT THE AUTHOR

Steven M. Wise, J.D. has practiced animal protection law for 20 years and teaches "Animal Rights Law" at the Harvard Law School, Vermont Law School, and John Marshall Law School. Former president of the Animal Legal Defense Fund, he is the author of numerous scholarly articles, and has collaborated and communicated for years with well-known scientists, and the animals with whom they work and share their lives. He is the founder and president of the Center for the Expansion of Fundamental Rights, Inc., and lives in Needham, Massachusetts, with his wife and law partner Debra Slater-Wise and three children.

ABOUT THE CENTER FOR THE EXPANSION OF FUNDAMENTAL RIGHTS, INC.

Since 1995, the Center for the Expansion of Fundamental Rights, Inc. (CEFR) has been the only nonprofit tax-exempt organization in the world with the primary purpose of obtaining fundamental legal rights for nonhuman animals, beginning with chimpanzees and bonobos. If you would like information about CEFR, please visit our website at www.cefr.org. If you would like to make a tax-deductible contribution, CEFR's address is 896 Beacon Street, Suite 303, Boston, Massachusetts 02215.

INDEX